THE MECHANICAL ADAPTATIONS OF BONES

John Currey

THE MECHANICAL ADAPTATIONS OF BONES

PRINCETON
UNIVERSITY
PRESS

Published by Princeton University Press, 41 William Street, Princeton, New Jersey 08540
In the United Kingdom: Princeton University Press, Guildford, Surrey

Library of Congress Cataloging in Publication Data will be found on the last printed page of this book

ISBN 0-691-08342-8

Clothbound editions of Princeton University Press books are printed on acid-free paper, and binding materials are chosen for strength and durability

Printed in the United States of America by Princeton University Press,
Princeton, New Jersey

Contents

Preface

Commenting on an excellent book on vertebrates, a colleague of mine remarked, unfairly, that it gave him the feeling that "animals were just a collection of levers held together by mathematical formulae." There is another class of book that leaves one with the feeling that animals are just "oohs" and "aahs" separated by a great deal of hand-waving. This book will occasionally lurch to one extreme or the other, but I have tried to make it travel down the middle.

As the title implies, I have written this book from the point of view of adaptation, and mechanical adaptation at that. To me the organic world makes no sense unless thought of as the product of evolution, and our skeletons have been as much part of that process as the rest of us. I regard bones as the best possible solutions to selective problems and, even if they are not, it is better to think of them as such rather than as having particular attributes through historical accident.

I hope this book will be intelligible and interesting to people who have little mechanics, or who have little biology, or whose thoughts about animals have been restricted to mammals. All these people will have to take some things on trust.

Many people have influenced my thinking about bones, but there are a few I wish particularly to mention. Harold Pusey taught me vertebrates at Oxford, and imbued me with the feeling that everything, in the end, was explicable. Arthur Cain, companion on many field trips, convinced me of the optimality of animals practically before the word had been invented. Al Burstein, a generous friend, taught me what little formal biomechanics I know, but failed to engender in me his love of skeet shooting. Neill Alexander, the master biomechanic, has the unsettling tendency to think of my bright ideas two years before I do. Both David White and Julian Bryant critically read much of the manuscript of this book, and those equations that are dimensionally correct are probably the result of Julian's disciplining. Margaret Britton introduced me to the dubious pleasures of word processing. Finally, my wife Jillian has taught me little about bones but, by telling me occasionally, in the evenings, about her job as a social worker, has reminded me that there is a real world outside.

THE MECHANICAL ADAPTATIONS OF BONES

1. The Mechanical Properties of Materials and the Structure of Bone

1.1 What Is Bone for?

Any biological material has an enormous number of mechanical properties that it is possible for scientists to investigate and that may also be tested by natural selection. Not all are likely to be of importance. I shall here discuss a question that will recur, by implication, frequently in this book: what are bones for? It is not worthwhile discussing at length the philosophical question of whether bone can be said to be designed for anything at all. As a convinced Darwinist, I believe that all living organisms are nicely designed for the conditions that their ancestors of the last few thousand generations have lived in. Organisms have evolved organs that help them to survive and to pass on their genes to the next generation. It is one of the jobs of biologists to find out what these functions are, and how the organs perform them. For a contrary view of the perfection of organisms see Gould and Lewontin (1979).

Often the function seems obvious, and discovering it seems trivial. However, this apparent triviality may be misleading. It would usually be possible, with the materials available, to design an organ that would perform the apparent function better. A heart would pump blood more smoothly, and with more reserves of power, if it were larger. Therefore, the fact that hearts are the size they are, and not larger, must imply to a Darwinist that there is some disadvantage in having a larger heart that would outweigh the hydraulic advantage. It may be, for instance, that the reserve of power is, in selective terms, not worth the metabolic cost of keeping a larger heart healthy.

We shall particularly attack this problem in chapter 4 when we discuss minimum mass analysis. However, it will be seen in that chapter that we have to assume certain functions, and then see whether bones seem to be performing them efficiently. What are these functions? Baldly, they are to be stiff enough, and not to break under either static or dynamic loading. There are some other functions, which we shall touch on later, but these are usually less important. It is perhaps natural to think that the *strength* of a bone, that is, the load it can bear before breaking, is its most important mechanical attribute. But it must surely be true that bones function

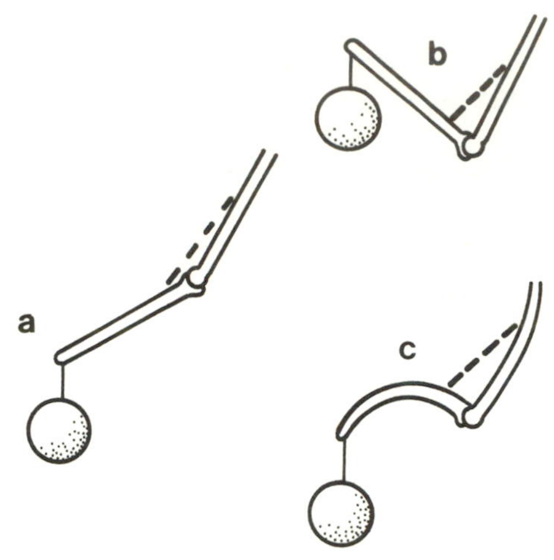

Fig. 1.1. The function of bone requires it to be stiff. (*a*) The arm is extended. (*b*) The muscle (interrupted line) shortens; the weight is lifted. (*c*) Flexible bones do not raise the weight.

mainly by not deforming appreciably under load. It is customary and correct to think of bones, especially limb bones, as acting as levers. The muscles, by their contraction and relaxation, alter the distance between fixed points on different bones, and so the bones are constrained to move. If the bones were floppy, they would not be constrained in the same way, and the movements of the muscles would be futile (figure 1.1). On the other hand, if the bone is stiff, but breaks, it becomes useless, and so the strength is of great but secondary importance.

The stiffness of a bone and its strength depend on two factors: the stiffness and strength of the bone material itself and also the build of the whole bone. By "build" I mean the amount of the bone material and how it is distributed in space. The question of the build of bones will be discussed in chapter 4. In this chapter we shall be concerned with the basis of the whole book: the mechanical properties of materials in general and the structure of bone itself. In the second chapter we can then talk about the mechanical properties of bone material.

1.2 Mechanical Properties of Materials

Because many readers may not be too clear about the mechanical matters to be discussed, I shall run through various basics now. Later, slightly more recondite mechanical concepts will be introduced. Only the minimum necessary for the understanding of this book will be given. Readers wishing to go further should read Gordon (1976) or Wainwright et al. (1982), and a good book on strength of materials.

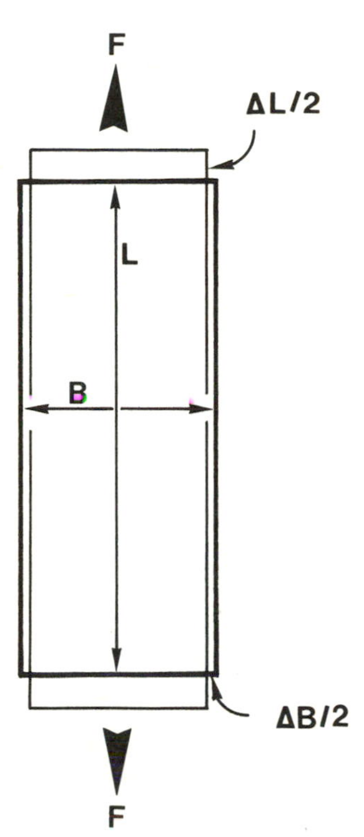

Fig. 1.2. Deformations produced by the application of load F to a block of material of original length L and original breadth B. The undistorted shape is shown by thick lines. The block gets longer by an amount ΔL, and narrower by an amount ΔB.

1.2.1 Stress, Strain, and Their Relationship

First: stress and strain. Consider a bar acted on by a force tending to stretch it (figure 1.2). If its original length is L, it will undergo some change in length, ΔL. It will also get thinner, and its breadth B will decrease by an amount ΔB. The proportional changes in length, $\Delta L/L$ and $\Delta B/B$, are called *normal strains*. They are often given the Greek letter ε. Note first that strains refer to changes in length in particular directions, and second that the strains tell us nothing, directly, about the forces causing them.

There is another way in which a material can distort: in such a way as to cause changes in *angles* between imaginary lines in the material (figure 1.3). A cubical block of material, which is part of a larger block, is acted on by a shearing force S. (For reasons of equilibrium, if S acts on the top and bottom surfaces, it must also act, in the way shown, on the right and left surfaces. This does not mean that forces have to be applied

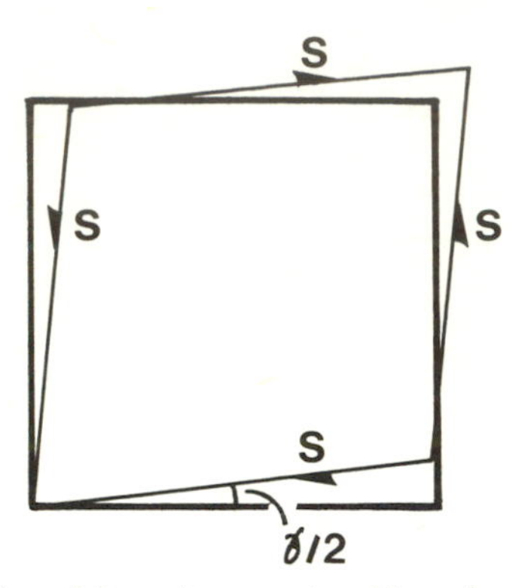

Fig. 1.3. Shear deformation γ produced by a shearing force S.

externally; if the cube is in a mass of material, the material takes care of it itself.) The shearing force will tend to distort the material in the manner shown in figure 1.3. The material is said to be showing *shear strain.* Shear strain is quantified as the change in angle undergone by two lines originally at right angles. The angle is measured in radians. It is frequently denoted by γ. In bone we are usually dealing with rather small strains, ε less than 0.005, γ less than 0.1. It may be helpful to show some loading situations in bones together with the kinds of strains that are produced (figure 1.4).

The basic idea of strain is fairly simple, that of stress less so. Stress is best thought of as the intensity of a force acting across a particular plane. Imagine an area in some plane in a body that is sufficiently small for the forces acting across it to be considered uniform, and to be represented by a vector F (figure 1.5). This vector can be resolved into three mutually perpendicular vectors, each in the direction of one of the three axes we have set up. The component normal to the plane, and parallel to the Z axis is F_z. The other two components are in the X–Y plane and are shear forces F_{zx} and F_{zy}. There is no necessary relation between these forces, their relative magnitudes depending on the angle F makes with the X, Y, and Z axes. If our little area has an area A, then we define the stresses acting on it thus: a normal stress F_z/A, and two shear stresses F_{zx}/A and F_{zy}/A.

It is most important to be clear that the stresses acting in a small region of a body depend upon the orientation of the plane we consider. To show this, look at a very simple loading system: a long bar loaded in tension by a force F. The bar has a square section of area A. The vector of the force acting on the cross section has no shear components, and so there are no shear stresses across the section. The only stress is the normal stress F/A (figure 1.6, middle). However, there is nothing in the situation

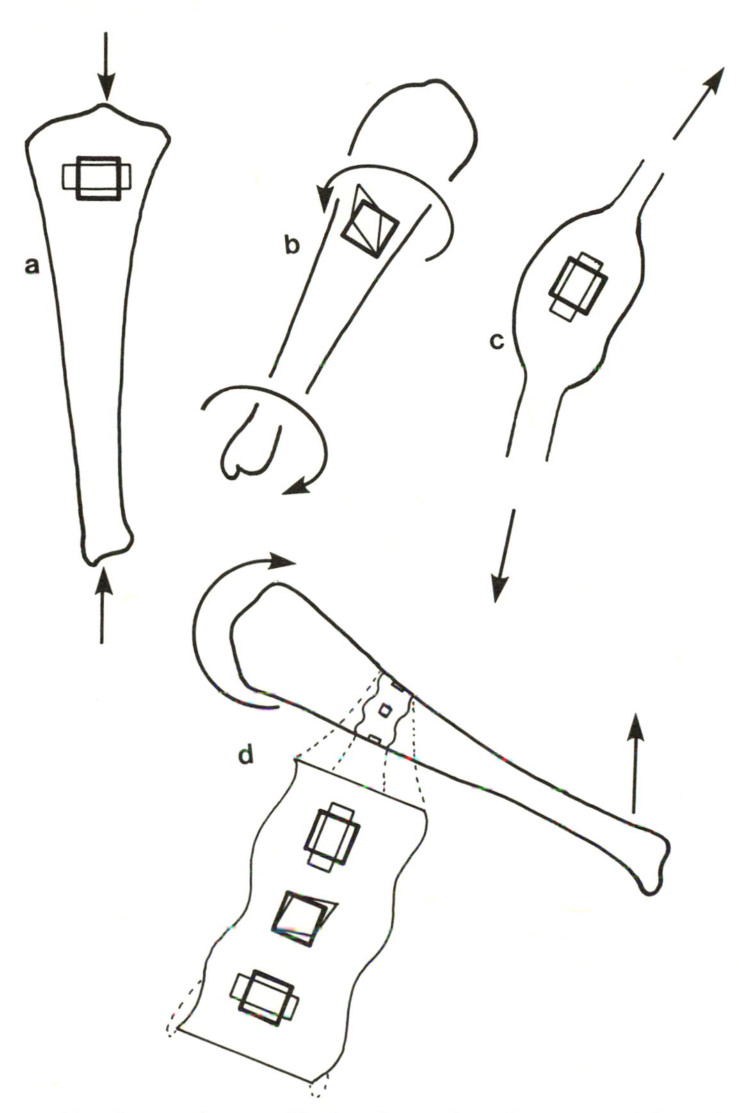

Fig. 1.4. Highly schematic diagrams of loads on bones, shown by arrows, and the resulting deformations. Thick lines are the undistorted shapes, thin lines the distorted ones. The deformations, much exaggerated, are: (*a*) compression; (*b*) torsion, producing shear; (*c*) tension, as in the patella; (*d*) bending. In (*d*), the deformations are shown on a piece of bone unwrapped from the whole bone. The lower part shows tension, the middle shear, and the top compression.

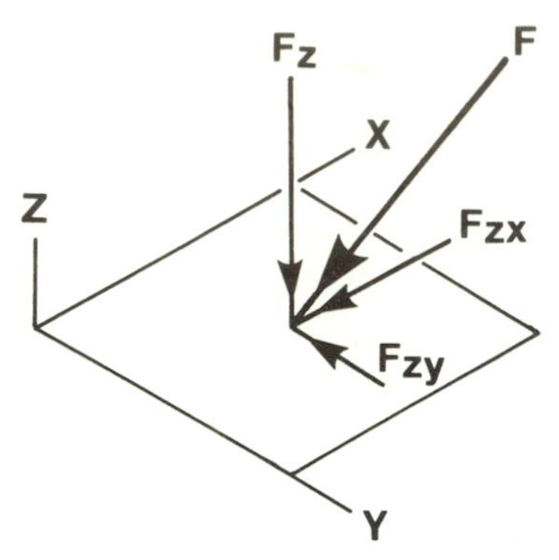

Fig. 1.5. A small area in the X–Y plane is acted on by a force F, which can be resolved into three forces at right angles, corresponding to one normal and two shear stresses acting on the area. The subscripts of the shear stresses refer to the axis normal to the plane and the direction in the plane in which the force is acting, respectively.

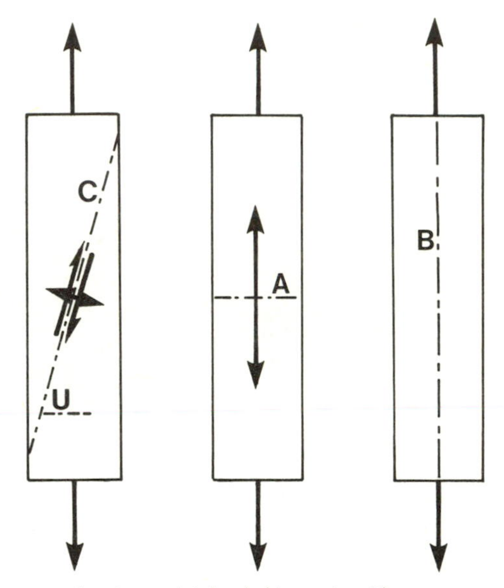

Fig. 1.6. A bar of cross-sectional area A is loaded in tension. The stresses across different planes are different. The different planes considered in the text are shown by the hatched lines.

that we have set up that necessarily makes the cross-sectional plane the important one. Consider instead a plane B lying right along the length of the bar (figure 1.6, right). There are no normal forces and so no normal stresses. Nor are there any shear stresses, because the forces do not act upward on one side of the plane and downward on the other, which is necessary if shear stresses are to occur. Therefore, there are no stresses acting across the B plane. What about planes at intermediate angles? Suppose the plane C is inclined at an angle U to A (figure 1.6, left). The force F can be resolved into a normal force $F \cdot \cos U$ and a shear force $F \cdot \sin U$. The area of C is greater than that of A; it is $A/\cos U$. The normal stress will be $(F \cdot \cos U)/(A/\cos U) = F \cdot \cos^2 U/A$. The shear stress will be $(F \cdot \sin U)/(A/\cos U) = F \cdot \cos U \cdot \sin U/A$. The normal stress is greatest when $U = 0°$, as seems obvious, and declines rapidly as U increases beyond about $U = 30°$. The shear stress is, less obviously, greatest when $U = 45°$, where it has a value half that of the maximum normal stress.

We have envisaged here a simple loading system. In fact, a small block of material in a body can be subjected to complex loading situations: tension on some faces, compression on others, and shear in various directions on all faces. Nevertheless, the stress across any plane in the blocks is always resolvable into just three, at right angles.

Now that we have some idea of stress and strain, we can consider the relationship between them. Suppose we load a small specimen of bone in tension until it breaks. We can do this by gripping it at the ends and pulling the grips apart at some constant rate, meanwhile measuring the load on the grips, which will be the same as the load on the specimen. There are devices for measuring the extension of a portion of the length of the test specimen, and the output can be displayed on an oscilloscope or similar instrument. The output will show a curve of load as a function of deformation. It will look something like figure 1.7. Starting from the origin, there is a part where the load varies linearly with the deformation. At the so-called yield point the curve flattens considerably, and now increasing deformation involves little extra load. Eventually the bone specimen will break.

A load-deformation curve is all very well for giving a rough idea of what is going on, but does not allow us to put any values on the variables that will tell us anything about the mechanical properties of the test piece. However, in a simple tensile test this is easily remedied. The force/(cross-sectional area) is the normal stress acting across the cross section, on which there are no shear stresses. Similarly, the increase in length of the part of the test piece being measured (the gage length), divided by the original length, is the strain. So the load-deformation curve can, with minimal arithmetic, be turned into a stress-strain curve. (Such simplicity does not hold for such loading systems as bending or torsion.)

The stress-strain curve gives a considerable amount of information. The linear part shows that there is a region where stress and strain are proportional to each other. For many biological materials this is not true. The stress-strain curve for cartilage, for instance, has no linear region; the curve is always curved, right from the origin. The less the deformation produced by a particular load, the stiffer the material. (Many people worry

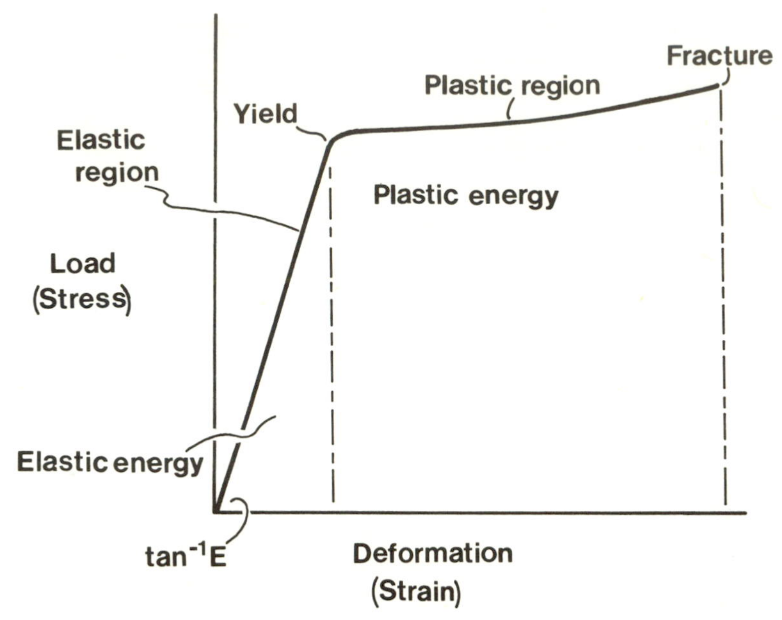

Fig. 1.7. A load-deformation curve of a bone specimen loaded in tension at a middling strain rate (0.01 s^{-1}). For a tensile specimen this is almost equivalent to a stress-strain curve. The area under the curve from the origin to the vertical interrupted line dropping from the yield point gives the amount of energy absorbed elastically. Plastic energy, between the two interrupted lines, is usually much greater. The angle made by the elastic part of the curve and the abscissa is the angle whose tangent is Young's modulus of elasticity.

that stress appears on the *Y* axis, and strain on the *X* axis, because they feel that the stress "causes" the strain. In fact, it is the strain that makes things happen in bone, as in other materials. For instance, one can show that bone breaks when the strain gets too big, not when the stress becomes too big.) The *modulus of elasticity*, or Young's modulus, is defined as stress/strain in the linear region of the curve of an ordinary tensile or compressive test. It is expressed in pascals (Pa), which are newtons per square meter, and in bone it has a value of roughly 2×10^9 Pa, or 20 GPa. It is usually denoted by *E*. The steeper the initial part of the curve, the greater the modulus of elasticity. In fact, the modulus of elasticity is the tangent of the angle made by the curve to the abscissa. Although Young's modulus of elasticity is often referred to as "the" modulus, we shall see that a number of independent moduli are needed to define the elastic behavior of a material. There is no reason why Young's modulus should have preference except that it is the most straightforward to measure and the easiest to understand.

It is found, near enough, that if a specimen of bone has been loaded only into the linear part of the curve, and the strain is then reduced, the force falls to zero when the strain falls to zero. If a material returns to its original state when loads are removed, it is said to behave *elastically*. If the material is also like bone in that the first part of the stress-strain curve is straight, it is said to be *linearly* elastic. Beyond the place where the curve bends over, something happens to the bone because beyond this point it is found that the stress will fall to zero before the strain falls to zero (figure 2.9). This residual strain is called plastic strain, and the material is said to be behaving *plastically*. As we shall see later, in chapter 2, so-called plastic behavior can be caused by a great variety of phenomena at the microscopic level; plasticity in bone has a different cause from plasticity in aluminum, for instance.

The point, or more properly the region, where the deformation changes from being elastic to being at least partially plastic is the *yield* point. Although a bone reaching this point still has far to go before it breaks, it is damaged to some extent once it enters the plastic region, and healing will have to take place if it is to be as it was before. It is, therefore, not surprising that what information we have shows that the loads placed on bones in life normally load them only into the elastic region. In the plastic region the bone extends with little extra stress. Therefore, the difference between the fracture stress, which is usually considered to be the strength of the bone, and the yield stress is usually rather small. So, if a bone were subjected to an ever-increasing load, it would make little difference to the load at failure whether the bone broke at the end of the linear part of the curve rather than at the end of the long plastic region. (Although this is true for tension, if the bone is loaded in bending, the plastic region does help to increase the load necessary to cause fracture.)

A material that breaks without showing any plastic deformation is said to be brittle. Characterizing a material as brittle gives no information about whether it is weak or strong in the sense of what load it can bear. However, the presence or absence of a reasonable amount of plastic deformation is an extremely significant feature of the mechanical properties of a material. Materials that show a reasonable amount of plastic deformation are very often "tough." A tough material is not weakened by small scratches on its surface, as brittle materials are. Bone may behave more or less brittlely according, among other things, to the direction in which it is loaded (figure 2.16). We shall see more of this later. In bone, toughness is important in determining how it breaks in life. What usually happens is that the bone is subjected to violence. This can be direct, as when a small boy falls off a wall onto his head, or it can be indirect, as when a galloping horse puts its hoof down a rabbit hole. In the latter case the bone is not

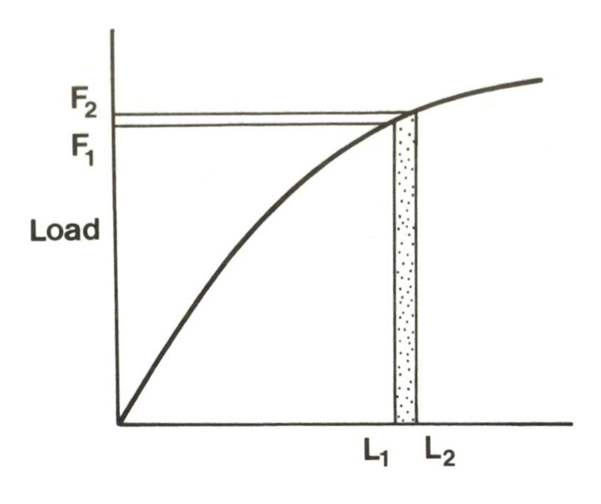

Fig. 1.8. A load-deformation curve. A load F_1 produces a deformation L_1; a load F_2, a deformation L_2. If the dotted area is narrow enough its area $\sim F_1(L_2 - L_1)$, and this respresents the work done in deforming the specimen from L_1 to L_2.

struck directly, like the boy's skull. Instead, the mass of the horse's body, moving on while the leg is stuck fast, produces a violent bending load on the leg. Sometimes, as in the spiral fracture of the tibia in a skiing fall, the bone is loaded by the ligaments that attach it to its neighboring bones, which may not themselves break. What happens in these cases is that the bone, and the surrounding tissues, are given a quantity of mechanical energy. This energy has to be dissipated in the bone and surrounding tissues without anything breaking. Now the energy absorbed by a material per unit volume is proportional to the area under the load-deformation curve up to the point considered. The reason for this is as follows: look at the load-deformation curve in figure 1.8. If we take a very narrow strip, its area is nearly $F_1(L_2 - L_1)$. In this part of the curve a force F_1 has acted over a distance $(L_2 - L_1)$, and therefore, the amount of work done is $F_1(L_2 - L_1)$. If the work done on the material is summed over the whole curve, the work done, or energy absorbed, per unit volume is the area under the stress-strain curve. So, the flat top of the curve of bone, although having little effect on the load at which the material breaks, can greatly increase the work that has to be done on it to break it.

1.2.2 Anisotropy

The information given by a stress-strain curve is considerable, but by no means exhausts what can be found out. Bone, like most real materials,

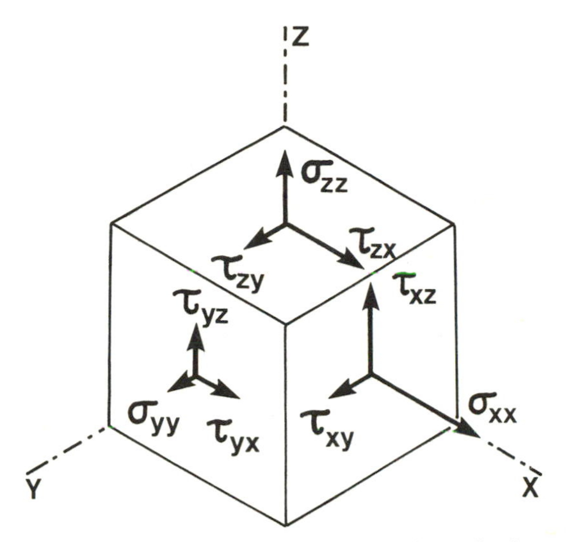

Fig. 1.9. Stresses that can act on the faces of a small cube.

is anisotropic. A material is anisotropic if its properties are different when measured in different directions. In this book we are really only interested in mechanical anisotropy, though anisotropy can occur in many physical properties. The stress-strain curve of an anisotropic material gives the value of Young's modulus in one direction, but it could be different in a different direction.

Consider a small cube of bone, in equilibrium under the forces acting on it. The forces on each face, divided by the area of the face, give the stresses (figure 1.9). As the cube is in equilibrium, we can ignore the equal and opposite normal forces on the opposite faces. For each force the first subscript refers to the face on which the force is acting, and the second subscript refers to the axis, parallel to which the force acts. Thus τ_{zy} refers to the stress caused by the force acting on the face *normal* to the Z axis that is acting in the y direction. It is a shear stress. σ_{xx}, σ_{yy}, and σ_{zz} act normally to the faces and are normal tensile or compressive stresses. Conventionally normal stresses are represented by σ, shear stresses by τ. There are notionally nine stresses that can act on the cube. In fact, the stresses are not all independent because, for reasons of equilibrium, $\tau_{yz} = \tau_{zy}$; $\tau_{zx} = \tau_{xz}$; and $\tau_{xy} = \tau_{yx}$. Nevertheless, there remain six independent stresses acting across the faces of the cube. Similarly, we can consider the strains in the cube. There will be three normal strains: ε_{xx}, ε_{yy}, and ε_{zz}. ε_{xx}, for instance, says how movement of a point in the x direction is dependent upon its position on the X axis. There are also shear strains such as γ_{xy}. γ_{xy} tells how movement of a point in the x direction depends upon its position on the Y axis. Fortunately, as with stress, not all the shear strains are independent; there are only six independent strains.

If the material is isotropic, the relationships between stress and strain are fairly simple. Young's modulus is given by $E = \sigma/\varepsilon$; the shear modulus is given by $G = \tau/\gamma$. If a stress σ acts in one direction and is the only stress acting on the specimen, the strain in this direction will be $\varepsilon = \sigma/E$, and there will be a strain $-\nu\varepsilon$ in directions at right angles to this. If the specimen is, say, a cube which is being stretched in one direction, it will shrink in others, and ν, Poisson's ratio, shows the ratio of the shrinking to the stretching. The shear modulus and Young's modulus are related by

$$E = 2G(1 + \nu) \text{ or } \nu = (E/2G) - 1.$$

For an isotropic material we need only determine two out of E, G, and ν to be able to characterize its elastic behavior completely. It is worth writing down the equations just to show how strains are, or are not, related to stresses in various directions.

$$\begin{aligned}
\varepsilon_{xx} &= (\sigma_{xx} - \nu(\sigma_{yy} + \sigma_{zz}))/E \\
\varepsilon_{yy} &= (\sigma_{yy} - \nu(\sigma_{xx} + \sigma_{zz}))/E \\
\varepsilon_{zz} &= (\sigma_{zz} - \nu(\sigma_{xx} + \sigma_{yy}))/E \\
\gamma_{yz} &= \tau_{yz}/G \\
\gamma_{zx} &= \tau_{zx}/G \\
\gamma_{xy} &= \tau_{xy}/G
\end{aligned}$$

Note that the shear strains depend only on the appropriate shear stresses, but that the normal strains depend on all three normal stresses. Crudely, if a cube is being pulled in the x direction, it will tend to elongate in the x direction, and shrink in the y and z directions. However, if it is also being pulled in the y and z directions, these will not shrink so much, and so the cube will not elongate so much in the x direction.

Most materials, including bone, are not isotropic. That is, their elastic behavior will vary according to the direction of loading. More information is needed before the elastic behavior can be described completely. For anisotropic materials a more complex notation than that used above is needed. A compliance, s, and a stiffness, c, are defined by $\varepsilon = s\sigma$ and $\sigma = c\varepsilon$.

If a stress σ_{ij} is applied to a material, the resulting strain ε_{ij} has several components which are related to the stress via the compliance (s) constants. A material loaded in compression in one direction will shorten in that direction but may squeeze out sideways to different extents in the other directions; it may also shear in various ways, so that a loaded body that started as a cube may have a very complex shape after loading.

However, with the small strains assumed by theory, the sides of the cube will remain plane, and its edges straight.

The state of strain in a material after stresses are applied or the state of stress in a material after undergoing strains can be determined by matrix algebra if enough values (c_{ij} or s_{ij}) of the stiffness of compliance matrix are known (Nye, 1957).

For a material that has the most complicated symmetry possible, there are twenty-one independent elastic properties that need to be evaluated. Recently it has been suggested that Haversian bone at least can be considered to be transversely orthotropic, or to show hexagonal symmetry. For such a material there are five independent stiffnesses or compliances to be evaluated. The matrix relations between stresses and strains are shown here:

$$\begin{bmatrix} \varepsilon_a \\ \varepsilon_b \\ \varepsilon_c \\ \gamma_d \\ \gamma_e \\ \gamma_f \end{bmatrix} = \begin{bmatrix} S_h & S_i & S_j & 0 & 0 & 0 \\ S_i & S_h & S_j & 0 & 0 & 0 \\ S_j & S_j & S_k & 0 & 0 & 0 \\ 0 & 0 & 0 & S_l & 0 & 0 \\ 0 & 0 & 0 & 0 & S_l & 0 \\ 0 & 0 & 0 & 0 & 0 & \frac{1}{2}(S_h - S_i) \end{bmatrix} \times \begin{bmatrix} \sigma_a \\ \sigma_b \\ \sigma_c \\ \tau_d \\ \tau_e \\ \tau_f \end{bmatrix}$$

I have given arbitrary subscripts here, because the subscripts used in the literature follow a logical but complicated convention, which is unnecessary for the present simple discussion. The matter is discussed in general by Nye (1957), and its application to bone is discussed very clearly by Yoon and Katz (1976a,b).

The point is that, knowing the values of the six possible independent stresses on a sample of bone (three normal and three shear), one can calculate the six possible strains resulting, if the compliance values S_h, S_i, S_j, S_k, and S_l are known. This can be done by matrix multiplication. (Computationally there are complications if the stresses or strains are in arbitrary directions, but the methods of dealing with this are well known.) An equivalent matrix can be constructed for determining stresses if the strains are known (Yoon and Katz, 1976a).

1.2.3 Technical Moduli

As I shall discuss briefly below, the values of the stiffness or compliance matrices can be determined by observing the behavior of sound waves in bone. However, the great majority of the mechanical testing of bone in the past, and at present, has used standard load-deformation curves. There are advantages and disadvantages in straightforward mechanical testing. Mechanical testing is fairly easy and allows one to compare the elastic, postyield, and fracture behavior of a specimen in a few tenths of a second. On the other hand, it is impossible to determine all the elastic properties on one specimen as can be done, given careful experimentation, by measuring sound waves. So, the question arises, what is the relationship between the "technical moduli" (the values for Young's moduli measured in various directions, shear moduli, and the bulk modulus) and the values in the compliance and stiffness matrices? They are, of course, related and, if bone can be considered to behave with hexagonal symmetry, the relationships are fairly straightforward. I quote them here, as taken from Yoon and Katz (1976a), but using the arbitrary values for the suffixes. Z is the cosine of the angle between the axis of symmetry, roughly along the long axis of the bone, and the direction in which the property is measured.

Young's modulus: $1/E = (1 - Z^3)S_h + Z^4 S_k + Z^2(1 - Z^2)(S_j + S_l)$

Shear modulus: $1/G = S_l + (1 - Z^2)(S_h - S_i - \frac{1}{2}S_l)$
$\qquad + SZ^2(1 - Z^2)(S_h + S_k - S_y - S_i)$

Bulk modulus: $1/K = S_k + 2(S_h + S_i + 2S_j)$

(Note that the bulk modulus does not depend on the direction of loading, and therefore there are no Z terms.) If the direction of loading is along the axis of symmetry, $Z = 1$, and so $1/E = S_k$, and $1/G = S_t$, and the relationship between the compliance matrix and the technical moduli becomes pleasingly simple.

It is certainly true that some types of bone do *not* show hexagonal symmetry. Reilly and Burstein have shown that the values of Young's modulus are different in three orthogonal directions in a type of bone known as "fibrolamellar." This type of bone must be described by more than five compliance or stiffness properties. I shall discuss the implications of this below.

1.2.4 Fracture and Toughness

If a material is stiff enough, the next question is whether it is strong enough to stand up to the conditions of service. In the last century materials testing had advanced sufficiently for it to be possible to find out, with great repeatability, the tensile and compressive strengths of materials. It nevertheless was apparent that something was wrong. Bridges and, more particularly, ships were breaking up under conditions that, it could be shown, imposed stresses in the material that could easily be borne by the material when it was tested in the laboratory. The reason

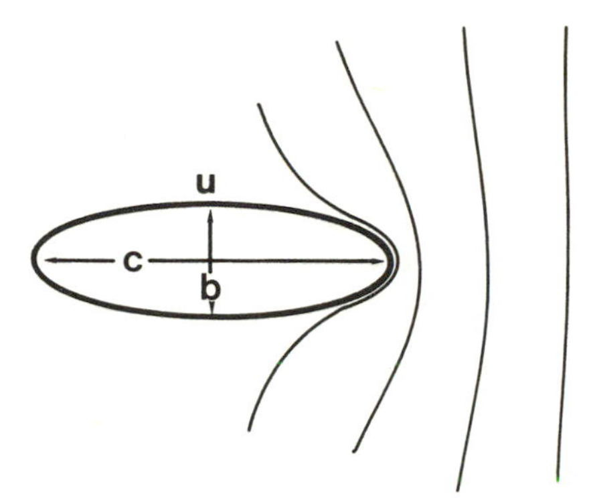

Fig. 1.10. Stresses near the tip of an elliptical cavity. The inverse of the distance between the lines of force is a measure of the stress. It is very high near the tip. The region marked *u* is nearly unstressed.

for this was almost completely obscure until great light was shone on the matter in 1920. From a classic paper written by A. A. Griffith a whole subscience, called "fracture mechanics," has grown.

The problem is this. It is possible, though difficult, to calculate for simple crystals the force theoretically needed to separate two planes of atoms from each other. However, compared with the results of laboratory tests the answers appear to be too high, by some orders of magnitude. Inglis (1913) supposed, as had others, that cracks developed from the corners of hatches and rivet holes because of stress concentrations. Figure 1.10 shows an elliptical hole in a large plate loaded in tension. The lines represent, roughly, the way the force is distributed around the hole. The force is not distributed evenly; it is more concentrated near the ends of the hole. There is, therefore, a higher stress and strain nearer the ends of the hole than farther away from it. Inglis took this relatively simple case and showed that, if the general stress was σ, then in a very large plate the stress in the material at the ends of the hole is $\sigma(1 + 2c/b)$, where c is the length of the ellipse normal to the stress and b is the length of the ellipse in line with the stress (figure 1.10). There is a "stress concentration" of $(1 + 2c/b)$. If the ratio c/b is very large, that is, if the ellipse is like a crack, the maximum stress $\sigma = (2c/b)$. It is clear that cracks are very potent producers of stress concentrations. However, these results were not taken seriously for some time for two reasons. The first was that the formula for the stress-concentrating effect of a hole was merely a ratio; there were no sizes in it. Therefore, the length of the ellipse or crack should make no difference. But it was a matter of vague experience that long cracks were more dangerous than short ones. The second was that if, on the other hand, one assumed that a sharp crack had a length c, which might be a few millimeters, and b was roughly the interatomic spacing, then the values of stress concentration would be so high that everything that was not perfectly smooth should fall apart under its own weight. So the analysis of Inglis, though in fact correct, seemed to be of no practical help.

This situation was remedied in 1920 by Griffith. He made the following analysis, in a more rigorous form, of course. Suppose we have a small crack in a large plate, which is subjected to a general stress σ. Suppose also, that the material is linearly elastic and completely brittle. Consider the energy changes that must occur if the crack is to spread. The surfaces of the crack have a certain amount of energy, U_S. Surface energy is related to surface tension, being caused by the surface atoms having fewer near neighbors and, therefore, less negative bond energy. Any increase in the length of the crack must result in an increase in the total surface energy of the plate, directly proportional to the increase in crack length. The plate also has strain energy, U_ε. Strain energy is the potential energy a material possesses because its interatomic bonds are strained. It can be released to do work, as in the uncoiling of a watch spring. In a linearly elastic material the amount of strain energy at any strain is equivalent to the area under the stress-strain curve up to that strain. Finally, in this list of energies, any displacement of the edges of the plate, caused by the load on them, will make the loads do work W_L, which will be negative if the work is done on the plate. The total energy of the plate, U, can be expressed as $U = -W_L + U_\varepsilon + U_S$.

Under what conditions will the crack spread? Let us suppose that it does, by some tiny amount. The relevant load-deformation curves are shown in figure 1.11. The plate has been loaded along OA up to a load F, equivalent to a general stress σ, and to a deformation D. The crack extends a little. This will make the plate more compliant, because there is less material at the level of the crack to bear the load, and so the plate will extend a little amount DC. The force remains the same, so the work done on the plate is force times distance $= F \times DC$, given by the area of the rectangle $ABCD$. The elastic strain energy of the plate *was* OAD, and is now OBC. If DC is small enough:

$$OBC - OAD \text{ (the change in strain energy)} = ABCD - ABO = ABCD/2$$

Therefore, the work done on the plate $= ABCD$, the extra strain energy is $ABCD/2$, and so there is $ABCD/2$ available for the surface energy of the crack.

Similar geometrical arguments show that if the crack extends under the "fixed-grip" condition, that is, without any increase in the value of OD, then the strain energy of the plate decreases, rather than increases. However, the same amount of strain energy is made available for the surface energy of the crack.

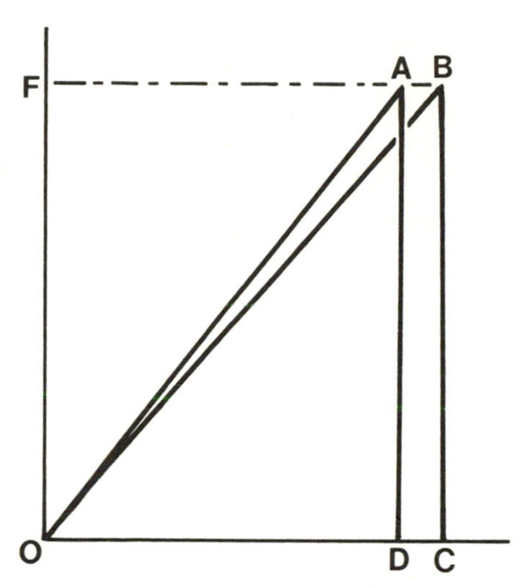

Fig. 1.11. Changes in the strain energy of a specimen with a spreading crack. For explanation, see text.

The critical question is: how can the strain energy released be expressed in terms of the surface energy needed? So far we have merely said that the extension of the crack makes the plate more compliant, but not how much more. It turns out that we do not necessarily need to know. Consider the state of stress near the tip of a crack (figure 1.12). At the tip the stress is very high; far from the tip it is σ; behind the crack tip there is a region that is effectively relieved from stress. As long as the crack is small relative to the plate, the *shape* of the area relieved from stress, which has consequently given up its strain energy, is not affected by the *length* of the crack. Therefore, the area of this region, or volume if we consider the plate to have thickness, is proportional to the *square* of the crack length.

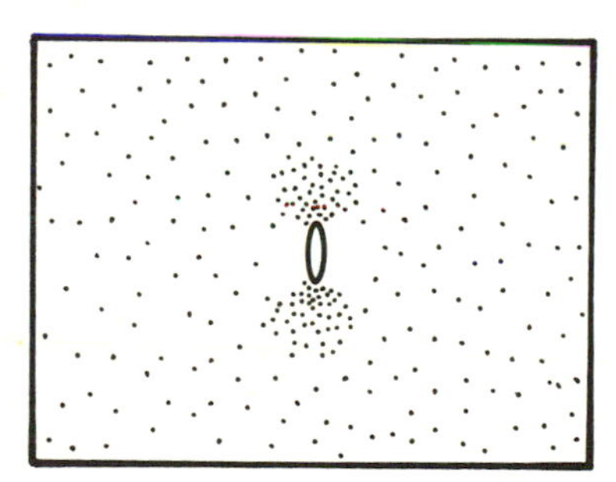

Fig. 1.12. General state of stress around a crack in a large plate. The spacing of the dots indicates the level of stress.

Griffith calculated the loss of strain energy caused by the presence of a crack of length $2c$ to be: $\pi c^2\sigma^2/E$, where σ is the general state of stress, and E is the modulus of elasticity of the material. What is the surface energy of the crack? If the surface energy per unit length is γ, the total surface energy is $4\gamma c$. (The factor of four comes in because the crack length is $2c$, and it forms two new surfaces as it extends.) Surface energy is conventionally given the letter γ; this is not the same as shear strain, which is also called γ, of course. Therefore, when the crack is just on the point of spreading, the strain energy released by a differentially small increase in crack length must just equal the surface energy needed:

$$d4\gamma c/dc = d\pi\sigma^2c^2/Edc$$

Notice that the release of strain energy is proportional to the square of the crack length, whereas the surface energy needed is proportional only to its length. From the equation above $\sigma = \sqrt{2\gamma E/\pi c}$. Since there are some uncertainties in the analysis, it can be taken that $\sigma = \sqrt{\gamma E/c}$. The value of stress given by this analysis is called the Griffith stress. The important point about it is that the stress needed to cause a crack to spread is less if the crack is longer. If we rearrange the equation, this point is brought out: $c = \gamma E/\sigma^2$. Thus, for a given general stress σ, there will be some length of crack c_{crit}, which will be sufficiently long to be able to spread.

We have used here a particular loading situation—tension on a plate—but the basic theory is applicable to all kinds of tensile loadings. Griffith tested his ideas using glass fibers, for which it was easy for him to get a reasonable idea of γ, the surface energy, and the tests worked out well. However, glass is rather a special material, just because it is so brittle. For glass, and materials like it, it is almost true that the only energy that has to be fed into the material near the crack tip is that required for surface energy. However, many materials, including bone, show plastic deformation, and we need to consider this.

Plastic deformation, whose presence is shown by a flattened region in the stress-strain curve, can be caused by various kinds of events at the molecular or microscopic level. In metals it is often caused by the movement of dislocations. These are faults in the uniform crystalline structure of the metallic lattice, which will move if the material is stressed sufficiently. In other materials plastic deformation may be produced by the appearance of tiny ruptures in the material, which allow much more deformation than would be possible if the movement were caused by the increasing of interatomic distances. These ruptures may heal spontaneously, the bonds reforming with new partners, though this is not usually the case.

These processes require a force, and as they involve movement, work has to be done to bring them about. They absorb energy. The plastic, energy-absorbing deformation that occurs when a material is deformed has two components. One is the plastic deformation that occurs generally in the body of the material and has nothing especially to do with the crack that may eventually break it in two. The other part is specifically associated with the crack and is a necessary concomitant of its travel. Even in glass, which is remarkably brittle and shows no general plastic deformation, a few layers of atoms under the fracture surface become irreversibly reordered, although some materials, such as alumina Al_2O_3, seem to develop cracks that are atomically sharp (Lawn, Hockey, and Wiederhorn, 1980).

Consider first the plastic deformation associated with the crack tip itself. Every differentially small increase in the crack length will require surface energy and the energy required to bring about plastic deformation. These are both linearly related to crack length, and can be combined as a term W. The extent to which W exceeds γ depends on the material and may be very large. γ is roughly 1 $J\,m^{-2}$ for the great majority of brittle materials. The value for W in bone is in the region of $2 \times 10^3\ J\,m^{-2}$. It is clear that the surface energy term in bone is trivial compared with the energy-absorbing events going on just below the surface.

Another effect of plastic deformation may be to blunt the crack. In a material with a low Young's modulus, crack blunting will take place even though the material remains elastic. One can see this easily by cutting a small slit in the cuff of a rubber kitchen glove and then pulling at right angles to the slit. The very sharp crack becomes completely rounded. However, in a stiff material like bone this macroscopic blunting does not happen. Nevertheless, the crack may be blunted by local plastic deformation and cause the local stress to fall significantly.

For a crack to spread, two conditions must be met. The thermodynamics must be right; that is to say, the strain energy released by the spread of the crack must be sufficient to produce enough surface and subsurface energy per unit length. Furthermore, the crack must have a sufficiently high stress at the tip to break the bond right at the crack tip. Plastic deformation has an effect on both these conditions, and tends to make it difficult for a preexisting crack in a material to spread. This is of the utmost importance in any real material, which is likely to have a host of tiny cracks on the surface.

Another effect of the plastic behavior of a material is that if the "worst" crack in the material can be prevented from spreading until the stress at which general yielding takes place is reached, then the material may undergo much more plastic deformation. Figure 1.13 shows this. Suppose

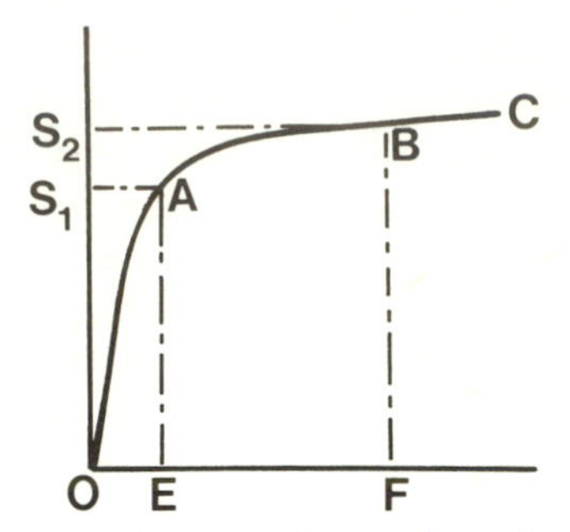

Fig. 1.13. A small increase in general stress at fracture (from S_1 to S_2) may produce a great increase in the energy absorbed (from *OAE* to *OABF*).

the stress-strain curve of the material has the shape $OABC$ and the specimen has a crack in it which will spread at a general stress in the specimen of S_1. The energy absorbed by the material, per unit volume, before the crack spreads, is OAE. If local plastic deformation near the crack tip, or just beneath the fracture surface, can raise the stress at fracture to S_2, the amount of energy absorbed before failure will be proportional to the area $OABF$. Thus a rather small increase in the general *stress* at failure will cause a great increase in the energy absorbed. This will be of great help during impact loading.

A further complication must be mentioned before these arguments can be applied to bone. We have so far seen that the energy-*absorbing* processes are associated with the growth of the fracture surface. The energy-*releasing* process is the release of strain energy. The Griffith equation assumes that all the energy put into the material in straining it is recovered when it becomes unloaded. But this is, in general, not the case. Some of the strain may be plastic strain, and some may be viscoelastic strain. In the former case some of the energy used up in straining the material is irrecoverably locked up in it, and cannot be released. Some of it will have been lost as heat. In the latter case, although the strain energy is eventually recoverable, it is so only if the unloading takes place slowly relative to the loading. Often, of course, the unloading tends to take place much more rapidly than the loading. In such cases, the energy put into the material is not all available to help the crack spread. The importance of accounting for energy recovery has been emphasized by Andrews (1980).

1.2.5 Fracture Mechanics

The situation of fracture in the real world looks so complicated, with so many apparent variables, both of the material and of the geometry of the situation, to account for that it might seem difficult to make any predictions about it. However, the new area of research "fracture mechanics" attempts to accomplish just this, by doing two things. One is to characterize the tendency of materials to fracture, in terms independent of the actual geometry of the loading system. The other is to determine how the actual geometries will affect the fracture behavior. (These are two sides of the same coin, of course.) For instance, if we are carrying out a tensile test on a sheet of length L, breadth B, and thickness T, with a transverse flaw in it of length f, the length of the flaw and the ratio of B to T are both important in determining at what load the sheet will rupture. The standard strength of materials approach would be merely to consider the cross-sectional area of the sheet at the level of the crack: $T \times (B - f)$. The length of the flaw appears only in its effect on the cross-sectional area, which may be trivial compared with its actual effect in reducing the strength of the specimen. The fracture mechanics approach, on the other hand, if successful, allows the engineer to know how dangerous will be flaws of particular size and orientation in any structure he may design.

There are currently two main ways of characterizing materials and cracks in them: (*a*) the crack will spread when the energy released from the material as the crack spreads exceeds the energy needed to extend the crack (the approach discussed above), which is called the critical strain energy release rate G_c; and (*b*) for a given geometry, the crack will spread when the stress at the crack tip reaches a critical value that overcomes the cohesive strength of the atoms just ahead of the crack, which is called the critical stress intensity factor, K_c.

These two properties might not seem to be related to each other, but for fairly brittle materials they are quite simply related. For instance, for a thick plate loaded in tension, $G_c = K_c^2(1 - \nu^2)/E$, where ν is Poisson's ratio.

The ways in which these properties are determined are fairly complex and are not suitable for small specimens. This makes them inappropriate for many bones. The applicability of fracture mechanics to bone is, as yet, rather limited, and it is not, I think, appropriate to say much about the methods used, although I shall discuss some results below. It is certain that the fracture mechanics approach will become important in the next ten years. The interested reader may care to consult Lawn and Wilshaw (1975) for a clear and suitably rigorous introduction to the subject.

There is another method of analysis that gives some idea of the toughness of bone. This is merely to prepare a specimen in such a way that it does not fail catastrophically when it breaks. Such a specimen is a beam loaded in three-point bending, with a deep notch in the middle of the beam (Tattersall and Tappin, 1966). The area under the load-deformation curve shows how much work was done in breaking the specimen in two, and the fracture surface area can be measured. In this way the "work of fracture," W, can be calculated. The calculated work of fracture of a material does vary somewhat with specimen size and geometry, but nevertheless, the value can be quite useful in comparisons between types of bone that vary considerably in their toughness. The critical point is that the higher the values of G_c, K_c, and W, the tougher the material is, and the less sensitive it is to the presence of flaws and cracks.

After this introduction to the mechanical properties of solids, we turn to the structure of bone itself.

1.3 The Structure of Bone

Throughout this book I shall be showing that the structure of bone tissue, and of whole bones, makes sense only if its function, particularly its mechanical function, is known or guessed. However, in this first chapter I shall deal only with the structure of bone, leaving all discussion of function until later. Much of the subject matter will be familiar to many readers, but I suspect not all to everyone. Only when the basic structure is known can we start to assess the significance of this structure.

Bone is such a complex structure that there is no level of organization at which one can truly be said to be looking at "bone" as such. I shall discuss bone starting at the lowest level, and working up to a brief description of the variety of shapes one sees in whole bones.

1.3.1 Bone at the Molecular Level

At the lowest level bone can be considered to be a composite material consisting of a fibrous protein, collagen, stiffened by an extremely dense filling of calcium phosphate. There are other constituents, notably water, some extremely ill-understood amorphous polysaccharides and proteins and, in many types of bone, living cells and blood vessels.

Collagen is a structural protein found in probably all metazoan animal phyla, but only in the vertebrates does it undergo a wholehearted transformation into a mineralized skeletal structure, although some soft corals have traveled some way along the road. Unmineralized collagen is also found in the vertebrates in skin, tendon, ligament, blood vessel walls, cartilage, basement membrane, and in connective tissue generally, in those circumstances where the material is not required to be extremely extensible. Collagen from different sites often has different amino acid compositions. The collagens of skin, tendon, dentine, and bone share the same type of composition. The protein molecule "tropocollagen," which aggregates to form the microfibrils of collagen, consists of three polypeptides of the same length—two the same, one different, in amino acid composition. These form on ribosomes, are connected by means of disulfide cysteine links, and leave the cell. Outside the cell the ends of the joined polypeptides are split off, the lost part containing the disulfide bonds. The three chains are by now held together by hydrogen bonds in a characteristic left-handed triple helix.

The primary structure of the polypeptides in the tropocollagen molecule is unusual, great stretches of it being repeats of glycine-proline-X, with X being hydroxyproline or some amino acid. The imino acids proline and hydroxyproline are unlike ordinary amino acids in that the nitrogen atom is included in the side chain as part of a five-membered ring. The effect of this is to reduce the amount of rotation possible between units of the polypeptide. It also prevents α-helix formation and limits hydrogen bond formation. These constraints result in a rather inflexible polypeptide (Dickerson and Geis, 1969).

The tropocollagen molecules have an inherent tendency to combine together to form microfibrils, by bonding head to tail, not with molecules in the same file, but with molecules in neighboring files. The tropocollagen molecules are 260 nm long, and the molecules alongside each other are staggered by about 1/4 of their length. There is a gap between the head of one molecule and the tail of the next, the "hole region," and, because many tropocollagen molecules are stacked side by side, these gaps and other features of the molecules produce a characteristic 67 nm periodicity. The whole microfibril becomes stabilized by intermolecular crosslinks. Microfibrils aggregate to form fibrils. Woodhead-Galloway (1980) gives a short, clear introduction to the structure of collagen.

Impregnating the collagen, in fully mineralized bone, is the bone salt, which is some variety of calcium phosphate. The precise nature of the mineral of bone, both its chemistry and its morphology, is still a matter of some dispute. The problem is that the mineral in bone comes in very small lumps having a very high surface-area-to-volume ratio. This makes it reactive, and so most preparative techniques used for investigating it, e.g., drying under vacuum for electron microscopy, are likely to cause alterations from the living state. There is agreement that some of the bone mineral is the version of calcium phosphate called hydroxyapatite, whose unit cell (the smallest part of a crystal that is repeated uniformly throughout a crystal) contains $Ca_{10}(PO_4)_6(OH)_2$. However, there is some evidence that initially the calcium and phosphate are deposited as an "amorphous" calcium phosphate, that is, as mineral whose short range order is so limited that it does not show as a coherent pattern in X-ray crystallographs. If the mineral is deposited in such an amorphous form, it quickly becomes transformed more and more into a form like apatite. Wheeler and Lewis (1977), for example, suggest that the X-ray diffraction pattern, which does not indicate a very long range order, should be interpreted as showing that the mineral is paracrystalline; it is neither amorphous nor perfectly crystalline over any long distance. They suggest that the distance over which a block of mineral can be considered to be a single crystal may vary from about 7 nm to 22 nm. However, the mineral will be in a larger lump than this. Driessens, van Dijk, and Borggreven (1978) suggest that the core of the lumps of mineral is apatite but that this is surrounded by a layer of octocalcium phosphate (unit cell $Ca_8H_2(PO_4)_6 \cdot 5H_2O$), which is transformed, autocatalytically, into hydroxyapatite. The question of the composition of bone mineral and of its mode of deposition is reviewed comprehensively by Glimcher (1976).

The position of the mineral relative to the collagen fibrils, as well as its shape, is becoming clearer, though there is still a fair amount of controversy. The main argument is whether the apparently crystalline mineral, which can be seen in electron micrographs, is needle shaped or plate shaped. Ascenzi et al. (1978) claim that the mineralization process starts off with small granules, about 4.5 nm across. These coalesce or grow into needles about 40 nm long. Jackson, Cartwright, and Lewis (1978), by using dark field imaging in the electron microscope, show, on the other hand, that the mineral is probably in platelike lumps about 35 nm across and 4 nm or so thick. It seems that the view that the mineral is platey is gaining general acceptance.

The mineral is apparently initially deposited in the holes between the heads and the tails of the tropocollagen molecules. This results in the initial mineralization having a 67 nm periodicity (Berthet-Colominas, Miller, and White, 1979). It is highly probable that in some way the particular conformation of the collagen molecule allows it to act as a nucleation site, permitting the precipitation of lumps of mineral which, without the presence of the energetically favorable sites, could not come out of solution. (See Glimcher, 1976, for a review.) Later, the mineral is deposited all over the collagen fibrils, and to some extent within them. The precipitation is not random; one of the long axes of the mineral plates is always fairly well aligned with the collagen fibrils. Finally, mineral is deposited between the fibrils, in the amorphous and rather tenuous ground substance.

1.3.2 Woven and Lamellar Bone

Above the level of the collagen fibril and its associated mineral, mammalian bone exists in three usually fairly distinct forms: woven bone, lamellar bone, and parallel-fibered bone.

Woven bone is usually laid down very quickly, most characteristically in the fetus and in the callus which is produced during fracture repair. The collagen in woven bone is fine fibered, 0.1 μm or so in diameter, and is oriented almost randomly, so it is difficult to make out any preferred direction over distances greater than a few μm. As in most bone, woven bone contains cells (osteocytes) and blood vessels. Rather frequently the spaces surrounding the blood vessels are quite extensive and differ in this way from those in lamellar bone. The osteocytes are imprisoned in cavities (lacunae) and connect, via delicate processes in channels (canaliculi), to neighboring osteocytes and ultimately to blood channels.

Lamellar bone is more precisely arranged, and is laid down much more slowly than woven bone (Boyde, 1980). The collagen and its associated mineral are arranged in sheets (lamellae) about 5 μm thick. The final degree of mineralization of lamellar bone is less than that of woven bone. The collagen in any particular lamella has an orientation almost normal to the short axis of the lamella; that is, the fibrils lie within the plane of the lamella, rather than passing from one to the next. Furthermore, the fibrils tend to be oriented in one direction within the plane of the lamella. Indeed, some workers suggest that the collagen fibrils in a particular lamella are all oriented in the same direction (Ascenzi et al., 1978). However, this is probably not the case: in many lamellae the fibrils are in small "domains" about 30–100 μm across. Within a domain the fibril orientation is constant, but it changes, within one lamella, from one domain to the next (Boyde and Hobdell, 1969; Frasca, Harper, and Katz, 1977). The collagen in lamellar bone forms branching bundles, 2–3 μm in diameter, much thicker than in woven bone. The osteocyte lacuna in lamellar bone is an oblate spheroid, about five times longer than it is across. The shorter axis of each lacuna is oriented within the short axis of the lamella. The division between one lamella and the next is abrupt, and there seems to be a sheet of "interlamellar bone," about 0.1 μm thick, with a rather high mineral content and little collagen, between the pairs of lamellae. Although all the collagen fibrils in one lamella do not all point in the same direction, it does seem, on passing from one lamella to the next, that the preferred direction of the fibrils changes. This is seen very clearly in fracture surfaces that cut across lamellae. As we shall see in the next chapter, this change in direction has mechanical consequences.

Parallel-fibered bone is structurally intermediate between woven bone and lamellar bone (described by Enlow [1969] and by Ascenzi, Bonucci, and Bocciarelli [1967]). It is quite highly calcified, but the collagen fiber bundles are not so totally random in their arrangement as those in woven bone. It is found in particular bones in particular situations.

1.3.3 Fibrolamellar and Haversian Bone

In mammals there are, at higher levels of structure, four main types of bone.

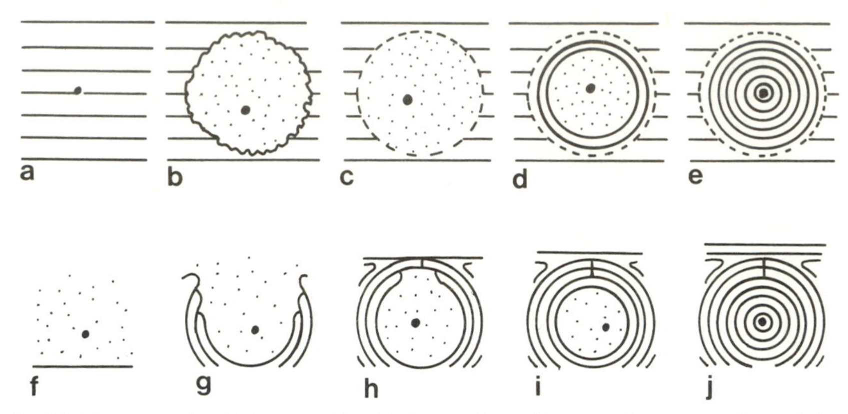

Fig. 1.14. Two ways in which concentric cylinders are formed in bone, shown very schematically. (*a–e*) the formation of a secondary osteon, or Haversian system. (*a*) A blood vessel is surrounded by preexisting bone. (*b*) Osteoclasts resorb bone around the blood vessel (tissue fluid indicated by dots). (*c*) The edges of the cavity are neatened off, and a cement sheath (shown by interrupted line) is laid down. (*d*) The cavity begins to be filled in by lamellar bone. (*e*) The mature secondary osteon. Note that the course of the preexisting lamellae is interrupted by the osteon. (*f–j*) The formation of a primary osteon. This can take place only on a growing surface. The lamellae are not interrupted by the osteon, and there is no cement sheath.

Woven bone can extend uniformly for many millimeters in all directions. Such a large block is found only in very young bone (of rather large mammals) and in large fracture calluses.

Lamellar bone may also extend for some distance. Usually, in mammals it does so in circumferential lamellae, initially wrapped around the outside or the inside of bones. There are blood channels in such bone, but they do not much disturb the general arrangement of the lamellae.

Lamellar bone also exists in a quite separate form: *Haversian systems*, or secondary osteons. In the English-speaking world the British incline toward the usage "Haversian systems," whereas Americans prefer "secondary osteons." It matters not which is used, except that it is critically important to distinguish primary osteons, which will shortly be described, from secondary osteons.

Haversian systems form like this: the bone around a blood vessel is eroded away by the action of bone-destroying cells, the osteoclasts. This leaves a cavity of diameter about 100 μm. The walls of the cavity are made smooth, and bone is deposited on the internal surface in concentric lamellae (figure 1.14). The end result is arranged like a leek, *Allium porrum*, with clearly distinguishable cylindrical layers, except that there is a central cavity in the Haversian system, which contains one, or sometimes two, blood vessels. There is an outer sheath to the Haversian system, called the cement sheath, which seems to be made of highly calcified mucopolysaccharide with virtually no collagen (Ortner and von Endt, 1971; Frasca, 1981). Hardly and canaliculi cross the cement sheath, so any cells outside are cut off metabolically from the blood vessel in the middle of the Haversian system.

A fourth characteristic type of mammalian bone is *plexiform*, or *laminar*, or *fibrolamellar* bone (Enlow and Brown, 1957, 1958; Currey, 1960; de Ricqlès, 1977). It is found particularly in large mammals, whose bones have to grow in diameter rather quickly. Lamellar bone cannot be laid down as fast as woven bone. If a bone has to grow in diameter faster than lamellar bone can be laid down, other bone must be laid down instead. For reasons that will be discussed in the next chapter, woven bone is almost certainly inferior to lamellar bone in its mechanical properties. The undesirable mechanical results of having a bone made from woven bone are obviated by the production of laminar bone. Essentially, an insubstantial scaffolding of parallel-fibered bone is laid down quickly to be filled in more leisurely with lamellar bone (figure 1.15).

In cattle, each lamina is about 200 μm thick. In the middle there is a two-dimensional network of blood vessels sandwiched between layers of lamellar bone. On each side of this layer is a layer of parallel-fibered bone, which is more heavily calcified than the lamellar bone (figure 1.16). The way fibrolamellar bone is laid down means that there are, in effect, alternating layers of parallel-fibered bone and lamellar bone tissue wrapped around the whole bone.

This description is of particularly neatly arranged laminar bone. Frequently the blood channels are more irregularly disposed or do not form a network, and the laminar arrangement gives way to one in which each blood channel is surrounded by more or less concentric layers of lamellar bone (figure 1.14). This produces an appearance like that of Haversian systems, and the structures around the blood vessels are called "primary" osteons. However, there is a most important difference between primary osteons and secondary osteons, or Haversian systems. For Haversian systems are *secondary*, that is, they replace bone that has existed previously. There are differences that enable one to distinguish Haversian systems from primary osteons histologically. In particular, Haversian systems are surrounded by a cement sheath, primary osteons are not. The distinction between the two types of osteon is not mere semantic hairsplitting, because differences between primary osteonal bone and Haversian bone correlate with very clear differences in mechanical behavior, as we shall see. Haversian bone is weaker than primary osteonal bone.

The kind of primary bone laid down depends on the rate of accretion. I examined the humerus of cats at different stages of growth. At birth the bone was almost entirely woven, with many large cavities. However, some fibrolamellar bone was forming on some parts of the periphery. At six

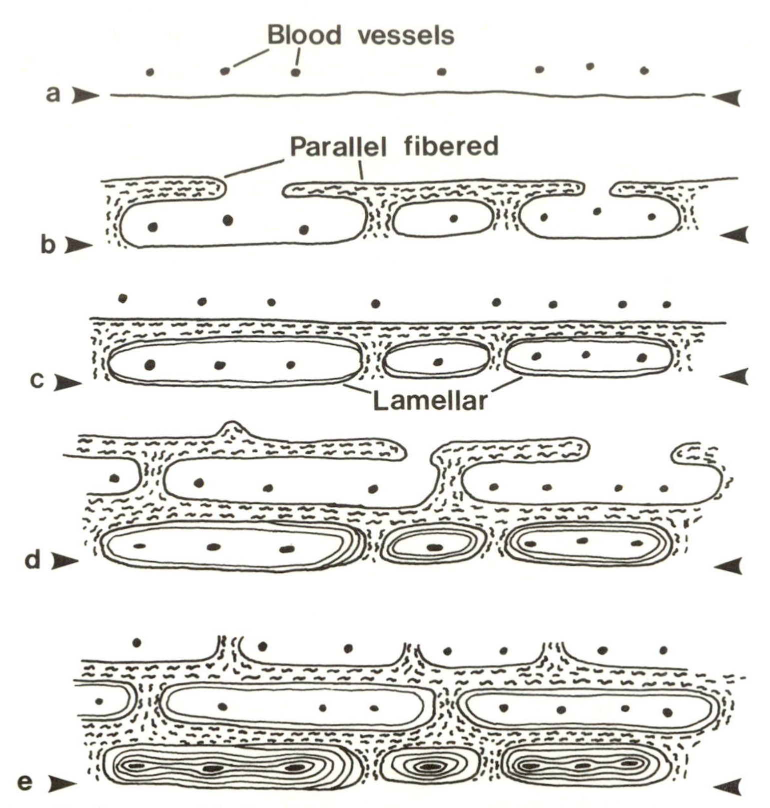

Fig. 1.15. The formation of fibrolamellar bone. These are cross sections of the outer surface of a rapidly growing bone. The arrowheads show the position of the original surface. Blood vessels are shown by black spots. (*a*) The original position. (*b*) Parallel-fibered bone, shown by squiggly lines, grows very quickly to form a scaffolding clear of the original surface. (*c*) Lamellar bone, shown by fine lines, starts to fill in the cavities left by the parallel-fibered bone. (*d*) As more lamellar bone is laid down, so is another scaffolding of parallel-fibered bone. (*e*) By the time the first row of cavities is filled in, the outer surface of the bone is far away.

weeks the growth is by well-formed laminae, with a fair amount of lamellar bone being formed. At six months the laminar appearance is much less obvious, as the pace of growth slows, and the blood channels are surrounded by primary osteons. At a year growth is by the laying down of circumferential lamellar bone, with no obvious relationship between the lamellae and the blood channels they enclose.

1.3.4 Primary and Secondary Bone

Primary bone is replaced by secondary bone in two ways: the bone can be eroded away at its surface, and then new bone can be laid down, or else

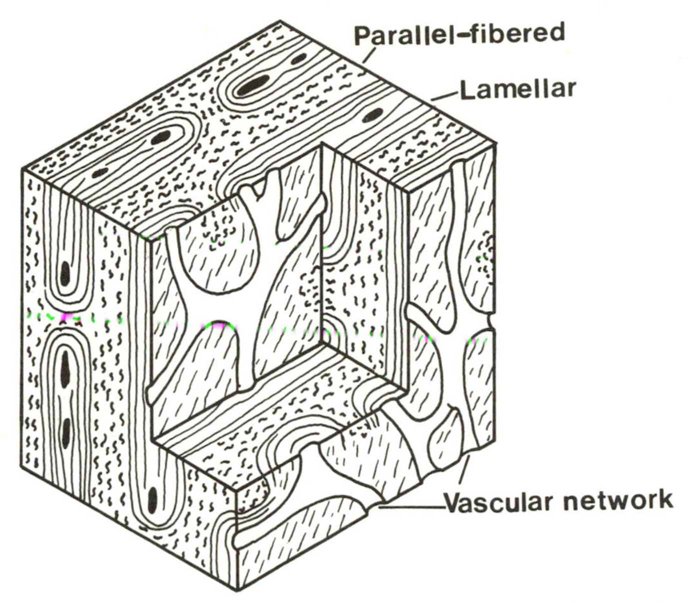

Fig. 1.16. Block diagram of fibrolamellar bone. Conventions concerning histological type as in Fig. 1.15. Two-dimensional networks of blood channels, sheathed by lamellar bone, alternate with parallel-fibered bone.

Haversian systems can be formed. It is often quite difficult to tell when the former has happened, and the effects, if any, of such replacement on mechanical properties are uncertain. The adaptive reason (if it is adaptive) for the formation of Haversian systems, which have a deleterious effect on mechanical properties, is obscure. The common explanation halfheartedly held by the bone community is that Haversian systems form when the bone mineral has, from time to time, to be released into the blood system for purposes of mineral homeostasis (Hancox, 1972). There are various problems associated with this explanation which need not concern us yet, because we can accept Haversian remodeling as a fact and explore the mechanical consequences. However, I shall say more, although rather inconclusively, about the function of Haversian remodeling in chapter 8.

The formation of Haversian systems tends to lead to the production of more Haversian systems. Each Haversian system is bounded by its cement sheath, across which very few canaliculi pass. This frequently results in blood vessels being separated from most of their catchment area, and osteocytes in this area may find it difficult to obtain nutrients, and are more likely to die (Currey, 1960; 1964b). It is probably for this reason that the formation of a few Haversian systems in a region is often followed by the formation of many more in the immediate vicinity. Eventually, a region of bone may be completely occupied by Haversian systems and by luckless "interstitial lamellae," little bits of bone that are separated by cement sheaths from all blood vessels, and so tend to be dead. The death of bone cells by no means always leads to the formation of new bone to replace the old.

Human bone is like that of many primates and carnivores in that primary fibrolamellar bone is laid down initially, but this bone type is soon replaced by Haversian bone. However, this is not the case in many other mammalian groups. In most bovids (cattle, deer, etc.), for example, the long bones keep their primary, laminar structure all through life, with only small regions, usually under the insertion of strong muscles, becoming Haversian. Many smaller mammals show no remodeling at all (Enlow and Brown, 1958), the bone being fibrolamellar or, often, mainly composed of circumferential lamellae.

1.3.5 Compact and Cancellous Bone

At the next higher order of structure there is the mechanically extremely important distinction between compact and cancellous bone. Compact bone is solid, with the only spaces in it being for osteocytes, canaliculi, capillaries, and erosion sites. In cancellous bone there are large spaces. The difference between the two types of bone is visible to the naked eye. The material making up cancellous bone of adults is usually primary lamellar bone or fragments of Haversian bone. In young mammals it may be made of woven or parallel-fibered bone.

The structure of cancellous bone can be considered to vary in three ways: in its fine-scale anisotropy, in its large-scale anisotropy, and in its porosity. Singh (1978) has a convenient description of cancellous bone morphology. The simplest kind of cancellous bone consists of randomly oriented cylindrical struts, about 0.1 mm in diameter, each extending for about 1 mm before making a connection with one or more other struts (figure 1.17). In a variation of this pattern the cylindrical struts are replaced by little plates. The amount of variation ranges from cancellous bone in which there is just the occasional plate among the struts to cancellous bone in which there is just the occasional strut among the plates. In other cancellous bone the plates may be considerably longer, up to several millimeters. When this happens there is a higher level of anisotropy: these longer plates are not randomly oriented but are preferentially aligned in one direction. The final form of such cancellous bone is shown in figure 1.17c, where there are parallel sheets of bone with fine struts joining them.

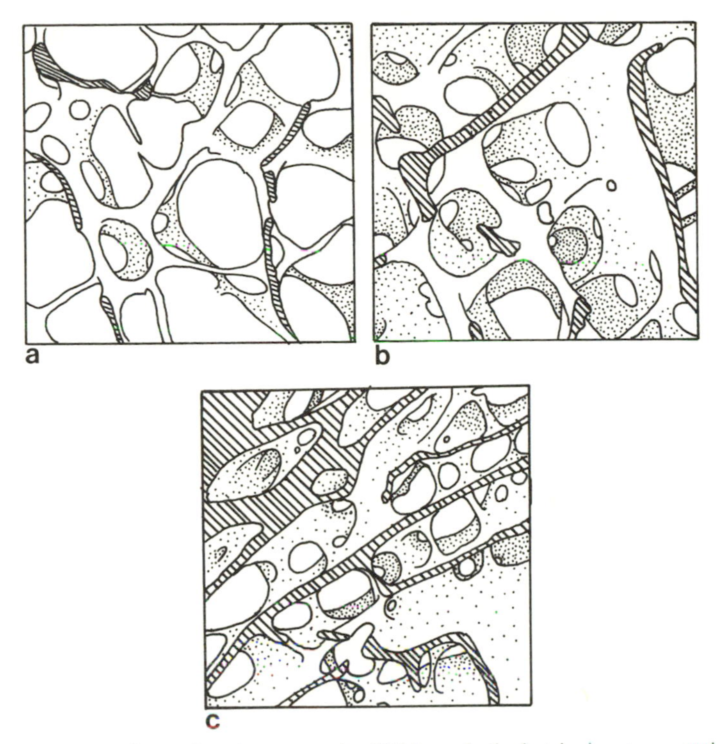

Fig. 1.17. Drawings of cancellous bone, seen by SEM. In each, the hatched parts are at the level of the top of the section. Everything else is below this, the farther away, the more closely spaced the dots. (*a*) From the middle of a human sternum. Rather fine, nearly random network of mainly cylindrical struts. Width of picture: 3.4 mm. (*b*) From a human greater trochanter. Many of the elements are plates. Width of picture: 3.4 mm. (*c*) From the human femoral neck. The longitudinal plates are very obvious. There are many struts and plates lying orthogonal to them. Width of picture: 8 mm (note smaller scale).
([*a*] Derived from Whitehouse, 1975; [*b*,*c*] derived from Whitehouse and Dyson, 1974.)

Another type of cancellous bone consists wholly of sheets, forming long tubular cavities which interconnect by means of fenestrae in the walls.

These different versions of cancellous bone are found in characteristically different places. Roughly, the type made of cylindrical struts, with no preferred orientation, is found deep in bones, well away from any loaded surface, while the more oriented types, made of many plates or completely of plates, are found just underneath loaded surfaces, particularly where the pattern of stress is reasonably constant.

The porosity of cancellous bone is the proportion of the total volume that is not occupied by bone tissue. Usually it will be occupied by marrow, but in birds there may be gas. The porosity varies from being effectively complete, where there is only the occasional tentative strut sticking into

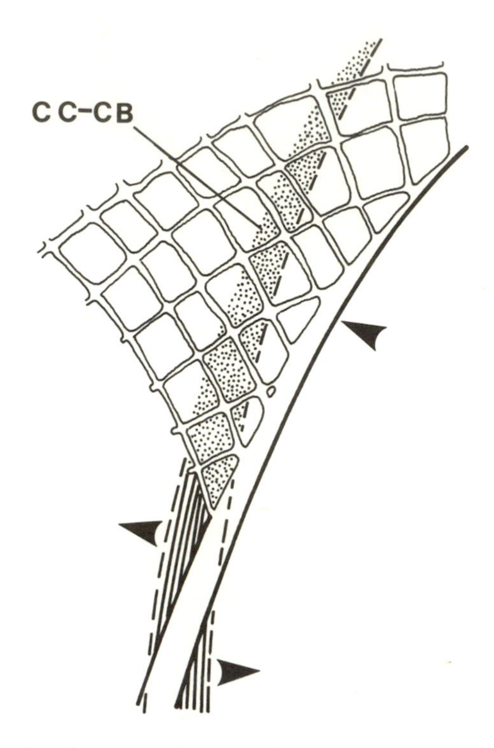

Fig. 1.18. The production of compact coarse-cancellous bone near the end of a long bone. The original bone is shown plain. It has a (highly schematic) cancellous network. As the bone grows in length, bone tissue is eroded, or deposited, in directions shown by arrowheads. At the bottom, ordinary lamellar bone is added, but among the trabeculae compact coarse-cancellous bone (CC–CB) is formed.

the marrow cavity, down to about 50%. If the porosity is less than about 50%, it becomes difficult to distinguish from compact bone with many holes in it. However, it is interesting that the change from compact to cancellous bone usually is clear and takes place over a small distance. The mechanical reasons for this are discussed in chapter 4.

Bone grows by accretion on preexisting surfaces. Long bones have cancellous bone at their ends for reasons discussed in chapter 4. The geometry of the situation is such that, quite often, compact bone has to be formed in a region where cancellous bone already exists. Where this is so, the old cancellous bone is not replaced; new bone is merely wrapped around the struts of the cancellous bone, producing an extremely confused structure with no obvious grain called "compact coarse-cancellous bone" (figure 1.18). The effect of bone growth on bone histology and many other aspects of bone growth and structure (but not fine structure) are very clearly discussed by Enlow (1963, 1975).

1.3.6 Nonmammalian Bone

The bone I have discussed so far is mammalian bone. Rather little is known about the mechanical properties of nonmammalian bone, but I shall say something about its structure because of its interesting similarities and dissimilarities with mammalian bone. A generation ago Enlow and Brown wrote a useful summary of fossil and recent bone of all the vertebrates (1956, 1957, 1958). De Ricqlès has produced a massive survey in ten parts of the histology of tetrapod bone, mainly that of reptiles. These papers are all cited in a comprehensive bibliography of 618 references, chiefly on bone histology, in de Ricqlès (1977). A shorter summary is in de Ricqlès (1979).

The bone tissue of birds is like that of mammals, although at the naked-eye level there are important differences in the proportions of wall thickness to overall diameter, which we shall discuss later. Some reptile bone is like mammalian bone, the dinosaurs in particular often having had well-developed fibrolamellar bone, Haversian systems, and a rich blood supply. However, in many reptiles the bone is poorly vascularized, and indeed is often avascular, although it does contain living bone cells. This poor vascularization is presumably possible because of the relatively low metabolic rate of many reptiles.

The modern amphibia tend to have a rather simple, often avascular, bone structure. However, the earlier amphibia, such as the Embolomeri, which were quite large, show ill-developed laminar bone, and also Haversian systems.

The bone of most modern bony fish—the advanced teleosts—has no bone cells. This is a remarkable fact whose significance, physiological or mechanical, is totally obscure. It seems that the bone cells form in the ordinary way, are incorporated into the bone, and then die, the lacunae they leave being filled up with mineral (Moss, 1961). In the lower teleosts and lungfish there are bone cells, although there is a tendency for bone to be replaced by cartilage in these groups.

The mineralized tissue of lower vertebrates, including extinct groups, is discussed by Halstead (1974), and in greater detail by Ørvig (1967). The considerable range of histological structures seen in the nonmammalian vertebrates is a challenge, because so little is known of their mechanical properties. Undoubtedly, when fully investigated they will turn out to have similarities to, and differences from, mammalian bone that will be most instructive.

1.3.7 The Shapes of Whole Bones

The shapes of whole bones are marvelously varied, but a large proportion of them fall into a few groups. I shall describe them briefly, mainly as a reminder of facts the reader is no doubt aware of. In chapter 4 we shall consider the mechanical reasons for these shapes.

Tubular bones are elongated in one direction and the section is often roughly circular. These bones are hollow; in their midsection the wall thickness is about one-fifth of the overall diameter. They are expanded at their ends and capped with a layer of synovial cartilage, forming part of the synovial joint. Near the ends the central lumen becomes filled with cancellous bone. In places the plain tubular shape may be distorted where the bone is drawn out into flanges and tubercles for the attachment of muscles and ligaments. Tubular bones include most of the long bones and the ribs.

Tabular bones are bones or, often, parts of bones that are rather flat, so that one dimension is much less than the other two. Examples are the scapula, the pelvis, many bones in the vault of the skull, the carapace of chelonians, and the opercular bones of fish. They usually consist of two thin sheets of cortical bone separated by some cancellous bone, but sometimes, as in some scapulae and the vault of the skull of many small mammals, the two cortices are not separate.

Short bones are bones or parts of bones that are roughly the same size in all directions. Examples are wrist and ankle bones and the centra of vertebrae. Many phalanges are intermediate between short bones and tubular bones in their structure. Short bones tend to have very thin cortices and to be almost filled with cancellous bone.

Often the extensions to bones for muscle and ligament insertions are so large that they take on the mechanical characteristics of tabular bones. This is seen particularly clearly in the transverse and spinous processes of vertebrae.

1.3.8 Epiphyses

Many bones of birds and mammals, and some reptiles, grow in length by means of epiphyseal plates. The problem with growing long bones is that the ends have to articulate with other bones, and it would clearly be awkward to have growth taking place actually at a sliding surface, exposed to quite large loads. The main growth takes place well away from the articulating surfaces. In many anamniote vertebrates, such as fish and amphibia, there is a large cartilage epiphysis between the line of bone growth and the articulation. This, however, means that if large forces are

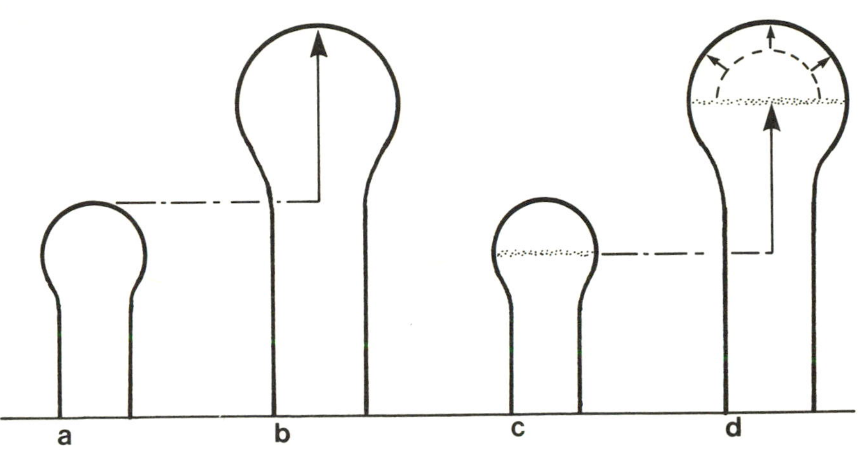

Fig. 1.19. The function of epiphyses. (*a*) The small bone has to become longer. (*b*) The amount of growth in length is shown by the arrow. (*c*) A bone with an epiphyseal plate (shown dotted) divides its growth into two parts. (*d*) The main growth is produced by the epiphyseal plate. The growth of the bone underneath the lubricating synovial cartilage can take place much more slowly.

applied across the joint the ends of the bones, being cartilaginous, will deform considerably. The epiphyseal plate is a device to get around this difficulty. The epiphyseal cartilage develops one or more secondary centers of ossification, and the end result is that the cartilaginous epiphyseal plate is sandwiched between two bones, and the whole system is fairly rigid. The epiphyseal bone, too, of course grows, but not at the frenetic pace often necessary for the bone as a whole (figure 1.19). Epiphyseal plates are an excellent solution to the problems of growth, but they are mechanically weak, and this produces design problems, whose partial solution is discussed in chapter 4.

2. The Mechanical Properties of Bone

This chapter is concerned with the basic mechanical properties of bone and how they are measured. Such a great amount of material has been gathered in the last twenty years or so about the mechanical properties of bone that it would be futile to try to summarize it. Fortunately, there is much agreement about most of the properties, so I shall take as datum points a few papers, and mention later the main variants. The mechanical properties I shall deal with are: elastic properties (mainly Young's modulus), strength (in tension, compression, and shear), and fracture mechanics properties. I shall discuss fatigue strength later in the chapter. These sections will be mainly bald statements of results; I shall discuss their implications in the next section. A point to bear in mind when looking at these results is that strain *rate* affects many of the properties to some extent. For practical reasons tests are often carried out at strain rates lower than those encountered in life, and this tends to produce values for Young's modulus, and for strength, that are less than those that will obtain in a more lifelike situation.

2.1 Elastic Properties

There are two main ways of measuring the elastic properties of bone: (*a*) by applying a load to a specimen and calculating the elastic properties from the resulting deformations (or, frequently, by applying a deformation and calculating the elastic properties from the load necessary to produce the deformation) and (*b*) by measuring the velocity of sound waves in bone. The velocity of sound in a medium is obtained from $V = \sqrt{E/\rho}$ where E is Young's modulus and ρ is the density of the medium. However, there are in reality considerable complications. The sound used is of high frequency, and I shall call A "mechanical" and B "ultrasonic," although both are at root mechanical, of course.

Mechanical testing has various advantages: it is relatively straightforward; Young's modulus can be determined in a variety of directions quite simply; cancellous bone and bone full of cavities can be tested; and the effect of strain rate on mechanical properties can be investigated.

Ultrasonic testing is much less straightforward. Indeed, judging by my experiences at fairly recent conferences (in 1983), there are still considerable arguments as to whether particular methods are valid even in principle. Furthermore, there are some difficulties in testing wet specimens, and indeed most tests have been carried out on dry specimens. The method cannot cope with cancellous bone. On the other hand, it should make possible the derivation of all the stiffness coefficients and, with difficulty, their determination all from the same specimen.

Table 2.1 gives values for technical moduli as determined by mechanical testing by Reilly, Burstein, and Frankel (1974), and by Reilly and Burstein (1975). These values are for man and for cattle. The table also gives values for technical moduli derived from ultrasonic tests by Yoon and Katz (1976b), Bonfield and Grynpas (1977), and Van Buskirk and Ashman (1981).

2.2 Strength

In most respects the strength of a material is fairly easy to measure. One can measure the load at which a specimen breaks and calculate the strength directly; there is no need to measure strain as well as stress. Nowadays many people would argue that the "tensile strength" of bone is not a useful concept, because such a measure gives no idea of how sensitive bone is to the presence of the little cracks and flaws that it must have in life and that will affect how it actually behaves. Such people say that fracture mechanics parameters give one a much better idea of what bone is really like. A fervent proponent of this view is Bonfield (Bonfield, 1981; Behiri and Bonfield, 1980). Although the fracture mechanics approach is obviously important, and some results will be discussed below, it nevertheless remains true that relatively simple determinations of strength do have several advantages.

First is simplicity. Fracture mechanics techniques usually require rather large specimens with precut notches, or various other awkward geometries. These requirements are difficult enough, but added to them is the requirement of measuring deformations and, often, the lengths of cracks. Second, although the usual fracture mechanics tests will measure the behavior of the bone when the crack surfaces are pulling apart, they have little to say about how bone behaves in compression. Third, *pace* Bonfield, the standard deviations achieved in well-conducted ordinary fracture tests are not large, and allow one, for example, to find fairly easily differences

Table 2.1 Technical Elastic Moduli for Compact Bone

		Reilly, Burstein, & Frankel	*Knets, Krauya, & Vilks*	*Van Buskirk & Ashman*	*Yoon & Katz*	*Reilly & Burstein*	*Currey*	*Reilly & Burstein*	*Currey*	*Van Buskirk & Ashman*	*Bonfield & Grynpas*
Species		Man	Man	Man	Man	Cow	Cow	Cow	Cow	Cow	Cow
Histology		*HS*	*HS*	*HS*	*HS* (Dry)	*HS*	*HS*	*FL*	*FL*	?	?
Method		Mech	Ultra	Ultra	Ultra	Mech	Mech	Mech	Mech	Ultra	Ultra
Property:											
Young's Modulus	3	17.0	18.4	21.5	27.4	22.6	20.3	26.5	25.9	21.9	17
	2	11.5}	8.5	14.4	18.8}	10.2}		11.0}		14.6	11}
	1	11.5}	6.9	13.0	18.8}	10.2}		11.0}		11.6	11}
Shear Modulus	23	3.3}	4.9	6.6		3.6}		5.1}		7.0	
	13	3.3}	3.6	5.8		3.6}		5.1}		6.3	
	12		2.4	4.7						5.3	
Poisson's Ratio	31		.32	.40		.36		.41		.21	
	32	.41}	.31	.33		.36		.41		.31	
	21	.41}	.62	.42		.51				.38	

Sources: Bonfield and Grynpas (1977); Currey (1975); Knets, Krauya, and Vilks (1975); Reilly and Burstein (1975); Van Buskirk and Ashman (1981); Yoon and Katz (1976b); Reilly, Burstein, and Frankel (1974).

Note: HS: Haversian systems; *FL*: Fibrolamellar bone. Young's modulus and shear modulus in GPa. The bracketed numbers refer to values that, because of assumption of symmetry, were not independently determined. 1, 2, and 3 refer to the radial, tangential, and longitudinal orientations in a long bone.

Table 2.2 Failure Properties of Compact Bone

Species	*Man*		*Cow*		*Cow*	
Histology	*HS*		*HS*		*FL*	
Direction of loading relative to long axis	Parallel	Normal	Parallel	Normal	Parallel	Normal
Tensile Strength	148	49	144	46	167	55
Yield strain	0.007	0.004	0.006	0.004	0.006	0.005
Ultimate strain	0.031	0.007	0.016	0.009	0.033	0.007
Compressive Strength	193	133	254	146	294	
Yield strain	0.010	0.011	0.011	0.014	0.010	
Ultimate strain	0.026	0.028	0.016	0.031	0.014	

Sources: Taken directly from, or from information in, Reilly, Burstein, and Frankel (1974) and Reilly and Burstein (1975). Strength in MPa.

between specimens with different histology, or between bones from individuals of different age.

Table 2.2 gives values for various fracture-related properties of human and bovine bone. They are taken from two papers (Reilly, Burstein, and Frankel, 1974; and Reilly and Burstein, 1975). I chose these two papers because they report carefully carried out work in which the compression tests and tensile tests are readily comparable, being made on the same-shaped specimens loaded at the same strain rate. It is fair to say that most other values in the literature, in which one has some confidence, are near these values.

2.3 Fracture Mechanics Properties

As I mentioned above, fracture mechanics properties are derived by rather complex testing methods. The values G_c, K_c, and W in table 2.3 are, respectively, the critical strain energy release rate, the critical stress intensity, and the work of fracture. The values are mainly reported from Behiri and Bonfield (1980).

Table 2.3 Fracture Mechanics Values for Compact Bone

Author	*Method*	*Bone*	*Crack Travel Relative to Grain*	*Crack Velocity* $10^{-5}\ m s^{-1}$	G_c $J m^{-2}$	K_c $MN m^{-3/2}$	W $J m^{-2}$
Melvin and Evans (1973)	Single edge notch	Bovine femur	Parallel		1,970	3.2	
			Normal		4,330	5.6	
Margel Robertson (1973)	Three-point bending	Bovine femur	Normal			6.6	
Wright and Hayes (1976)	Compact tension	Bovine femur	Parallel		1,200	3.5	
Bonfield et al. (1978)	Compact tension	Bovine femur	Parallel		1,850	3.8	
Currey (1979)	Three-point bending	Bovine femur	Normal				1,710
		Deer antler	Normal				6,190
		Whale bulla	Isotropic				200
Behiri and Bonfield (1980)	Compact tension	Bovine tibia	Parallel	1.75	1,740	4.5	760
				3.6	1,810	5.0	1,340
				12.6	2,250	5.2	1,900
				23.6	2,800	5.4	2,120
				Catastrophic	<1,500	<4	125

Sources: All are given in Behiri and Bonfield (1980) or Bonfield (1981) except Currey (1979).

2.4 Modeling and Explaining Mechanical Properties of Bone

One of the never-failing pleasures of being a biologist is finding out how well organisms are designed. It will become apparent from later chapters that whole bones and skeletons are, indeed, designed well for their functions. However, before being able to say that natural selection is wonderful and has done it again, one has to find out what exactly has been done. Doing this for bone *material*, as opposed to whole bones, is difficult, because materials scientists have not yet developed a comprehensive framework for explaining the mechanical properties of a material like bone. In this section I shall try to show how we are beginning to be able to model some of the mechanical properties of bone. It will become clear that we still have far to go.

2.4.1 Modeling Elastic Behavior

As I said at the beginning of the book, I think the most important feature of bone material is its stiffness. The operations of the skeleton cannot be described without at least the implication that these functions could not be performed by flexible structures. On the other hand, there are no particular implications that these structures should be strong, or tough, or good stores of calcium ions. I shall therefore start consideration of the properties of bone with stiffness.

Bone is essentially a collagen framework packed with calcium phosphate mineral. Collagen is widespread in the animal kingdom, and it is surprising that no other phylum has adopted this method of producing a stiff skeletal material. In fact, no other animals, except some crustaceans, seem to have such an intimate association of organic material and mineral. Lowenstam (1981) lists all minerals used by living organisms, and it is apparent from this that although the number is considerable, easily the most widely used ones are calcium carbonate, silica and, a poor third, calcium phosphate and its relatives. Calcium carbonate is nearly always found in a composite of organic and mineral parts, but the volume fraction of organic material is very low, nearly always less than 5%, usually much less. Calcium carbonate may also occur as isolated spicules. These spicules are usually large, visible to the naked eye. Muzik and Wainwright (1977) provide an account of the spiculation found in a number of soft corals. That paper gives a good idea of the kind of variation one finds in animals, and how different mechanical properties are produced by different volume fractions and arrangements of spicules.

Silica usually occurs as isolated spicules, though it may form the entire skeleton of some protozoans. It never behaves as a composite in which there is tight bonding between the organic and mineral parts. Most of the animals with calcium carbonate skeletons have collagen, but do not deposit mineral in it. Stiffening in the invertebrates is produced by having very high volume fractions of fairly large mineral crystals, or spicules, or by tanning (cross-linking) organic materials. This last method is adopted by the arthropods in particular, whose cuticle consists of fibrils of chitin bound together by a more or less highly cross-linked protein. The resulting composite is probably superior to bone in every respect on a per-weight basis, an embarrassing fact I shall not pursue further, except to say that this mechanical superiority is bought at the cost of considerable developmental complexity. During growth the cross-linked cuticle cannot be resorbed, and so must be shed periodically. Vincent (1982) has written an excellent account of the devices used by organisms to produce adaptive mechanical properties in their structural materials.

The Young's modulus of tendon, which we can take to represent collagen in a fairly uniformly aligned arrangement, is about 2 GPa. The Young's modulus of hydroxyapatite is about 110 GPa. The problem, which has been attacked by several workers over the last dozen years or so, is to obtain a model that will explain the elastic behavior of bone using this and other information, such as the volume fraction of mineral, anisotropy, and the detailed arrangement of the microscopic constituents of bone. Katz and his co-workers have been particularly concerned with this (Katz, 1971), and although a long way from complete success, are making good progress. Katz started by modeling bone as if the mineral and the collagen formed a multilayer sandwich (figure 2.1). If such a sandwich is loaded parallel to the layers, the *stiffness* of the two components are additive; this is the so-called Voigt model. If the sandwich is loaded normal to the layers, the *compliances* are additive. (Compliances can be considered to be simply the inverse of the stiffnesses.) This is the so-called Reuss model. If the bone is considered to be a multilayered composite material of modulus E_b, made of layers of collagen and layers of mineral of modulus E_c and E_m, and of relative volumes V_c and V_m respectively, then in the case of loading parallel to the layers $E_b = E_c \cdot V_c + E_m \cdot V_m$. This is the Voigt model. For the Reuss model $1/E_b = V_c/E_c + V_m/E_m$. Although bone and enamel do fit within the upper and lower bounds, the bounds themselves are so far apart that this model of Katz can hardly be considered to be satisfactory (figure 2.2). Piekarski (1973) came to a similar conclusion. Of course, by using some arbitrary mixture of the Voigt and the Reuss models, it would be possible to get quite a good fit. However, it is then found that the modulus for loading at angles to the long axis is not well predicted.

Later, Katz refined his model by assuming (which is almost certainly true) that the apatite crystals are oriented along the collagen fibrils and

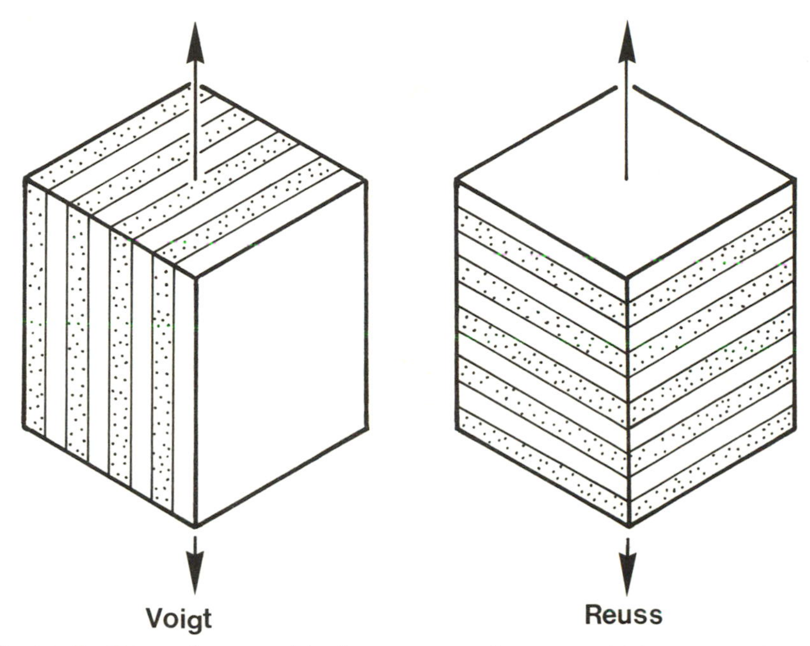

Fig. 2.1. The Voigt and Reuss models of composites. In the Voigt model the strains in the two components are equal; the stiffness of the composite is given by adding the stiffnesses of the components. In the Reuss model the stresses in the two components are equal; the compliance of the composite is given by adding the compliances of the components.

tightly bonded to them. The collagen fibrils, however, will be in various orientations in various lamellae. Following Krenchel (1964), Katz (1981) developed the equation:

$$E_{\text{bone}} = E_c V_c (1 - v_{c\ \text{bone}})/(1 - v_c^2) + \sum E_{ha} V_{ha} \alpha_n (\cos^4 \phi_n = v_{\text{bone}} \cos^2 \phi_n \sin \phi_n).$$

Subscript c refers to collagen, ha to apatite, v is Poisson's ratio, α_n is the fraction of the apatite crystallites that lie at an angle ϕ_n from the direction of stressing. The first part of the equation refers to collagen, which is assumed to have the same modulus in all directions. This, though not true, does not matter, because the modulus of collagen is so low. The second part gives the effects of apatite. The important point to note here is the $\cos^4 \phi_n$ term, which implies that the effectiveness of the mineral as a stiffener will fall off very rapidly as it becomes misaligned. ($\cos^4 10°$ is 0.94, $\cos^4 20°$ is 0.78, $\cos^4 45°$ is 0.25.) Making some reasonable assumptions about the orientations of the crystals, Katz was able to compare theory with experiment, using his and other people's data (Reilly and Burstein, 1975; Bonfield and Grynpas, 1977). The fit was not very good, particularly when the grain of the bone was at a large angle to the

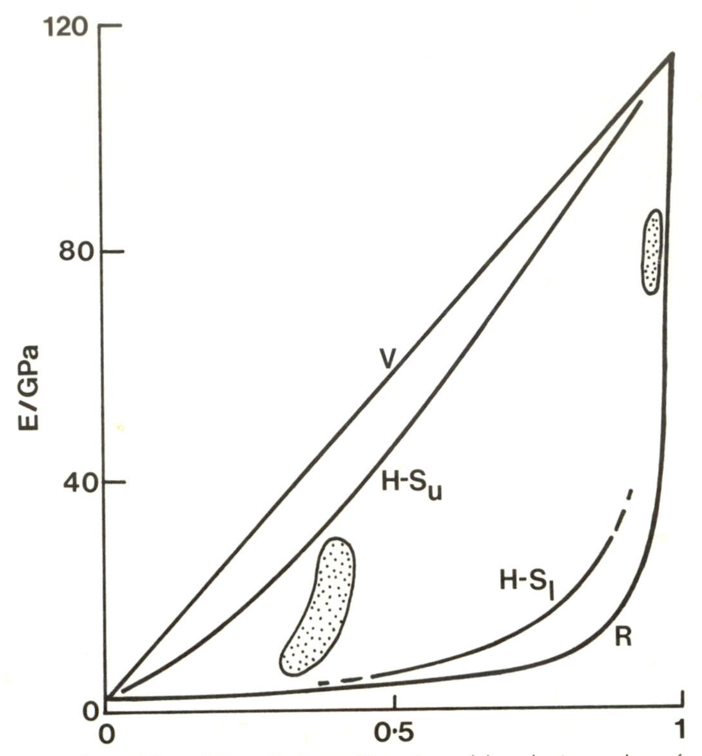

Fig. 2.2. The early modeling of Katz. Ordinate: Young's modulus; abscissa: volume fraction of mineral. The upper and lower curves are the predictions of the Voigt (*V*) and Reuss (*R*) models. The inner lines ($H\text{-}S_u$ and $H\text{-}S_l$) are the upper and lower bounds of another model studied by Katz, the Hashin-Shtrikman model. The lower stippled area represents the actual values for dentine and bone; the upper stippled area represents enamel. (From Katz, 1971.)

loading direction. At large angles the observed values for Young's modulus were considerably greater than the model predicted.

Katz and his co-workers have developed a more complex model to account for the behavior of Haversian bone. They suggest that the Haversian systems are arranged roughly in hexagonal packing, and that the moduli of the systems themselves and of the interstitial lamellae are different. They include in the effective modulus of the interstitial lamellae the stiffness of the cement lines surrounding the Haversian systems, which they claim is rather low. It would, I think, be surprising if cement lines really did have a low modulus, because they are highly calcified, usually more so than the systems they enclose. However, Frasca, Harper, and Katz (1981) have tested isolated groups of secondary osteons from human bone in shear and find a marked size effect: the greater the number of osteons in the sample (from one to eleven), the more viscous the sample becomes and the lower its shear modulus of elasticity. Although the scatter of the data is rather large, and the specimen was immature (from a twelve year old), these results do show the possibility that the cement lines are allowing relative movement of the osteons, and that they are more viscous than the rest of the bone. I admit to being puzzled by this result.

Katz adduces an observation of Lakes and Saha (1979) as a further evidence for the viscous nature of cement lines. These workers showed, by scribing a line on a specimen of fibrolamellar bone and then loading the bone in shear for a long time, that most irreversible strain took place at the "cement line." This is an extremely interesting observation, but I think it is irrelevant to the present discussion. The line is not a cement line around a secondary osteon, but a line produced in the growth of fibrolamellar bone that is certainly not homologous with a cement line, and may well not have a similar composition. Gottesman and Hashin (1980) have applied Katz's model to the viscoelastic properties of Haversian bone. Remarks about the lack of experimentally determined data apply with even greater force to this analysis.

Even if these models do seem to account accurately for the behavior of Haversian bone, they have two shortcomings. One is that they seem, if anything, to lead us even further away from an understanding of fibrolamellar bone. For in fibrolamellar bone there are no cement lines, yet this type of bone is more anisotropic in its elasticity than Haversian bone. Of course, the observation of Lakes and Saha will come into its own when fibrolamellar bone is analyzed. The other shortcoming is that the basic model does not account for the observation that the Young's modulus of bone increases very markedly with mineral content over a small range of mineral values. The equation above shows that, if the modulus of the mineral is much greater than that of the organic material, then the modulus should be roughly proportional to the volume fraction of mineral. Yet there have been a number of studies that show that the modulus increases much more rapidly than the volume fraction of mineral (Currey, 1969; 1979; Ramaekers, 1977). Katz assumes that the apatite needles are bound rigidly to the collagen, but makes no allowance for their length; the equations he uses are for very high aspect ratio fibers. Making allowances for the length of fibers is difficult, and the problem has not been completely solved yet except for very simple cases.

The analyses of Katz and his co-workers have all assumed values for the Young's modulus of collagen tested on its own. However, McCutchen (1975) proposed that the mineral phase of bone "straightjackets" the collagen and prevents it from straightening under stress. Lees and Davidson (1977) proposed a similar scheme (without mentioning McCutchen's paper). Hukins (1978) proposed a modification of McCutchen's model, in which the collagen fibrils are considered to be arranged as liquid crystals whose habitual reorientation under stress is prevented by the mineral. All these suggestions are ingenious, not to say amusing, but until they can be quantified it is probably best to think of them as suggestions that should be continually borne in mind when considering models such as those of Katz and his colleagues. Certainly, when there is such an intimate relationship between the collagen and the mineral at the nanometer level, it is dubious to what extent one should consider the collagen to be behaving just like collagen tested on its own.

We have seen that bone shows a certain amount of anisotropy of Young's modulus: it is about twice as stiff in the stiffest direction as in the less stiff direction. This is because bone is like wood in having a marked grain—the structural elements tend to be oriented preferentially in one direction, and it is in this direction that they will be stiff. However, if it were adaptive to have an isotropic material, isotropy could be achieved by the bone having the collagen fibrils and all other elements arranged randomly in space. This is effectively what happens in woven bone and, to a lesser extent, in parallel-fibered bone. Of course, there is a price to be paid for this isotropy; the greatest stiffness will be reduced. Unfortunately, there seem to be no studies on the mechanical properties of woven bone. (They would be difficult to carry out because woven bone usually comes in very small lumps.) We do not know, therefore, how stiff three dimensionally isotropic bone is. As the mineralization of woven bone is different from that of lamellar and parallel-fibered bone, such a comparison would not separate the effect of fiber direction from that of mineralization. Harris (1980) and Hull (1981) give good accounts of the effect of fiber direction on the Young's modulus of composites.

In general, one expects materials that are fairly stiff to have the same Young's modulus in tension and in compression. Originally, Reilly, Burstein, and Frankel (1974) showed this to be the case in a set of very careful experiments. More recent work from Burstein's laboratory (Torzilli *et al.*, 1982) shows that, at least in dog's bone, the compressive Young's modulus is greater than the tensile modulus. This result is disturbing and, if not caused by some subtle artifact, will have to be explained by people modeling the Young's modulus of bone.

2.4.2 Fracture of Bone in Tension

In order to understand the mechanisms in bone that tend to stop it from breaking, we must try to understand what goes on when it does break. Bone usually breaks in tension or shear, and tensile fracture is easier to understand than compressive fracture; we shall consider tensile fracture first.

Materials can break in tension in various ways. A truly brittle material will simply cleave catastrophically from some preexisting flaw when the energy balance is right. A very plastic material such as a mild steel will deform plastically by sending off dislocations from the highly stressed regions. These dislocations will distribute the strains over quite a large volume, far from the highly stressed region. The dislocations will be shear dislocations, and the result of their movement will be that in a conventional cylindrical test piece the highly stressed region will become thinner, tending to increase the stress on it. This leads to the phenomenon of "necking," in which a short region of a tensile test piece becomes quite narrow and bears an increasingly higher stress, even though the load on the specimen may be increasing little, if at all. Eventually, small cavities appear in the neck region; these coalesce and the material fails. The fracture surface is pitted, and the broken structure, if fitted back together, would be obviously highly deformed.

Bone deforms rather little before fracturing, and it cannot send off dislocations in the way metals do, so the fracture mode must be different from that of a ductile metal. On the other hand, it does not fracture with a smooth surface, as a really brittle material does, and so cannot be truly brittle. In fact, it turns out to be fairly illuminating to think of bone as being a fibrous composite material. So, a short discussion of fracture in fibrous composites follows.

2.4.2.1 *Fibrous composites with continuous fibers*

Ordinary fibrous composites can be idealized as consisting of indefinitely long fibers of tensile strength S_f and modulus E_f, embedded in a continuous matrix of tensile strength S_m and modulus E_m. The proportion of the volume occupied by fibers is V_f and by the matrix V_m: ($V_f + V_m = 1$). I shall use f, m, and c to denote fibers, matrix, and composite. Usually, in engineering composites the fibers are stiffer and much stronger than the matrix.

If $V_f = 1$, the composite consists entirely of fibers. A wire rope is in effect a composite of $V_f = 1$. Such a composite may be strong in tension but, like a wire rope, has little stiffness in bending or compression, because the individual fibers are free to slide past each other. If the fibers are embedded in a matrix of reasonable shear stiffness, the composite becomes much stiffer in bending, and the fibers cannot buckle so easily when loaded in compression.

Suppose $E_f/E_m > S_f/S_m$, that is to say, that compared with the matrix the fibers are stiffer than they are stronger, and that we have a composite consisting of a matrix containing a single stiff, slender, strong fiber. The value of V_f will be very low. When the composite is loaded in tension, the strains in the matrix and in the fiber embedded in it will be the same.

When strain = S_f/E_f, the fiber will break. Because $E_f/E_m > S_f/S_m$, the matrix will not have reached its failure stress. Also, because there is only a single fiber, the reduction in cross-sectional area of the matrix caused by the fracture of the single fiber is small. Therefore, even after the fiber has broken, the matrix will be below its failure stress. What, then, has been the effect on its strength of introducing the fiber? It has been to *weaken* the structure, because the matrix could have borne the load that caused the fiber to break, and after it broke, the cross-sectional area remaining to bear greater loads was less.

Suppose, on the other hand, that we had a very high value for V_f, approaching 1. Now the theoretical strength of the composite is nearly that of the fibers, and this is much stronger than the matrix. What is the value of V_f (called V_{crit}) at which the presence of the fibers leads to strengthening rather than weakening? If the matrix is sufficiently compliant never to break before the fibers, $V_{crit} = (S_m - S'_m)/(S_f - S'_m)$, where S'_m is the stress in the matrix when the fibers break. So, it is no good trying to get a modest improvement in tensile strength by introducing a modest amount of strong fibers; the fibers will strengthen the composite significantly only if V_f is well above V_{crit}.

According to figure 2.3 the greatest strength the composite can have is the strength of the fibers, when $V_f = 1$. There might not seem to be any advantage in having a matrix at all. In fact, there are great advantages. First, as discussed above, individual fibers must be bound together in order to have any bending or compressive stiffness. In addition, as we shall now see, there is an increase in strength and, most importantly, an increase in toughness.

The strengths of the reinforcing fibers will not, in reality, all be the same; there will be a spread about the mean value, caused by larger or smaller imperfections in the fibers at various places along their length. Imagine a specimen made of N continuous parallel fibers with no matrix. A load P is applied, which causes the weakest fiber to break. It bears no more of the load, so each of the remaining fibers now bears a load of $P/(N - 1)$. This may be sufficient to cause the second weakest fiber to break, in which case each remaining fiber would bear a load of $P/(N - 2)$, and so on. Even if one breaking fiber does not cause a cascade of further breaks, the load required to break the next fiber will be less than if the first fiber had not broken. The result of all this is that the load on the specimen required to break it is less than the value derived from the average strength of the fibers. Sooner or later the fibers will break in a cascade, each breaking at its weakest point. How much less the experimental value will be than the calculated one will depend on how variable the fibers are.

If there is a matrix, the situation is quite different. Figure 2.4 represents a simple composite consisting of five fibers in a matrix. The middle fiber,

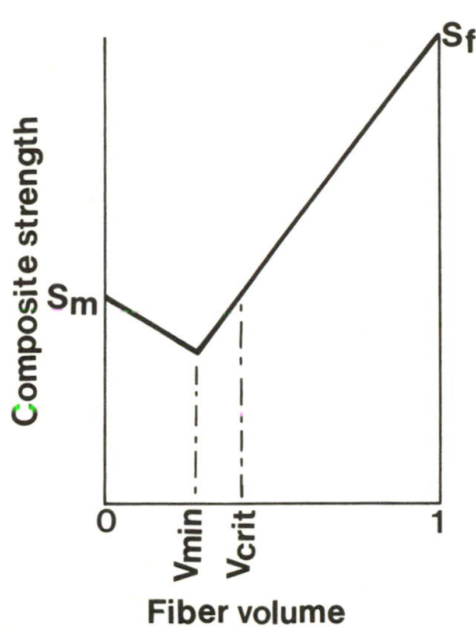

Fig. 2.3. Relationship between the strength of a composite (ordinate) and the proportion of the total volume occupied by fiber (abscissa). With no fibers, the strength is the strength of the matrix (S_m). With no matrix the strength is the strength of the fibers (S_f). If $E_f/E_m > S_f/S_m$, then the addition of fibers will weaken the composite until V_{min} is reached. Not until V_{crit} is reached will the addition of fibers strengthen the composite. (The particular value of V_{crit} will depend on the relationships of E_f/E_m and S_f/S_m. If $E_f/E_m < S_f/S_m$, which is unlikely to be the case, the addition of fibers will always strengthen the composite.)

c, has broken. At the level of the break the strain and the stress on the other fibers are increased by nearly one-quarter. (In fact, the compliant matrix will take a small share of the extra load.) At levels away from the point of break the extra load on the unbroken fibers is transferred back to the broken fiber via shear stresses in the matrix (figure 2.5). A short distance beyond this point, then, the situation is as if the fiber had not broken. The distance over which the load can be transferred back to the fiber is called the "transfer length." For reasons that will become clear in a moment, we are interested in how great the transfer length is when the stress in the fiber is to be the breaking stress S_f. The load is transferred over a shorter distance if the interfacial shear stress τ is high. Increasing the radius of the fiber, r, increases the transfer length. The transfer length $L_c = rS_f/\tau$. Only if the other fibers have flaws *within* the transfer length on each side of the fiber break will the composite be weakened by the presence of the break. Therefore, the shorter the transfer length, the less the likelihood of a second fiber failing because the first failed.

The toughness of composites is brought about by several mechanisms. It is important to be clear that a composite can be tough even though both fiber and matrix may be completely brittle. First let us assume that

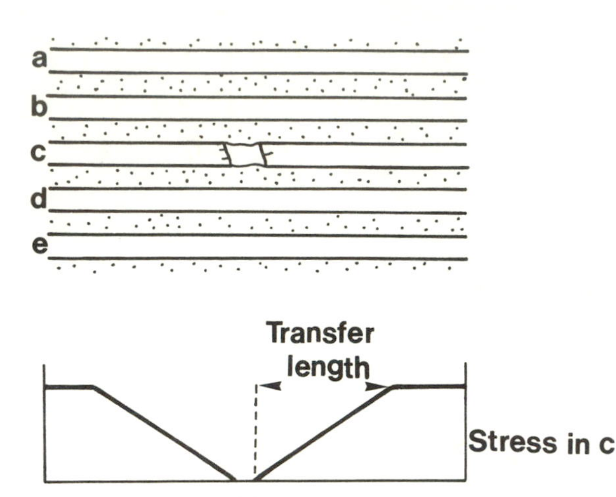

Fig. 2.4. *Above:* a composite of five fibers in a matrix. The middle fiber, *c*, has broken. The fibers *a, b, d,* and *e* must be imagined as being arranged around fiber *c*. *Below:* the stresses in the broken and the unbroken fibers. The stress in the broken fiber is zero at the break but returns to the average value over the transfer length.

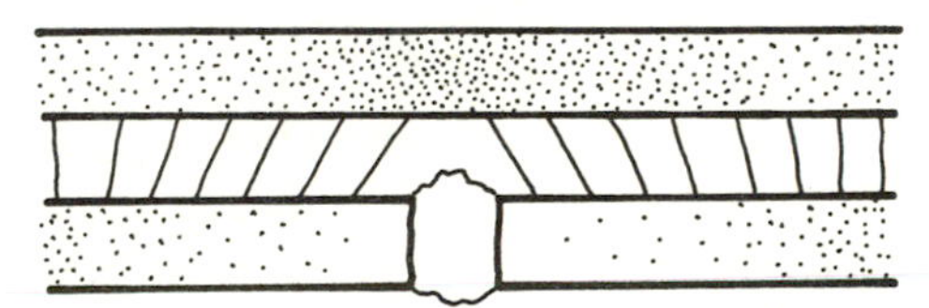

Fig. 2.5. Diagram showing how stresses are transferred back from an unbroken fiber to the broken one by shear stresses in the matrix. The intensity of stress and strain in the fibers is indicated by the intensity of stippling.

both are, indeed, brittle. When the composite has broken, there will be a fairly flat surface of broken matrix. We can call this growing surface, before the composite has completely broken, the "fracture path." Three factors will tend to increase the toughness of a fibrous composite.

1. Suppose the fibers break fairly near the line of the fracture path. For the material to break in two, the two surfaces have to pull apart. But the fracture levels of the fibers are not *exactly* on the line of the main fracture. To be pulled out from the matrix, energy has to be supplied to overcome the interfacial shear strength and the frictional forces tending to keep the fibers in place in the matrix (figure 2.6).

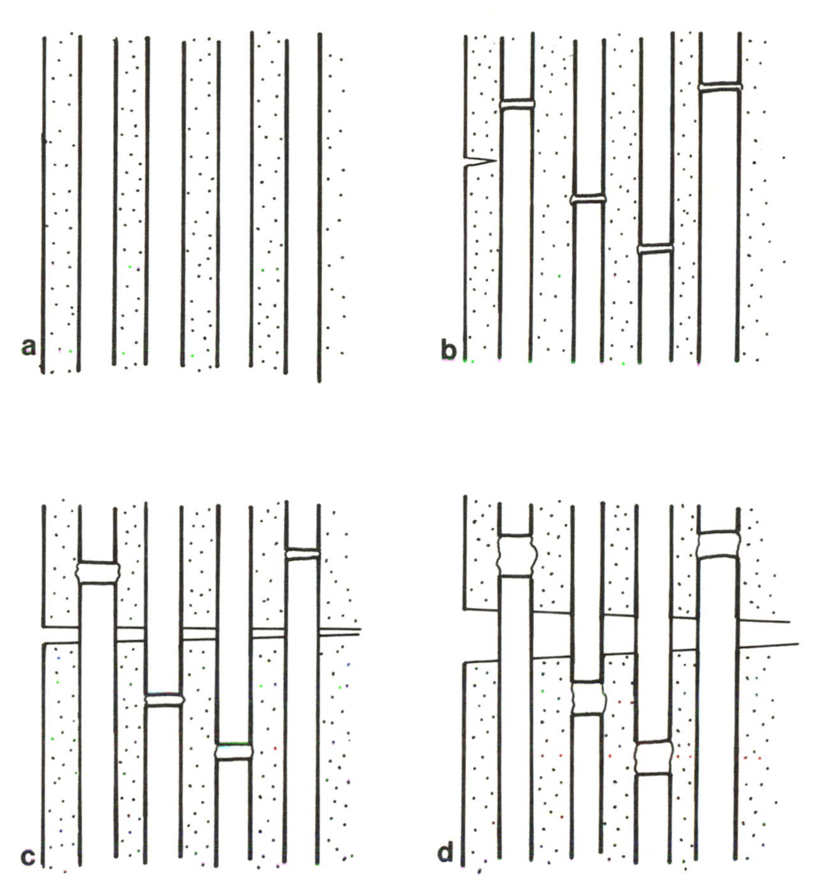

Fig. 2.6. The spread of a crack through a fibrous composite. (*a*) Unstressed. (*b*) The crack starts to travel, and the fibers have broken. (*c,d*) As the crack travels farther, the surfaces must pull apart near the origin of the crack. The fibers have to be pulled through the matrix to allow this.

2. Suppose, again, that the fibers break fairly near the line of the fracture path. As the crack approaches a fiber, the matrix pulls away from the fiber. This peeling off, or delamination, requires energy. The reason the matrix pulls away is that there is a tensile stress, just ahead of the crack tip, which is *in line* with the direction of crack travel. If the matrix is isotropic, this tensile stress is about 1/5 of the tensile stress normal to the line of crack travel (Cook and Gordon, 1964). As the crack approaches the fiber the "in line" stress will tend to pull the matrix away from the fiber. If the strength of this interface is low, it will separate, and in so doing will require energy.

This delamination may be particularly important in composites made of alternating sheets of material of different strength, because in this case the crack will be *blunted* all along its front. Not only will energy be absorbed in separating the two components from each other, but it will also be difficult for the crack to continue because the crack tip will be blunted; there will be no great stress concentrations to produce local high stresses. The situation is less favorable in a fibrous composite, because the crack front remains sharp between the fibers. The availability of strain energy to continue the progress of the crack will, of course, be strongly reduced by the energy being used up in delaminating the matrix from the fibers.

3. If the matrix is ductile, rather than brittle, then energy will be used up in deforming it. Further, the high stresses at the tip of the crack will cause a ductile material to deform so much that the crack will be blunted.

Composites, then, have mechanisms that make it energetically difficult for a crack to travel and that also blunt the crack, reducing stress concentrations at the leading edge.

2.4.2.2 *Discontinuous fibers*

So far we have considered composites made from fibers that run from one end of the structure to the other. Many real composites have fibers that are much shorter than the length of the structure. In these composites it is obvious that, as load cannot be transferred from one end to the other by uninterrupted fibers, it must be transferred from fiber to fiber by the matrix. Indeed, the situation is like that in a composite made of continuous fibers, all of which have broken in many places, except that, before load is applied, the matrix around the ends of the fibers is not greatly strained. The importance of the transfer length will now become clear.

We have seen that the stress increases roughly linearly from the end of a fiber. Assuming a cylindrical fiber, radius r, with the shear stress present at the fiber/matrix interface being τ, then the total force on the fiber up to a distance l from the end will be $2\pi rl\tau$. The stress at that section will be $2\pi rl\tau/\pi r^2 = 2l\tau/r$. To make full use of the strength of the fiber it should break before the interface cracks; it should satisfy the condition $\pi r^2 S_{\mathrm{fmax}} < 2\pi rl\tau_{\mathrm{max}}$, S_{fmax} being the failure stress of the fiber. $(S_{\mathrm{fmax}}/4\tau_{\mathrm{max}}) < l/d$, where d is the diameter of the fiber.

So the greater the length of the fiber in relation to its diameter (the greater its aspect ratio), the better will the composite be able to make use of the strength of its fibers. A short fiber cannot be loaded to its failure strength; it pulls away from the matrix first. The transfer length is the length, at each end of the fiber, required for the stress in the fiber to reach a value determined by the general state of stress in the composite. The critical length of the fiber, l_c, is twice the transfer length and is given by: $l_c = dS_{\mathrm{fmax}}/\tau_{\mathrm{max}}$.

If the matrix does pull away from the fibers, it may not necessarily be a bad thing. This debonding is an energy-absorbing process, and the toughness of the composite may be increased, even though the static strength is decreased; it depends on the particular values of strength and interfacial friction.

2.4.2.3 *Orientation effects*

So far we have considered only composites in which the fibers are oriented in the direction of the principal tensile stress. If the principal stress is at an angle to the fiber direction, the effectiveness of the fiber falls precipitously. When the fibers are normal to the direction of the principal tensile stress, the strength is determined almost entirely by the matrix. The theory of composite materials has not yet produced analytical solutions to what happens when the fibers are neither at 0° or 90° to the principal tensile stress. However, there are various more or less empirical solutions. A simple one whose predictions are often near to experimental results is that of Stowell and Liu (1961). This supposes that if the strength of the composite loaded parallel to the fibers is S_p, the strength when loaded normal to the fibers is S_n, and the strength in shear parallel to the fibers is τ, the failure stress at an angle θ in each of the three modes is: $S_p/\cos^2 \theta$; $\tau/\sin \theta \cdot \cos \theta$; $S_n/\sin^2 \theta$, for the parallel, shear or normal modes respectively. To find the composite strength at any particular angle, one merely calculates the lowest stress of the three possible modes.

We have seen in the case of modulus of elasticity that it is possible to reduce the anisotropy of a composite by arranging the fibers at different angles throughout the composite. Often, in engineering practice, the composite consists of a series of laminae, within each of which the fibers have all approximately the same, preferred, direction.

The strength of a composite can also be made more or less isotropic in a similar way. Figure 2.7 shows different arrangements of fibers. Associated with each is an "efficiency factor" η. The strength of the composite is given by $S_c = \eta S_f V_f + S'_m(1 - Vf)$ (Krenchel, 1964). This simple analysis ignores, for instance, problems caused by differences in Poisson's ratio between the fibers and the matrix, and the value of the interfacial shear strength. However, the lesson is clear: there is a very obvious tradeoff between maximum strength and isotropy. For instance, if we assume that the matrix has no strength, then fibers arranged isotropically produce a composite of about 3/8 the strength of fibers arranged all in the direction of loading. If the fibers are arranged isotropically in *three* dimensions, the strength is about 1/5 that of a unidirectional composite.

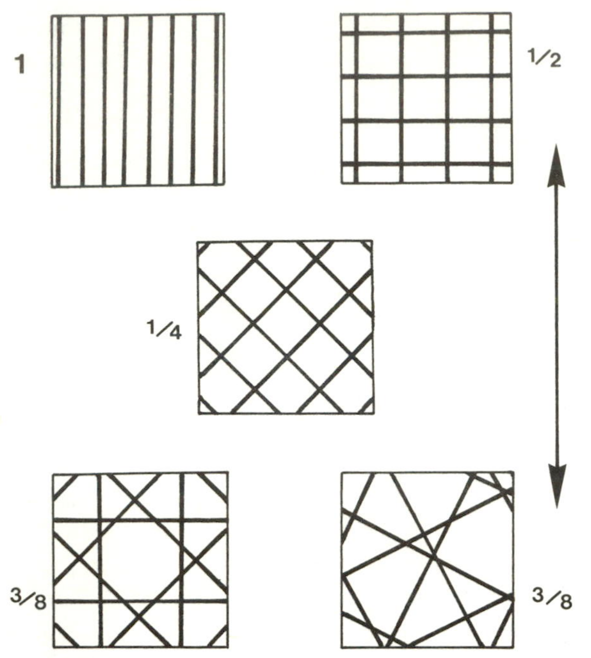

Fig. 2.7. Krenchel's model of how misorientation of fibers affects the strength. The loading direction is shown by the double-headed arrow. The vulgar fractions are the "efficiency factor" η in the equation $S_c = \eta S_f V_f + S'_m(1 - V_f)$.

2.4.2.4 *Levels of analogy*

Such a long introduction to the tensile strength properties of fibrous composites is necessary as much to show what bone is not doing as to show what it is doing. In trying to analyze bone as a composite, we meet the problem of the level at which the analysis should be made (figure 2.8).

At one level we could consider the collagen to be the matrix, in which the mineral "fibers" are embedded. At the next level we could consider the collagen fibrils, with their associated mineral, to be the fibers, or plates in the case of laminae, each more or less loosely connected to its neighboring fibers, all of them making lamellae. At the next level the lamellae are separated by interlamellar regions, which are relatively poor in collagen and which have more ground substance, of dubious chemical constitution, than the lamellae themselves.

At the highest level we could consider the osteons or laminae to be the fibers, again more or less loosely connected to each other by cement lines, "bright lines," and other structures.

The early analogies between composite materials and bone (Currey, 1964a; Mack, 1964) concentrated on the lowest level, that of collagen as

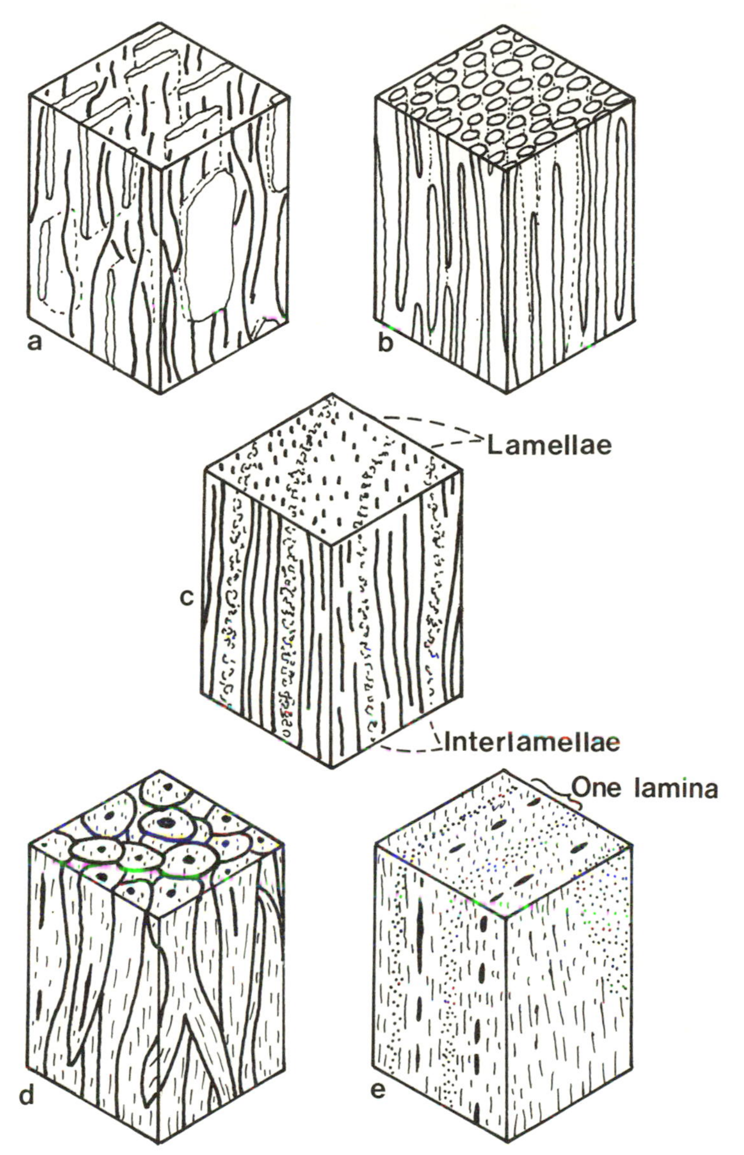

Fig. 2.8: At what level should we try to model bone as a composite? The figure shows, highly schematically, various possible levels. (*a*) Apatite plates in a matrix of collagen fibrils. (*b*) Collagen fibrils, with their associated mineral, loosely connected to their neighbors. (*c*) Lamellae, separated by collagen-poor interlamellae. (*d*) Haversian systems in Haversian bone. (*e*) Laminae in fibrolamellar bone. Blood channels: solid; lamellar bone: fine lines; parallel-fibered bone: dots.

the matrix and the mineral as the fiber. However, this turns out not to have been very fruitful. The reason is that at such a low level, with such an intimate relationship between the mineral and the protein, it is impossible to have much idea of how the collagen is behaving. We have seen in the discussion on elasticity in section 2.4.1 that it has been suggested, and indeed there is some corroborating evidence, that the properties of collagen in bone may be radically different from those it has in, say, tendon (Lees and Davidson, 1977; Hukins, 1978; McCutchen, 1975). At the moment we cannot talk meaningfully about the critical aspect ratio and radius of the mineral because we know little about the mineral morphology, and less about the mode of attachment of the collagen to the mineral. Furthermore, at this level we cannot see the fracture surface clearly enough to make out, for instance, whether the minute mineral blocks pull out from the matrix. It seems to me likely that the composite nature of bone is important, at this level, for fracture as well as for elasticity, but that we shall have to wait a good while before we know much. Professor L. Katz remarked to me at Rensselaer Polytechnic in 1981 that he supposed the reason that the vertebrates had adopted apatite, rather than calcium carbonate, as the mineral for bone was that apatite was very reluctant to form large crystals, and therefore the vertebrates did not have to adopt elaborate mechanisms as, for instance, the mollusks do, to avoid growing large crystals. Large crystals are, of course, brittle and in general to be avoided in materials that need to be tough. Although this idea may not be correct, it struck me at the time as being such a classic case of the "blinding glimpse of the obvious" (Haldane, 1953), that I feel I should introduce it anecdotally.

At the next higher level things are better because the individual fibers and larger structures can be made out with the scanning electron microscope, which is particularly useful for examining fracture surfaces.

2.4.2.5 *Load-deformation curves*

When tested in tension, wet specimens of bone produce a stress-strain curve of characteristic shape: there is a straight elastic region; then a short transitional corner; then a longer, much flatter region, the "yield region" (figure 1.7). The bone is behaving elastically in the first region, as is shown by the fact that if the bone is loaded and unloaded cyclically the stress-strain curves are virtually superimposable. However, if the bone is unloaded and reloaded after the stress-strain curve has flattened out, the curves are no longer superimposable (figure 2.9). The ordinary stress-strain curve to failure and the set of loading and unloading curves have important things to tell us.

(*a*) Bone is to some extent viscoelastic, and it might be that the residual strains *a*, *b*, and *c* in figure 2.9 would recover with time. I have found

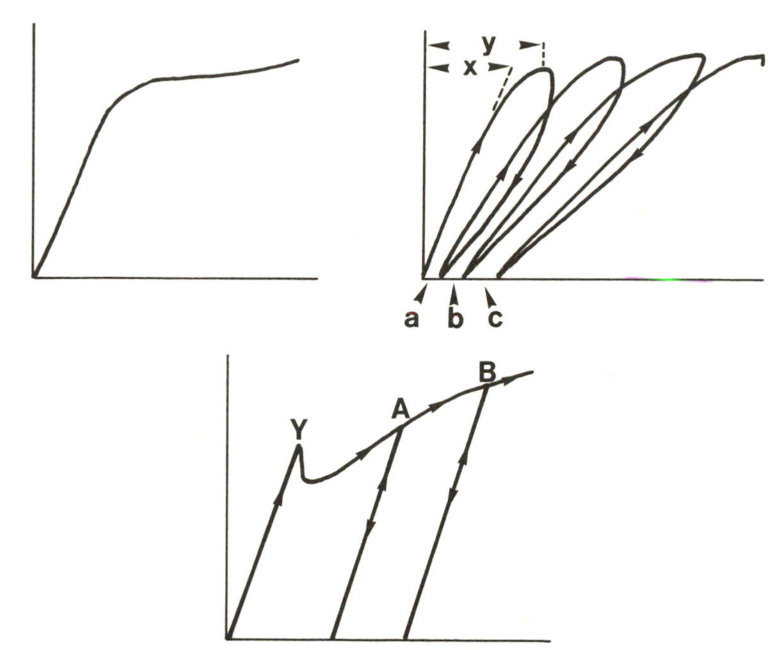

Fig. 2.9. *Top left:* stress-strain curve of bone loaded uniformly in tension until fracture. *Top right:* a bone specimen loaded and unloaded four times. Explanation of the letters in text. *Bottom:* mild steel loaded beyond the yield point *Y* and then unloaded twice, at *A* and *B*.

that some of this strain *is* recovered after a few minutes, but that most of it is not. This indicates that the strains are to a large extent irrecoverable. The related experiment of *bending* a specimen shows that this is correct. A specimen that is bent so that the load-deformation curve flattens out, and is then unloaded, keeps some of its distortion indefinitely. (*b*) The strain at y has two components, an elastic component x, and a component $y - x$, which in some way we can attribute to the yield behavior of the specimen. Because a is less than $y - x$, some of the strain occurring in the yield region is recoverable almost instantaneously on unloading, while some is not. (*c*) The unloading and reloading curves form loops. This shows that the specimen is indeed viscoelastic. (*d*) The slopes of the three loading curves are different, the specimen becoming more compliant each time it is loaded into the yield region. (*e*) The stress achieved does not increase much with strain, the envelope of the curves being rather flat-topped. This is also true of the loading curve that does not involve unloading before fracture.

Before discussing what these observations imply, it is helpful to look at a stress-strain curve for unloading and reloading in a typical metal such as mild steel (figure 2.9). There is a drop at *Y*, at the end of the elastic region, which is produced by a mechanism that need not concern us. On unloading at *A* and reloading there is no loop, showing that steel is barely viscoelastic. The unloading and reloading curves have the same slope as the initial part of the curve, before yield occurred. Steel behaves as if its yield behavior is in some way a function of strain (for instance, *B* is considerably higher than *A*, but its elastic behavior is unaffected by yielding). Basically, this is the case. When steel yields, shear dislocations travel through the material, allowing it to distort rather easily. For various reasons these dislocations become trapped and entangled with each other, and in so doing become much more difficult to move. That is why the stress in the yield region rises with strain. When the material is unloaded and reloaded again, the elasticity is unchanged, because any dislocations whose easy motion might make the steel more compliant have become trapped and cannot move until the unloading stress is again reached. In particular, there are no cracks in the steel; instead it has a population of dislocations, which will be able to move if stressed to various levels.

There is evidence that the yield behavior of bone is caused by tiny cracks forming but not growing very much. For the moment suppose that this is so. How would we account for the behavior shown in figure 2.9? Each crack in the specimen would increase the specimen's compliance; that is, the specimen would become more flexible. So the increased compliance with movement into the yield region can be accounted for, as can the fact that the residual strain on unloading is not as great as the strain that has occurred in the yield region. There is some residual strain, which perfectly neat cracks would not produce. However, in such an irregular material as bone many cracks, once formed, would not be able to close up completely when the specimen was unloaded.

The fact that the yield curve is fairly flat-topped is not explained ipso facto by the formation of cracks. I suggest that what is happening is this. At any cross section in the specimen we can imagine many elements (without specifying what level of organization we are imagining). Some will be weaker and some stronger. When a weak element cracks, the crack will spread a little way, but will then be stopped by some crack-arresting mechanism. This event will have two main consequences.

(*a*) There will be an increased tendency for nearby elements in the section to crack because of the local high stresses produced by the crack.

(*b*) The stress in the section as a whole will be increased because its effective area has been reduced by the crack. This effect will be trivial at first.

Consider first the events at the section that is going to break eventually. There is some distribution of strengths of the different elements. Usually the distribution will be of a few weak elements, many more somewhat stronger elements, and many, many more elements somewhat stronger again (figure 2.10). The first element to break is unlikely to have any

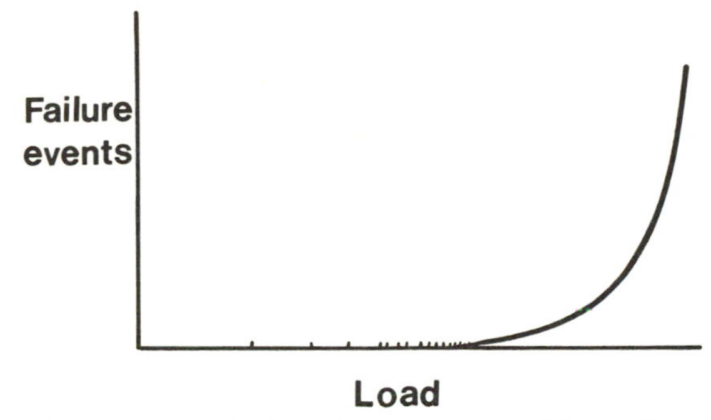

Fig. 2.10. Relationship between the load on a bone specimen and the number of elements failing. Each blip represents the failure of a single element. The solid rising line indicates the distribution when many elements fail at each increment of load.

nearly equally weak elements around it, and so the formation of one crack is unlikely to cause another. Now the load is increased slightly. Soon another element in the section will crack. The tendency of any element to crack will be related to its strength, the amount of the cross section that has already cracked (because this will raise the stress *generally* across the section), and the presence near it of any cracks (because these will modify the stress *locally*). A small increase in load will induce more and more elements of the section to crack, and sooner or later a cascade will occur, and the section will break.

Such a set of events would explain why the *stress* does not increase much after the yield point has been reached, but not why there is so much *strain* in the yield region. To help explain this we can make use here of an observation made by Burstein and his co-workers (Currey and Brear, 1974). When a wet bone section is loaded in tension and the curve bends over, showing that yielding is occurring, part of the specimen goes opaque. The appearance is very similar to crazing in fibrous composites, in which it is caused by the development of many tiny cracks (Harris, 1980). This opacity does not appear at different places all over the specimen; it starts at one level and spreads along the specimen. When the specimen breaks, it usually does so very close to the level from which the opacity started to spread.

Presuming the opacity to represent some kind of microcracking, we can explain the spread of the opacity as being caused by the effect of the section that is beginning to crack on the sections on each side of it (figure 2.11). The cracks in one section will have stress concentrations at their ends, which will locally raise the stress in the neighboring sections up and down the specimen. These volumes of greater-than-average stress will increase the likelihood of cracks appearing in the neighbors. Although the increase in compliance of the specimen as a whole produced

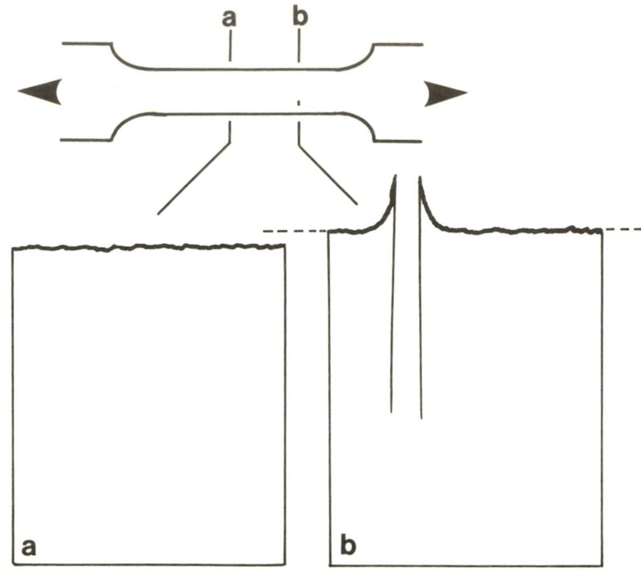

Fig. 2.11. A specimen loaded in tension. *Above:* At section *a* there are no flaws; there is one at section *b*. *Below:* the levels of stress across the two sections. The stress in *a* is not completely uniform because of slight variations in the stiffness and grain of the bone. The stress in *b*, shown by the interrupted line, is generally higher than in *a* because of the reduction in cross-sectional area caused by the flaw. Near the ends of the flaw the stress is much higher due to the stress-concentrating effect.

by a single crack will be tiny, their cumulative effect will be great. The theory behind these ideas of progressive increase in the number of failed elements in a test specimen is discussed by Zweben and Rosen (1970).

So far we have merely assumed that microcracks occur in bone when it yields in tension. We must now briefly review the evidence for this assumption.

Optical effects. I have already mentioned the opacity that appears and spreads in tensile specimens loaded into the yield region. This somewhat fleeting effect can be made permanent by loading a specimen in bending and then immersing it in stain (Currey and Brear, 1974). The stain settles on surfaces, particularly rough ones, and the yielded region on the tensile side, which had become opaque, is now picked out in color. The coloring is not uniform but extends in thin wavy lines part way across the breadth of the specimen. Similar diffuse lines can sometimes be made out at the ends of tensile cracks that have failed to break a specimen in two. These observations do not conclusively demonstrate cracking, but it is difficult to see how the stain could get into the interior of the yielded part of the specimen unless passageways were opened up for it.

Submicroscopic examination. There have been a few reports in the literature of the appearance of microcracks in bone that has yielded but not failed (Carter and Hayes, 1977c; Ascenzi and Bonucci, 1967; Frost, 1960). The observations of Carter and Hayes, who loaded bending specimens six or so times into the yield region, were particularly clear. They reported, "The damage . . . consisted primarily of separation (or debonding) at cement lines and interlamellar cement bands . . . . Occasional microcracking of interstitial bone was also observed. High magnifications of debonding around an osteon revealed significant fibrous tearing." (Carter and Hayes, 1977c, page 269).

It must be added, however, that often it is very difficult to see evidence of microfracture in bone that has yielded. Deer antler is very tough (see chapter 3), and a specimen loaded in impact will deform into a bow without breaking. Examination of sections from the part of the specimen that has gone opaque does not reveal any cracks at the light microscope level. What cracks there are must be very small indeed.

Acoustic emission. Thomas, Yoon, and Katz (1977), Yoon, Caraco, and Katz (1979), and Netz, Eriksson, and Strömberg (1980) loaded bones in bending and examined, electronically, the energy distribution of the noise they made. The noise made by the generation or spreading of cracks is distinguishable from the noise made by the movement of dislocations (Dunegan and Green, 1972). Essentially all the noises made by bone were characteristic of microcracking. Furthermore, Netz and his co-workers showed that they occurred in, or after, the part of the load-deformation curve where the bones started to yield. Knets, Krauya, and Vilks (1975) report similar results with an "avalanche like rise" in the number of cracks as the bone approached failure. However, as these workers probably tested their specimens dry (the paper is not clear on the point), their results are less relevant to the life situation than those of Netz and his co-workers.

It seems reasonable to conclude that the yield behavior of bone in tension is the result of the formation of innumerable cracks which, because of the way the various elements of bone, such as osteons, laminae, lamellae, and fibrils, are put together, cannot travel far and are soon brought to a halt. This ability to crack, yet not fail, accounts for the toughness of bone, such as it is.

Although progressive cracking seems reasonable as an explanation of the flat part of the load-deformation curve, some ingenious experiments of Burstein et al. (1975) permit a different explanation. Tensile specimens were progressively decalcified with dilute acid. Yield stress and ultimate stress declined progressively, but there was no significant change in yield or ultimate strain. Furthermore, the gentle slope of the postyield region remained constant. These workers suggest that the yield strain is determined by the mineral, but that the postyield stiffness is determined by the organic matrix alone (figures 2.12 and 2.13). This explanation is at variance with the idea of progressive cracking set out above. It seems to

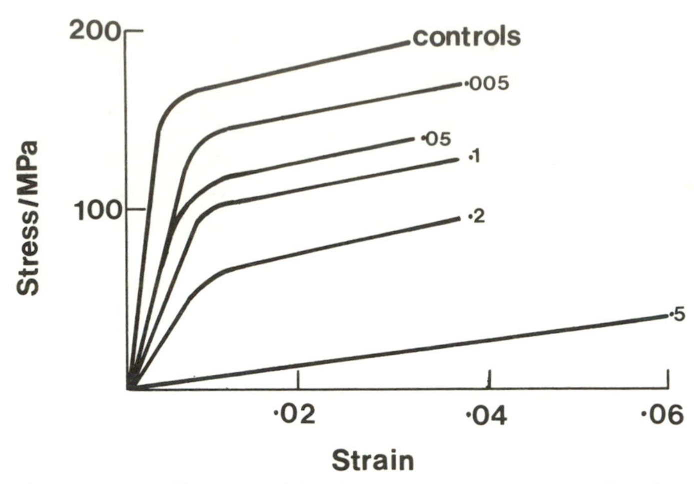

Fig. 2.12. The experiments of Burstein et al. (1975). Average stress-strain curves of tensile specimens of bone decalcified in different strengths of HCl whose normality is shown at the ends of the curves. The postyield slope is almost the same for all amounts of decalcification.

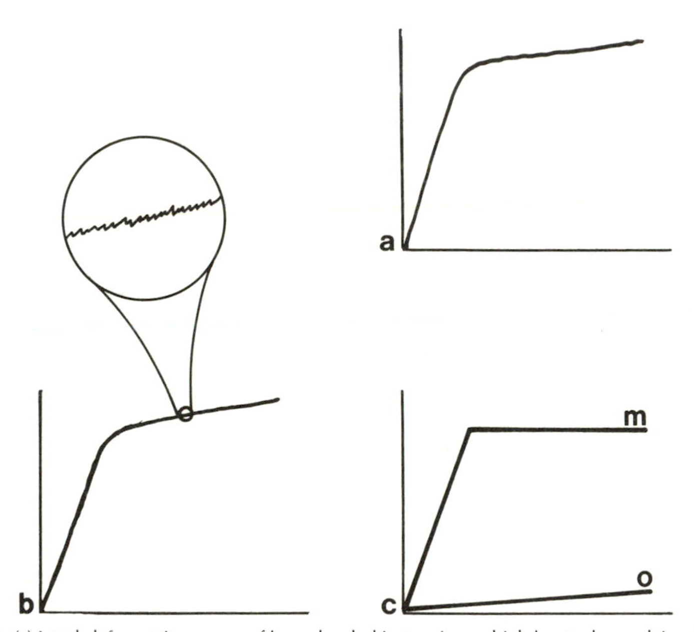

Fig. 2.13. (a) Load-deformation curve of bone loaded in tension, which has to be explained. (*b*) My explanation. After the yield point the bone suffers a very large number of tiny cracks (part of the curve is shown greatly enlarged). (*c*) The explanation of Burstein et al. The curve is the sum of the curves for mineral and for the organic component. The mineral, *m*, behaves as if it were perfectly plastic after the yield point. The organic component, *o*, is thereafter the sole contributor to the increasing load.

me that the experiments of Burstein et al. do not really distinguish between their own hypothesis and that of progressive cracking, both of which would, with a progressive decline in the mineralized bone/collagen ratio, predict a lower elastic modulus, a lower yield stress, unchanged yield strain, and a low postyield modulus. Intuitively it seems improbable that the collagen could free itself, as it were, from the mineral, and show its own mechanical properties once the mineral had yielded.

2.4.2.6 *Crack starting and stopping mechanisms*

In a truly brittle material, if high strength is to be achieved, there must be no flaws on the surface or within the body of the material. The only material I know of, produced by a living organism, that comes in reasonably large blocks and that is truly brittle is the skeleton of echinoderms (starfish, sand dollars, and sea urchins, etc.). This skeleton consists of blocks made of single crystals of calcite. The external surfaces are kept remarkably smooth by a layer of living tissue. Nevertheless, echinoderm skeletons are probably rather low stress structures (Nichols and Currey, 1968). All other skeletal materials seem to have at least some degree of toughness, and are usually rather flawed. This is certainly true of bone.

Bone, in fact, has a mass of potential stress-concentrating discontinuities. Most obvious are the cavities for blood vessels, osteocyte lacunae and canaliculi. It is fairly easy to calculate their stress-concentrating effect because they are, for all intents and purposes, voids, though filled by fluid. Currey (1962) showed that the stress-concentrating effects of the flattened osteocytes could be as great as $7S$, where S is the general level of stress, in the region of the stress concentrator. Even cylindrical blood channels could have a stress-concentrating effect as great as $3S$. Whether these effects are important depends on the size of the cavities and their orientation with respect to dangerous stresses.

Fracture mechanics theory shows that a sharp crack will not spread if it is below a certain size, the size depending on the intensity of the local stress field. A large stress concentrator in the region of a crack may allow it to run, even though the average stress in the material is low (figure 2.14). The crack on its own could not spread because the amount of strain energy available locally would not be great enough. (The thermodynamic criterion for crack spread would not be met [section 1.2.4].) The large stress concentrator on its own would not be dangerous because, although the stress in its neighborhood is higher than generally, it is still far below that required to initiate a spreading crack. (The stress criterion would not be met.) However, the large stress concentration will produce a stress field around the crack sufficiently intense and extensive for strain energy to be available to drive the crack forward.

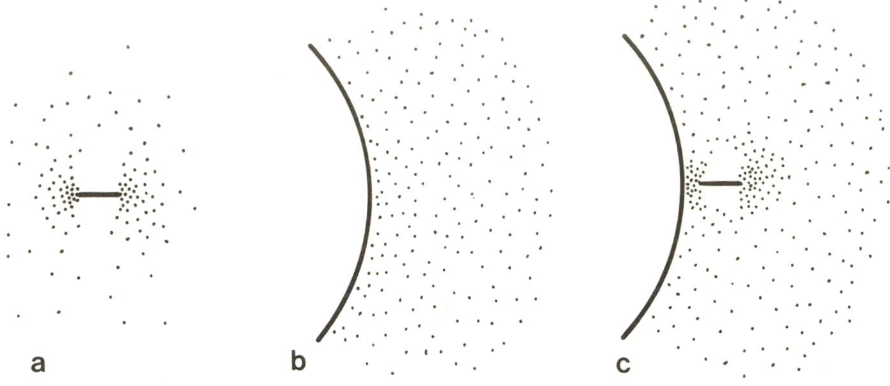

Fig. 2.14. (*a*) A short sharp crack does not strain a large enough volume of its surroundings for there to be enough strain energy to allow it to spread. (*b*) A large stress concentrator may not have a sharp enough end to allow it to spread. (*c*) Together they may be able to initiate crack travel.

The other potential stress concentrators in bone are morphological features, such as sharp notches or discontinuities on surfaces, and concentrations caused by large differences in modulus, such as may possibly occur at cement sheaths around osteons. Figure 2.15 shows the stress-concentrating effects of some simply shaped discontinuities.

Stress concentrations formed by cavities and discontinuities are, therefore, a potential hazard in bone. Does the overall or histological structure of bone give any indication that it is adapted to this state of affairs? Perhaps the most obvious feature, in long bones anyhow, is the way in which the blood channels are arranged. Blood channels are the largest potential stress concentrators. They run predominantly obliquely through long bones, at quite a small angle to the bone's long axis. They also tend to take a gently spiraling course around the long axis. In this arrangement their stress-concentrating effect on most ordinary stresses will be very small, though whether spiraling has any mechanical function is doubtful The obliquely arranged blood channels have a roughly circular cross section. However, in laminar bone there are two-dimensional networks of blood channels between each pair of laminae (figure 1.16). In these networks the blood channels run in all directions in the tangential plane. Nevertheless, there is a predominance of channels running nearly longitudinally. Also, the channels in the networks are elliptical in cross section, and this will reduce the stress-concentrating effect of those that run nearly horizontally (figure 2.15).

Given the necessity for blood channels in bone, their arrangement seems to be mechanically sound. But this might be an example of mechanical serendipity—the arrangement of blood channels being determined mainly by the requirements of an efficient blood supply. This may be so,

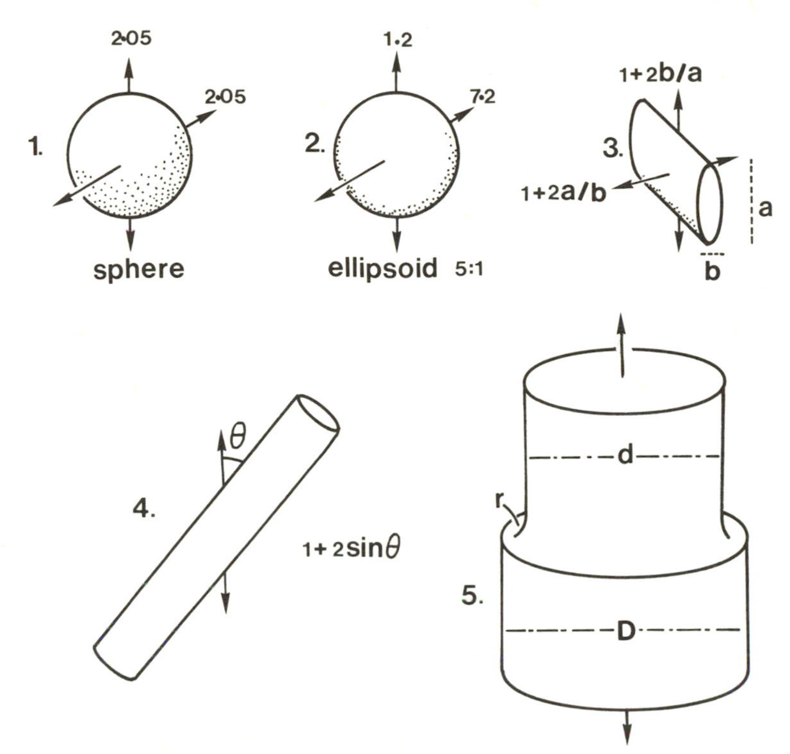

Fig. 2.15. Stress concentration factors of some simple discontinuities. (*1.*) A sphere. The stress concentration is 2.05 for all normal stresses. (*2.*) An ellipsoid of revolution with a ratio of 5:1 in the major and minor axes. If the stress is normal to the major axis, the stress concentration is 7.2, but only 1.2 if normal to the minor axis. (*3.*) An elongate cavity of elliptical cross section. (*4.*) A cylindrical cavity at an angle θ to the stress direction. (*5.*) A cylinder with a step, loaded in tension. If $r = D - d$, and $r = d/10$, the stress concentration is about 1.78.

yet the blood supply of long bones would seem a little bizarre if the mechanical effects of blood channels were not important. The efficiency of nutrient and waste transfer would be greatest if the very narrow vessels in the cortex of the bone were short, running directly from endosteum to periosteum. Yet few do.

Long bones usually have one or more quite large nutrient arteries that pierce the cortex and supply the marrow. These nearly always run obliquely through the cortex. There are many arguments about the mechanism producing this obliquity (Brookes, 1971), yet the mechanical advantages of it never seem to be mentioned by anatomists.

The lacunae of osteocytes show a variety of shapes, from nearly spherical in woven bone to flattened oblate spheroids, with a ratio of major to minor axes of 3.3:1, in lamellar bone (Currey, 1962). A tensile stress in the direction of the short axis of such an oblate spheroid would produce a stress concentration of 5.2, while in the direction of the long axis it would be only about 1.3 (Peterson, 1974). Lamellar bone is very, very rarely arranged so that the long axis of the bone cuts across the lamellae. The osteocyte lacunae lie in the plane of the lamellae, so they also lie with a long axis along the long axis of the bone, mechanically the most advantageous direction.

Canaliculi spread right through bone tissue at all angles and, therefore, theoretically must produce stress concentrations of times three throughout the tissue. However, they are very small, about 0.2 to 0.5 μm, even in relation to potentially dangerous cracks, and so will not be effective in helping cracks to spread.

Bones very rarely have sharp steps on their surfaces. All necessary changes in diameter are smooth. This is particularly evident in cancellous bone, where the holes in the plates and the junctions between longitudinal and transverse struts are well "radiused" (figure 1.17).

In woven bone the osteocyte lacunae are fairly spherical, and the blood channels are much more variously arranged. However, woven bone is adapted for speed of construction, not mechanical excellence.

It does seem, therefore, that the cavities in adult bone are arranged in such a way as to minimize their stress-concentrating effects, and the external form of the bone is also reasonably smooth. This apparently appropriate arrangement of potential stress concentrators is real only if the dangerous stresses run reasonably along the length of the bone which, of course, they usually do. If bone is loaded in other directions, its fracture behavior alters drastically.

2.4.2.7 *Anisotropy in fracture*

Reilly and Burstein (1975) studied the anisotropy of bone in some detail. Their findings about Young's modulus have already been discussed (section 2.4.1). They consider Haversian bone to be approximately transversely isotropic. That is to say, it has only one axis of symmetry, along the length of the bone. The behavior in all directions at right angles to this axis is the same. Their data do support this. Human and bovine Haversian bone are rather similar, except that the ultimate strain in bovine bone loaded along the axis is rather low (table 2.2). The contrast in the properties in the two directions is striking. The ultimate tensile strength is only about 30%–40% as great in the transverse direction as it is longitudinally. The ultimate strain is similarly reduced. This reduction in strain is brought about mainly by a reduction in length of the yield region.

The behavior of fibrolamellar bone is not transversely isotropic, the behavior in the radial direction being different from that in the transverse direction (figures 2.16, 2.17). The tensile strength in the radial direction

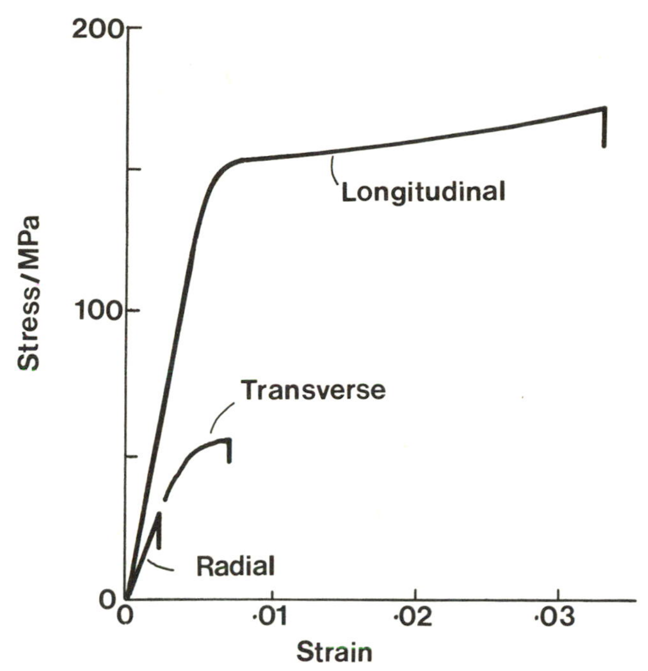

Fig. 2.16. Idealizations of stress-strain curves of bovine fibrolamellar bone loaded in tension. (Derived from data in Reilly and Burstein, 1975.)

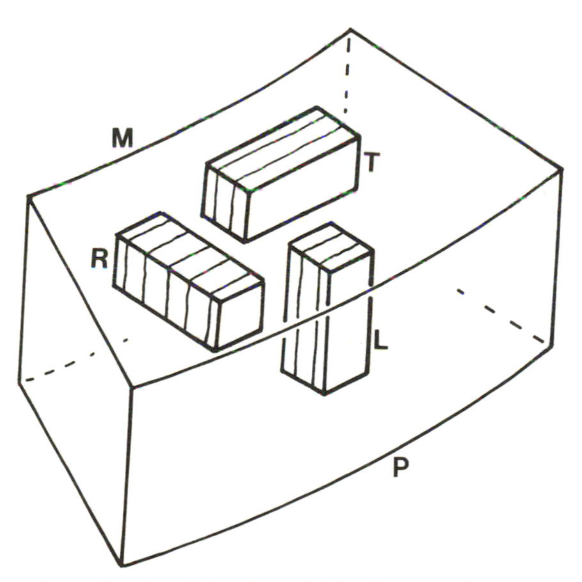

Fig. 2.17. Diagram of a small section of a long bone. *M*: the side next to the marrow cavity; *P*: the side next to the periosteum. Three differently oriented test specimens are shown, all loaded along their long axes. *L*: longitudinal; *T*: transverse; *R*: radial.

is pathetic. The amount of energy absorbed by radial specimens before fracture is only about 1/150 the energy absorbed by longitudinal specimens. This is brought about by the combination of low tensile strength and, more importantly, by the almost completely brittle behavior of the radial specimens. Of course, the poor performance of radial specimens usually would not matter in life, because it is difficult to arrange things so that bone is loaded radially. When bones are loaded radially, as may happen under muscle insertions (the tuberosity on the radius under the insertion of the biceps, for example), the histology of the bone is quite altered, its grain leading smoothly into the line of action of the tendon.

This anisotropic behavior is what we should expect. The collagen fibrils, the cement sheaths, the blood vessel cavities, and indeed, nearly all the structures in long bones are for the most part oriented along the long axis. They do not cause stress concentrations and, if a crack does start to spread, they may well be positioned to prevent its further spread, depending on the direction from which the crack comes. They will do this by forming a mass of relatively weak interfaces at a large angle to the direction of crack travel. The structures forming these interfaces will delaminate or pull out, increasing the energy needed to continue the crack spread. The crack will also be blunted. In fact, the situations shown in figure 2.16 are for an isolated test specimen. The longitudinal specimen will tend to break with the crack going in the transverse, easier direction *E* (figure 2.18), because the crack-interrupting effects of the laminae will not be met. In life, however, the crack will have to travel from the outside of the bone (where the stresses are greatest) inward, in the difficult direction *D*, and the laminae will be more effective.

When the crack is traveling along the line of the structures in the bone, as happens when the test specimen is radially oriented, the situation is trebly different. First, the structures in the bone are oriented with respect to the stresses so as to produce large stress concentrations. Second, the side-to-side connections between neighboring fibrils, between neighboring lamellae, between laminae, and between Haversian systems and their neighboring interstitial lamellae are relatively weak. As explained above, this is advantageous when the crack is spreading across these interfaces. The cement sheaths around Haversian systems are certainly rather weak—cracks have frequently been seen going around them (McElhaney, 1966; Carter and Hayes, 1977c; Behiri and Bonfield, 1980). Laminae are very weak at the level of the blood channel networks. There is considerable reduction of effective area caused by the presence of blood channels all in one plane, and laminar bone, loaded radially, nearly always breaks at the level of the blood channels. Third, when the crack is traveling along the length of the bony structures, there is little tendency for the crack to

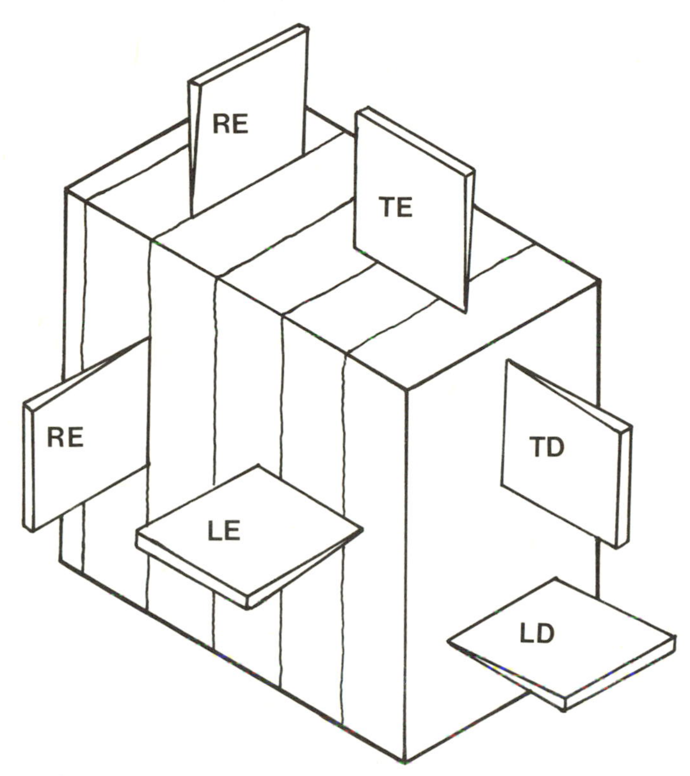

Fig. 2.18. The directions of crack travel in tensile specimens of fibrolamellar bone. *R*, *T*, and *D* refer to the direction of loading that would cause cracks (shown by wedges) to travel in these directions. *E* and *D* refer to more or less easy directions for crack travel. The radial crack can travel very easily in either direction. The difference between *TD* and *TE* and between *LD* and *LE* is probably not great, but in the difficult direction the periodic interlaminar vascular network would act as planes of weakness, tending to stop the crack.

become blunted, any delamination occurring in line with the direction of crack travel.

The effects of these factors are seen particularly clearly in fibrolamellar bone. When laminar bone is pulled longitudinally in life, all the factors cooperate to maximize the strength and crack-stopping ability of the material. When loaded in the transverse direction the general grain of the fibrils is at right angles to the stress. As a result strength and toughness are much reduced. In the radial direction the lamellae, the blood vessel networks, and the laminae are in the worst possible orientation, and strength and toughness are minuscule.

The description given so far of tensile fracture is somewhat idealized, because there are so many factors, ignored till now, that can alter the tensile strength of bone. Apart from the direction of loading, the strength can be influenced by the histological structure of the bone, its amount of mineralization, its age, and the strain rate at which it is loaded. Most of these factors are connected with each other, but their influence can be summarized.

2.4.2.8 *Histological structure*

In bovine bone, which is the only type of bone that has been thoroughly studied in this respect, it is now accepted that primary fibrolamellar (laminar) bone has a higher tensile strength than remodeled, Haversian bone (Currey, 1959; Evans and Bang, 1967; Reilly and Burstein, 1975; Carter and Hayes, 1976a; Saha and Hayes, 1976; Saha, 1982). This is true in both static and dynamic loading. The effect is quite marked, a completely laminar specimen being more than one and one-half times as strong as one made entirely of Haversian systems or their remnants (Currey, 1975).

I have recently carried out some experiments on unnotched bending specimens in impact, and have calculated work of fracture from sharply notched bovine femoral specimens loaded slowly, which have rather complicated the issue. Impact strength ranged from 3 to 10 $\mathrm{kJ\,m^{-2}}$. Work of fracture ranged from 1.3 to 2.8 $\mathrm{kJ\,m^{-2}}$. The highest impact strengths and highest works of fracture were from specimens with clear laminar structure, with no, or virtually no, Haversian remodeling. However, the weakest specimens tended to consist of compact coarse-cancellous bone, with a greater or lesser amount of Haversian remodeling. This compact coarse-cancellous bone is organized in a rather chaotic way; in particular, it is usually oriented in directions at an angle to the long axis of the bone from which the specimens came. This orientation is caused by the grain of the cancellous bone, which is the scaffolding on which the compact bone is deposited. The cancellous bone may be oriented in all sorts of ways relative to the long axis of the definitive bone. When this chaotically oriented bone is replaced by Haversian systems, these are oriented more nearly along the axis of the bone than the bone they replace, as if attempting to bring order out of chaos. So, the strongest bone is fibrolamellar bone, but the weakest is compact coarse-cancellous bone, which is also primary, but very badly oriented. I find it faintly disturbing that Haversian bone and compact coarse-cancellous bone are often difficult to distinguish. One hopes that the reports in the literature have not confused them, because they are very different tissues, one being primary, the other secondary. (On checking my old specimens I find there is, fortunately, effectively no compact coarse-cancellous bone among them.)

2.4.2.9 *Mineralization*

The greater the mineralization, the greater the tensile strength (Currey, 1975). This is not true, as we shall see in the next chapter, when the amount of mineralization becomes very great. Also, it may not be true of toughness at all. It is difficult to distinguish the effects of mineralization per se and of reconstruction, because Haversian systems, being necessarily younger than the bone in which they lie, tend to be less well mineralized, so reconstruction and low mineralization tend to go together. However, Carter, Hayes, and Schurman (1976) have shown, in investigating fatigue strength, that it is possible to distinguish the effects with a large sample size and careful experimentation.

The reasons for the weakening effect of reconstruction are not clear. There is a trivial effect caused by the area of bone tissue in Haversian bone being less than that in a similar-sized piece of well-developed fibrolamellar bone because of the presence of erosion cavities. Another possible factor is the difference in modulus caused by differences in mineralization. The less mineralized bone in Haversian systems will have a lower modulus than the surrounding bone. This modulus mismatch can have a stress-concentrating effect, but not an important one, because even if the Haversian system bone had a modulus half that of the surrounding tissue, the stress-concentrating effect when the bone was loaded in the worst possible orientation would be only times 1.5 (Peterson, 1974, figure 213). Cement sheaths around Haversian systems are strange things. Ortner and von Endt (1971) claim that they consist of highly calcified mucopoly-saccharide, with no collagen. This constitution would certainly seem to fit their strength, which apparently is rather low.

Nevertheless, for whatever reason, Haversian bone is definitely weaker in tension than primary fibrolamellar bone. Where Haversian systems replace compact coarse-cancellous bone, they probably have a beneficial effect on the strength of bone by improving the orientation of its grain. However, they do not have a good effect when they replace well-organized fibrolamellar bone. Haversian bone might seem to be well designed for toughness in that it has many potentially crack-stopping cement sheaths throughout the tissue. Presumably the deleterious effects of modulus mismatch and the boring of holes more or less at random through previously well-ordered tissue outweighs this possible advantage.

2.4.2.10 *Strain rate*

The tensile strength of bone increases with strain rate, at least up to rates producing fracture in a time not much less than that occurring in life, say twenty milliseconds. Torsional tests, in which the bone is loaded in shear, produced a 1/3 increase in ultimate strength over a 3,000 to 10,000 increase in strain rate (Sammarco et al., 1971). I found an increase of tensile strength and tensile yield stress of about 1/2 with a thousandfold increase in strain rate, from 0.0001 to 0.1 per second (Currey, 1975). The strain rate of 1/10 per second is rather low. More satisfactory, in that it incorporates strain rates including physiological rates, is the study of Wright and Hayes (1976). They tested femoral cortical bone over a series of strain rates from 0.001 to 100 per second. They found the stress at failure continued to increase at all the strain rates they investigated. The effect was not strong, tensile strength being proportional to strain rate to the 0.07 power.

Saha and Hayes (1976) have loaded bone in tension in impact at very high strain rates (about 1,000 per second, which took about 1/10,000 seconds to fracture). Although they found a weakening effect of Haversian systems, they were unable to obtain the very uniform load-deformation curves that, for instance, Burstein and his co-workers consistently obtained at lower strain rates. This may well be because of the shock waves that would be set up by such extremely rapid loading. These waves would bounce off interfaces, cavities, and the edges of the specimen, reinforcing and canceling each other out in unpredictable ways. Despite the fact that these tests were performed at unphysiologically high rates of loading, the disparity of the shapes of the curves they found and of those produced at lower strain rates is a warning: static tests are useful, but they may not give a clear idea of what goes on at higher loading rates. Utenkin (1975) loaded human femoral specimens in impact, using a falling pendulum, and found marked anisotropy. Those specimens loaded so that the tensile stresses were along the bone axis were eight times stronger than specimens loaded at right angles to this direction.

Bonfield and Behiri (1983) find that fracture toughness of bone increases with *crack velocity* at rather low velocities, up to 10^{-3} m s^{-1}, but is lower at the catastrophic rates seen when a bone breaks in life.

2.4.3 Fracture of Bone in Compression

So far, we have talked only about fracture in tension. Compressive fracture is certainly less important in life, but it can occur, and the mode of compressive fracture is rather different from tensile failure. Of course, bones are often loaded in compression and, indeed, as I shall discuss in Chapter 6, the skeleton seems to be designed so that tension and bending loading of bones are minimized. Nevertheless, most accidents that fracture bones produce loading that deforms the bone in bending, and this will produce tension on one side of the bone.

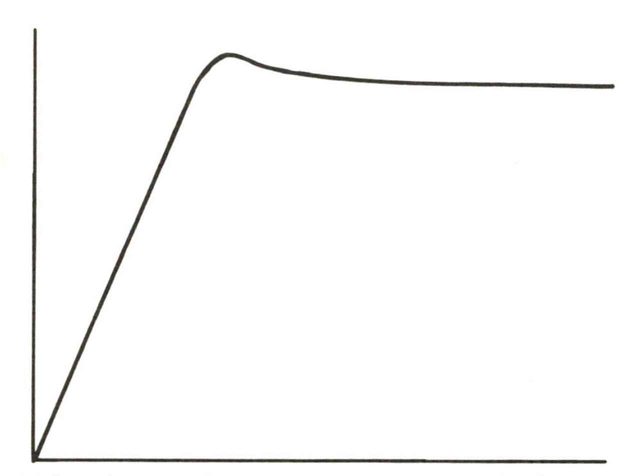

Fig. 2.19. Load-deformation curve of compact bone in compression. After the yield point the load drops somewhat, and then remains more or less constant for very variable amounts of deformation; ultimate strains may be as great as 5%.

The load-deformation curve of a compressive specimen shows a linear portion followed by a short yield region, the load soon dropping slightly and then continuing at roughly the same level until the specimen finally breaks up (figure 2.19). This final breakup may occur only after considerable deformation.

Yield stress is higher in compression than in tension. Yield in compression is microscopically quite different from yield in tension. What we can call "shear lines" appear. They are clearly described by Tschantz and Rutishauser (1967) and Chamay (1970). These shear lines show a strong tendency to initiate at large stress concentrators such as blood channels, which is not the case with the diffuse lines seen in tensile loading. When the specimen is loaded along the length of the bone, the lines form an angle of 30° to 40° with the long axis. These lines probably represent very small-scale buckling of lamellae or fibrils. The discontinuities initially formed soon coalesce and make small cracks. These spread through the specimen, which eventually disintegrates.

There have been a number of studies on the effect of strain rate and other variables on the compressive failure of bone; in general the same story emerges as for tensile failure.

2.4.3.1 *Anisotropy*

The anisotropy of bone in compression is not as marked as it is in tension (table 2.2). This is what we might expect. When a specimen is loaded laterally or radially, there are more stress concentrations in dangerous orientations than when it is loaded longitudinally. On the other hand, in longitudinal loading the rather weak side-to-side loading of fibrils, which is an advantage in tension, now becomes a disadvantage, allowing these structures to separate from each other by buckling. This effect is seen strikingly in wood in which, because of the presence of cavities into which the cell walls can buckle, the compressive strength is only about one-half to one-third the tensile strength (Jeronimides, 1980).

2.4.3.2 *Histological structure and mineralization*

Haversian reconstruction makes bone weaker in compression as well as in tension. This was first determined by Hĕrt et al. (1965), and has been confirmed since by Robertson and Smith (1978). There seems to be no work on mineral content and compressive strength of ordinary specimens, though Ascenzi and his co-workers have done many tests on isolated osteons, to which I shall refer shortly.

2.4.3.3 *Strain rate*

McElhaney (1966) tested small blocks of bovine and human bone in compression and found that, between strain rates of 0.001 to 1 per second, there was an increase in compressive fracture stress. Bovine bone increased from 180 to 250 MPa, human bone from 150 to 220 MPa. McElhaney's specimens showed reasonable plastic deformation, the ultimate strains at strain rates of one per second being 1.2% and 1.8% for bovine and human bone respectively.

Robertson and Smith (1978) have observed a slight increase in compressive strength in rather immature pigs' mandibles, with strain rates from $2.4 \times 10^{-4}\,s^{-1}$ to $2 \times 10^{-1}\,s^{-1}$. They also observed a change in the shape of the stress-strain curve. This had a long plastic region at $2 \times 10^{-4}\,s^{-1}$ and became linear and brittle by $10^{-1}\,s^{-1}$. They found a transition from ductile to brittle behavior at about $2.4 \times 10^{-3}\,s^{-1}$. It is difficult to reconcile these findings with those of Reilly and Burstein (1975). These latter workers observed very long plastic zones at compressive strain rates of $0.02–0.05\,s^{-1}$. McElhaney (1966) also found considerable plastic deformation. It may be that the findings of Robertson and Smith are affected by the age of the bone; they used pigs of six weeks to seven months old, and unfortunately do not report age-related effects, which surely there must be. The maximum strengths they report, of the order of 120 MPa, are certainly low.

2.4.4 Single Haversian Systems

Over many years Ascenzi and his co-workers have attempted to discover the mechanisms of the elastic and fracture behavior of bone by investigating single secondary osteons, mainly from human bone. This work has given much useful information, but it does suffer, as these workers acknowledge, from the fact that only one element in a composite is being studied. In particular the machining out of Haversian systems removes the cement sheath, by which the Haversian system is connected to the rest of the bone. Also, we must remember that Haversian systems are not found at all in many bones.

To test Haversian systems in tension, they grind longitudinal sections of bone about 20 to 50 μm thick. This is much less than the diameter of a Haversian system, which is characteristically about 200 μm, and so only longitudinal segments of Haversian systems remain. Where a Haversian system is sectioned so that its central canal is in the plane of the section, one side is isolated as a test specimen. For compressive tests they, in essence, mill out a cylindrical test specimen about 1/2 mm long. They tested specimens wet and dry. I shall describe tests on wet specimens only. They consider, usually, two types of Haversian system, apparently distinguishable from each other under the polarizing microscope. I call these "longitudinal" and "alternating." In the longitudinal specimens the fibrils in the lamellae have a very low angle to the long axis of the system, and there is a small difference in the direction between neighboring lamellae. In alternating systems the fibrils in one lamella will be nearly longitudinal, and almost at right angles to this in the neighboring lamellae. (It should be pointed out that some workers think that Haversian systems are not so neatly classifiable as this [Evans, 1973; Katz, 1981].) Even so, the photographs produced by Ascenzi and his co-workers do seem to fit their description.

The Ascenzi school also distinguishes between fully calcified systems and those that have just started to calcify. Initially, Haversian systems calcify very rapidly, and then the process runs more slowly, so the difference in the amount of mineralization between "partially" and "fully" calcified systems is not great.

The results of their studies on single systems are given in table 2.4, derived from Ascenzi and Bonucci (1967, 1968). The effect of completing calcification is always to increase the strength and modulus. Considering now only completely calcified systems, there is an indication that longitudinal systems are stronger in tension and weaker in compression than alternating specimens.

Two things are striking about these results. One is that the ultimate strains seem very high. In particular, a strain of 10% in failure in tension is unheard of in the testing of ordinary samples. I can think of no reason why the strain should be so large, while the strength is not unusual. The other striking thing about these results is that there are very large differences in the values for Young's modulus in tension and compression,

Table 2.4 Mechanical Properties of Isolated Haversian Systems

Species	Mineralization	Fiber Direction	Strength/MPa Tension	Strength/MPa Compression	Young's modulus/GPa Tension	Young's modulus/GPa Compression	Tensile Elongation at Break/%
Man	Starting	Alternating	91	100	4.5	7.4	9.8
		Longitudinal	110	90	6.1	4.9	9.4
	Complete	Alternating	96	167	5.6	9.5	10.3
		Longitudinal	116	112	11.9	6.5	6.8
Cow	Starting	Longitudinal	113	—	5.5	—	11.1
	Complete	Longitudinal	121	—	15.0	—	8.3

even though they are low. Admittedly, the values for tension and compression were not measured on the same samples, but it does seem unlikely that the tensile modulus of the longitudinal system should be almost double the compressive modulus, these differences, incidentally, being highly statistically significant.

The mode of failure in single systems is interesting. The compressive failures involved fissures traveling at 30° to 35° to the long axis of the system. They are very similar to the kind of thing seen in larger test specimens. The fissures show little tendency to deviate or interrupt when traveling from one lamella to the next. The behavior of alternate systems in tension is more complicated (Ascenzi and Bonucci, 1976). Above a load of about 30% of the failure load the transversely oriented lamellae break periodically along their length, but the structural integrity of the specimen is maintained by the longitudinally oriented lamellae. This is accompanied by a change in the slope of the load-deformation curve (figure 2.20). The transversely oriented lamellae crack frequently along their length and eventually, of course, the longitudinal lamellae break. The situation seems to be much as in some engineering laminates, in which alternate sheets are oriented at about 90° to each other. As we saw earlier in this chapter (figure 2.7) such a laminate would notionally have a strength, when loaded in the direction of one of the sets of fibers, not much greater than half that of a unidirectional laminate if the matrix is weak. In fact, according to Ascenzi and Bonucci's 1967 study (table 2.4) there is only a 15%–20% difference. Even so, these microscopic studies do show why, for unidirectional loading, the alternate arrangement may not be very efficient. The Young's modulus of the transverse lamellae is presumably less than that of the longitudinal lamellae. Also the sets of lamellae are in parallel. Therefore, for any given strain the stress in the transverse lamellae is less than in the longitudinal lamellae. However, if as seems to be the case, the

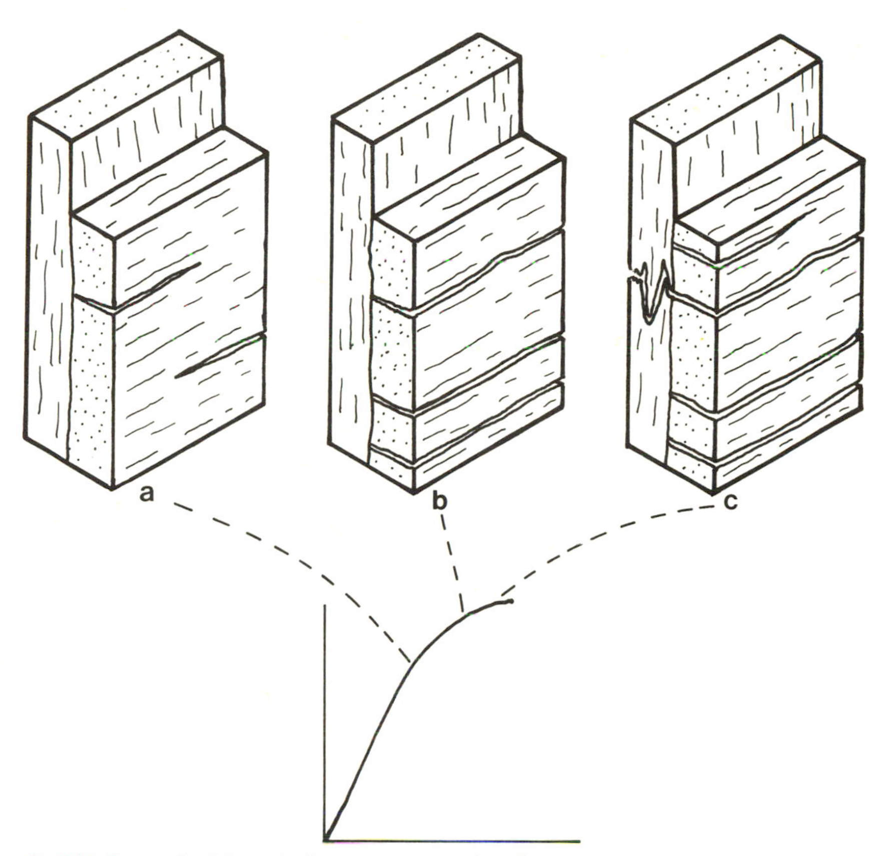

Fig. 2.20. Progressive failure of "alternate" Haversian lamellae. (*a*) Cracks appear in the transversely oriented lamellae. The load-deformation curve stops being linear. (*b*) More cracks; the curve flattens. (*c*) Longitudinal lamella fails; failure of the whole specimen is imminent. (Based on ideas in Ascenzi and Bonucci [1976] with modifications.)

difference in strength is greater than the difference in modulus, then the transverse lamellae will break first. As the transverse lamellae break, more and more of the load is thrown onto the longitudinal lamellae. The system is weakened, therefore, by the presence of the transverse lamellae.

However, the situation is not nearly so clear as has been made out so far. First, the spatial arrangement of the lamellae into longitudinal and transverse, neatly alternating, has been questioned. In particular, Frasca, Harper, and Katz (1977) have isolated and examined individual lamellae from Haversian systems. Systems the Ascenzi school would classify as longitudinal do indeed have a predominance of longitudinal fibers. However, alternate osteons still show a dominance of longitudinal fibers, according to Frasca et al., but now there are a number of "oblique fibers, circumferential fibers and coexisting fibers of various orientations." In general, the studies of Katz's group reveal a great predominance of fibers oriented in the direction of the bone axis and, in trabecular bone, along the axis of the trabeculae (Katz, 1981). The group could find no sets of lamellae in which longitudinal and transverse fibers alternated neatly and regularly. Yet, the photographs of Ascenzi and Bonucci clearly show regions where this alternation occurs, and in these regions the bone does fail like an alternating composite.

The studies of Ascenzi's school are very interesting. They have recently started to investigate the properties of single Haversian lamellae (Ascenzi, Benvenuti, and Bonucci, 1982) with intriguing, but inconclusive, results as yet. However, there are problems associated with them, which I have mentioned and, of course, the isolation of single systems precludes the incorporation of cement sheaths into any model of the behavior of bone.

2.4.5 Cancellous Bone

The mechanical properties of cancellous bone are not nearly as well known as those of compact bone. Because some of the ideas necessary to understand cancellous bone, such as buckling, are not introduced yet, and because cancellous bone never occurs on its own, but instead always forms structures with compact bone, I shall discuss it in chapter 4. Cancellous bone is always much less stiff, and is rather weaker, than compact bone.

2.5 Fatigue Fracture

In most materials, repeated loading of a specimen to stresses lower than the failure stress can result in fracture. Such a fracture is called a fatigue fracture. Fatigue fractures occur in bone. I shall consider the phenomenon in general before discussing the particular behavior of bone. The most useful way of testing the fatigue behavior of a material is to construct a so-called S–N diagram (figure 2.21). A set of specimens is subjected to various stress amplitudes, usually in a cyclical manner. This mode of loading is convenient, because one can load a specimen in bending and then rotate it about its long axis. The stress amplitude S is then plotted against the number of cycles required to cause the specimen to fracture, N. The resulting S–N diagram is usually plotted with S on a linear or a logarithmic scale and N on a logarithmic one. It has two characteristic features. One is that the lower the stress amplitude, the greater the number of cycles to failure. The other is that materials may show one of two behaviors: they either show a stress, the "endurance limit," at or below,

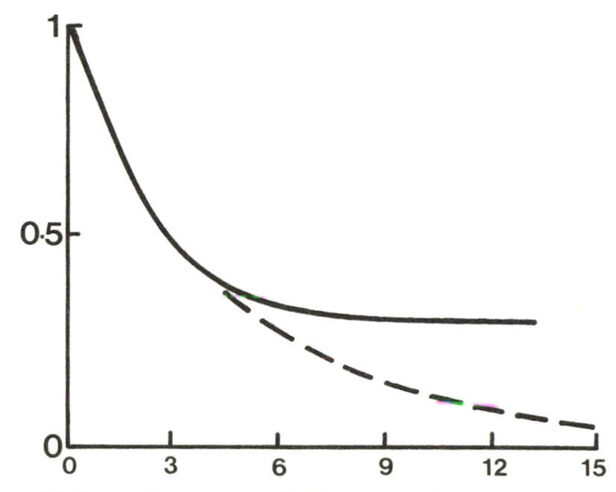

Fig. 2.21. S–N curves. Ordinate: the stress amplitude expressed as a proportion of the stress at failure in a single, static, tensile test. Abscissa: Log_{10} of the number of cycles before fracture occurs. Solid line: a material such as steel that has an endurance limit. Interrupted line: material that has no (known) endurance limit. The rate of decrease of strength is different between different materials.

which they can be stressed an indefinite number of times without fracture; or the S–N curve continues monotonically downward, and the specimen eventually fractures no matter how low the stress amplitude. Steel shows an endurance limit, aluminum does not. Of course, the behavior of materials with an endurance limit can be inferred only from the shape of the curve. The number of cycles tested must eventually increase beyond the budget, interest, patience, or life of the investigator.

The mechanisms of fatigue failure in metals have been the subject of intense study, particularly after the beautiful British Comet aircraft started showering themselves like confetti over the Mediterranean. It seems that the origin of fatigue in metals is that stress concentrations on the surface or, less commonly, within the body of the specimen cause the material to be loaded locally into the plastic region, even though most of the material is in the elastic range. The alternating size or sign of the stress makes the plastically deformed defect enlarge by the process of slip. I shall not discuss slip further—the subject is covered comprehensively by Frost, Marsh, and Pook (1974). Slip is of great interest to metallurgists, but not to us because there is no evidence that it is of importance in bone. Eventually, the slowly traveling defect so weakens the specimen that it fails catastrophically.

Fatigue in bone is best documented in humans, but it is generally a rather obscure phenomenon. Devas (1975) discusses the clinical aspects fully. Orthopedic surgeons rather confusingly call what would seem to be fatigue fractures "stress fractures." This is a strange usage, but is unfortunately well dug into the literature. It comes, I think, from the idea, which is correct, that such fractures can result from stresses imposed by the muscular system during locomotion. The clinical symptoms of fatigue fracture are usually that of pain developing in a bone when activity has been increased over the normal. This may happen, for instance, when a young man joins the army and is set marching, when an athlete restarts training, or when an old widow has to walk to the stores, having been previously driven there by her husband. Often the radiograph shows nothing, though radiodensity develops later, as repair starts. No obviously excessive loading has taken place, yet the repetition of somewhat higher than usual loads has resulted in imminent failure. Fatigue fracture may also develop in different circumstances, such as when the bone is loaded strongly by muscular action, and fracture takes place without premonitory pain. The difference between these two modes will be discussed below.

Evans and his co-workers were the first to study the fatigue properties of bone in the laboratory (Evans and Lebow, 1957; Lease and Evans, 1959; King and Evans, 1967). More recently, Swanson and others (1971) and Carter and his co-workers (Carter and Hayes, 1976a, 1977b, 1977c; Carter, Hayes, and Schurman, 1976; Carter, Harris, Vasu, and Caler, 1981; Carter et al. 1981a, 1981b) have carried things further.

The laboratory tests showed bone to have an *S–N* curve like that of other materials. Figure 2.22 shows that the results of most workers agree quite well, but that one set of results, from Carter et al. (1981a, 1981b), is very different indeed. This latter set of results is so different from the rest that I shall consider it separately. Several remarks made in the next few paragraphs should be kept in mind as rather provisional, until the validity of the other results is considered.

The results of the other workers agree well, despite considerable variation in mode of loading, type of bone, and so on. There is no evidence of an endurance limit, but the greatest number of cycles to failure, about 10^8 in the work of King and Evans, is huge. A person walking at a brisk 100 steps a minute, 10 hours a day, would take 9 years to reach this number of cycles on one leg bone. A migrating mammal might load each leg once a second for 12 hours a day. If it migrated for a month it would load the leg about 10^6 times. According to figure 2.22 a stress amplitude of 60 MPa would be sufficient to break about 50% of specimens after this number of cycles. Alexander et al. (1979a) have calculated the maximum stresses in various long bones of animals traveling fast (see chapter 7 for details). The calculated peak stresses vary from about 60 to 150 MPa. The rough correspondence between these two sets of figures is misleading, for various reasons. First, Alexander's animals were traveling fast; our hypothetical migrator would not be galloping, and the stresses in the bone would be correspondingly less. Second, the value of 60 MPa is a stress *amplitude*, that is it refers to movements away from zero stress

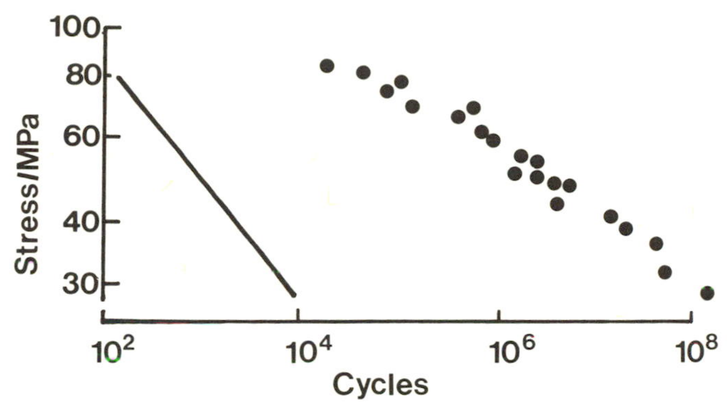

Fig. 2.22. *S–N* data for compact bone. (Note that in this diagram, both axes are on a log scale.) The circles represent values from various workers, including the earlier research of Carter and his co-workers. The continuous line represents the least squares regression for Carter's recent work (Carter, Caler, Spengler, and Frankel, 1981). The distributions are totally different.

equally in both directions. Although the maximum stresses calculated by Alexander's group will diminish and even change sign at other periods in the gait cycle, it is certain that, because of the geometry of the limb, the stress will not be so great on the other side of zero. Third, and very important, is the fact that we are here comparing situations in vivo with laboratory tests. Living bone has the ability to repair fractures, and so is presumably capable of repairing the tiny cracks caused by the fatigue process. Devas (1975) shows great bone reorganization taking place when a bone has not completely fractured. It is unknown, however, whether this repair can take place before gross bone damage occurs. When gross damage does occur, the repair process will be obvious, but we do not at the moment have any real idea of how the race between microdamage and microrepair goes. Martin and Burr (1982) have suggested, indeed, that Haversian systems have a function in preventing the spread of fatigue cracks. Their proposed mechanism is dealt with in chapter 8.

A number of factors affect the fatigue life of bone specimens in the laboratory, apart from the imposed stress. Increasing the frequency of loading increases the fatigue life. Lafferty (1978) showed that the increase took place only above about 30 Hz. (The fatigue life is about 2–3 times greater at 125 Hz than at 30 Hz.) It would be sensible, therefore, to test bones below this value and be reasonably confident that the results, from this point of view anyhow, would be like those in vivo, where the frequency of loading is about 1–2 Hz. Unfortunately, at low frequencies the tests take inordinately long, and Carter and Hayes (1976a), for instance, tested at 125 Hz. However, Lafferty and Raju (1979) have shown that an empirical equation can bring tests carried out at different frequencies into line.

Most studies on the mechanical properties of bone are carried out at room temperature. For tests lasting a few seconds or less the fact that the temperature of mammal bones may be 20° above room temperature (in English rooms, anyhow) is probably unimportant (Evans, 1973, chapter 6). However, for long-lasting tests temperature does become important. This is seen in tests on viscoelasticity (Currey, 1965) but also, strikingly, in fatigue. Carter and Hayes (1976a) showed that between 21° and 37° there was a fall of roughly one-half in the fatigue life (the number of cycles required to fracture a specimen at a particular stress amplitude). Carter, Hayes, and Schurman (1976) went on to show that two features of the specimen itself had an effect on the fatigue life of the bovine bone they tested: the density and the histological structure. The greater the density, the greater the fatigue life; the more of the specimen that was occupied by Haversian systems, the shorter the fatigue life. These two features are difficult to disentangle because the mineralization of the specimen is less, the more that Haversian remodeling has gone on. Nevertheless, by use of a large number of specimens the effects of these two variables were distinguished. The authors produced a formula relating experimental and specimen variables to the number of reversals before failure of the form: log (no. reversals) $= A \cdot \log(\text{stress}) - B \cdot \text{temp} + C \cdot \text{density} + D \cdot \text{histology}$ (histology having arbitrary values 1 to 4). This equation explains an extremely high proportion of the variance in fatigue life. The effect of histological structure is strong; specimens with 100% primary bone have a fatigue life about five times greater than those that are completely Haversian, even when the effect of mineralization has been removed.

Carter and Hayes (1977b) showed that when bone had been loaded sufficiently hard for it to be on the way to failing by fatigue, it had a lower modulus of elasticity and lower strength than unloaded bond. This is not the case with metals; their gross properties remain almost unaltered until just before fracture. However, bone's behavior is characteristic of many composite materials (Hancock, 1975).

It is probable that bone is undergoing diffuse damage, with many tiny cracks, rather than failure by slip along a particular plane. This suggestion of Carter and Hayes is to some extent backed up by their observations on fatiguing bone under the scanning electron microscope (1977c). Interestingly, they found more obvious damage on the compressive, rather than on the tensile, side of specimens, even though bone is considerably weaker in tension than in compression. The bone on the compression side showed tiny cracks all through the heavily stressed region, many cracks being longitudinal and associated with stress concentrators such as osteocyte lacunae and blood channels. On the tension side the cracks usually resulted from the debonding of cement sheaths around Haversian systems and the junctions between lamellae. Again, the damage was diffuse. Carter and Hayes loaded their specimens rather fiercely in these studies on microstructure, often so much so that the bone was loaded into the plastic region of the load-deformation curve in the first cycle. It may be that less severe loading, continued for more cycles, might produce a different picture of prefailure damage. Drs. Burstein and Frankel have suggested to me that fatigue fracture in bone may be of two types. One is "standard," the loads being fairly high, but a large number of cycles occurring before symptoms appear. The other type is when, through excitement, fear, or other intense emotion (and their equivalent in racehorses greyhounds, and, presumably, antelopes), the normal neuromuscular inhibitions to the overloading of bone are overridden. In such circumstances the bone may be loaded into the plastic region, and rather few reversals will cause failure. If such a distinction is real, it is probable that the submicroscopic results of Carter and Hayes refer to the latter type. We do not know.

I have mentioned above the newer results of Carter and his co-workers (Carter, Harris, Vasu, and Caler, 1981; Carter, Caler, Spengler, and Frankel, 1981b). These were tests done on human cortical bone in tension. These workers were concerned to show whether it was the range of strain that is important, or the maximum strain. They loaded at a low strain rate, 0.01 s^{-1}, which is characteristic of strain rates produced by locomotion during life. The fatigue life they found was about $3\frac{1}{2}$ decades less than other workers have found. They attribute this primarily "to the use of uniaxial rather than rotating bending and flexural tests." They also loaded at lower strain rates and at higher temperatures than others had used before. Even so, it is a little difficult to credit that these variables can have had such an extremely large effect on the fatigue life of the bones.

In summary, fatigue fractures occur in bone in life. Laboratory specimens show the characteristic S–N curve, and fatigue with loads near those experienced in life. There is, however, the evidence of Carter et al. that the fatigue life of bone may be much less than previously thought. If this evidence is corroborated, we shall have to reconsider drastically our ideas of the rate at which fatigue damage can be repaired in life. The mechanism of fracture is not clear, but it is more like that seen in fiber-reinforced materials than that seen in metals.

2.6 Fracture in Bone: Conclusions

There are innumerable influences on the strength that bone will actually show in life. In interpreting studies made in the laboratory we must remember, for instance, that static tests, though useful, are different from the situation in which bone usually breaks in life: in milliseconds. Strength increases with strain rate. On the other hand, evidence is beginning to appear that some of the variables concerning energy absorption decrease with strain rate. Certainly the energy absorbed during crack travel decreases with increase in crack velocity. This energy reduction is associated with a much smoother fracture surface. There has been very little work done on the effect of temperature on fracture behavior; indeed, most experiments are done at room temperature.

Bearing such reservations in mind, we can, nevertheless, begin to see that bone has a number of design features reducing its likelihood of fracture.

It is stronger in compression than in tension and, as we shall see in chapter 6, bones are rarely loaded in tension overall. However, if they are loaded in bending, then parts of them will be loaded in tension. Again, we shall see how the skeleton is designed to minimize bending, though it cannot be completely eliminated.

The microscopic structure of bone is designed to reduce the stress-concentrating effects of potential concentrators that must be present. Most bone, except woven bone, has a definite grain, produced by the co-oriented cementing together of collagen fibrils and their mineral. This gives bone a microstructure equivalent to that of a fibrous composite with a very high volume fraction of fibers. Between the lamellae there are the interlamellar regions, which seem to have more "ground substance" and less collagen. The cement sheaths around Haversian systems, and the blood channel networks in laminar bone, are again structures that, though quite capable of holding the tissue together, are relatively weak interfaces at which cracks may be stopped, diverted, or made to split up, and the general level of energy input needed to keep the crack going increased.

Perhaps the most problematical aspect of the microstructure of bone in relation to fracture is the presence of Haversian systems. In no respect does Haversian bone seem to have mechanical properties superior to those of fibrolamellar bone. The penalty for having Haversian systems seems high. There must be a countervailing advantage, and possibly it is related to making available to the body in general the phosphorous and calcium locked in the bone substance.

There are, however, two possible mechanical advantages in Haversian reconstruction. One is that such reconstruction will enable the repair of microcracks. It is very difficult to evaluate the importance of this. The other possible advantage is that it allows a reorientation of the grain of the bone if this orientation is wrong for some reason. This may be an important effect. If bone tissue within the substance of a bone is badly oriented, there is no way in which it can be altered except by Haversian remodeling. In the first chapter I discussed briefly how bone could become fossilized in the wrong direction. Bad orientation reduces the strength by about 60% to 80%. The effect of having completely Haversian bone rather than laminar bone is to reduce the strength by 30% to 40%. It would, therefore, be advantageous to replace badly oriented primary bone by well-oriented Haversian bone. Unfortunately for the mechanical effectiveness of bone, Haversian systems develop, often extraordinarily densely, in bone that was suitably oriented already. This happens particularly in primates and carnivores. Haversian systems remain an embarrassment for anyone trying to explain the structure of bone solely as a functional response to mechanical requirements: the physiological effects of remodeling must also be important.

3. The Limits of the Mechanical Properties of Bone

In general, the mechanical properties of bone taken from limb bones of mammals and birds seem to vary rather little between bones and between species. For instance, Biewener (1982) tested whole bones of animals as small as the mouse (0.04 kg body mass) and the painted quail (0.05 kg body mass). The strength of the bone of these animals, and of somewhat larger ones, was about the same as the bone of the horse, the cow, and man. However, as soon as one strays from testing the limb bone of adult, quadrupedal, terrestrial amniotes, it becomes apparent that bone can have quite a range of mechanical properties, and it is also clear that this variation is adaptive. This chapter explores this variation. First, let us look at the properties of three rather different types of bone of adult mammals: deer antlers, mammalian limb bones, and ear bones. This discussion is based mainly on Currey (1979a).

3.1 Properties of Bone with Different Functions

Ordinary limb bones can be taken as "standard." They must be fairly stiff and strong, but also quite good at resisting impact loads. Compared with them, antlers have rather different requirements. In the red deer, *Cervus elaphus*, as in all deer except the caribou, *Rangifer tarandus*, antlers are found in the males only. They are grown in the spring and summer, used in the rut in the fall, and are shed in the late winter. During the rutting season males compete to collect, maintain, and impregnate harems. The ability to maintain a harem depends on several factors; important among these is the ability to outface an opponent male. For this, antlers are to some extent important insofar as they signal the age and physical state of the bearer. During the display the antlers' mechanical properties are irrelevant—waterproof cardboard would do as well. However, if two opposing males appear to each other to be closely matched, they may fight. Fighting involves smashing the antlers together, fencing with them, and attempting to make the opponent lose his footing. During the smashing together and fencing, the impact properties of the antler are very important. Broken antlers are common; about 30% of deer on the Scottish island of Rhum have some fracture of the antlers by the end of the rut (Clutton-Brock et al., 1979). Broken antlers reduce fighting ability. During the pushing phase of the fight the antlers need to be reasonably stiff, but in this respect they are almost certainly well over-designed. The stresses imposed during the pushing phase will be much lower than during impact and fencing, and so the static strength will be more than adequate. Chapman (1981) mildly criticized the zoology in my 1979 paper and pointed out that the Muntjac deer, *Muntiacus muntiacus*, has small antlers that are not used in fighting. These antlers have a considerably greater density than those of red deer, and we have found that the mechanical properties are more like those of ordinary bone.

Ear bones in mammals have markedly different functions from most other bones. By "ear bones" I mean the auditory ossicles, the otic bone around the inner ear, and the tympanic bulla. The function of the auditory ossicles is to transmit the vibration of the tympanum to the fenestra ovalis. In doing this they usually increase the force of the vibrations and decrease their amplitude. However, in some sea mammals the amplitude is increased and the power decreased.

Consider first the auditory ossicles of a land mammal. Their function is to transmit the excursions of the tympanum to the oval window without distortion or loss of sound energy. They will do this best if they are very stiff, and if they have a low mass. If they are compliant, they will distort under the influence of the tympanum, rather than moving as a whole, and this will affect the movement of the oval window. If they are massive, it will be difficult for the tympanum to move them at high frequencies because of their inertia. Both these effects will reduce the amount of sound energy reaching the fluid of the middle ear.

The otic bone, surrounding the cochlea, should be stiff. As the oval window vibrates, the fluid in the cochlea must vibrate to stimulate the hair cells. It is important for the function of the cochlea that the energy of the vibrating window should remain in the cochlear fluid and not be dissipated into the otic bone. The amount of sound energy that is reflected from the interface between the fluid and the bone is determined by the impedance matching between the two media. For two media meeting at a flat surface the proportion of the total energy reflected is given by: $((Z1 - Z2)/(Z1 + Z2))^2$ where $Z1$and $Z2$ are the acoustic impedances of the two media. $Z = \sqrt{(D \cdot E)}$ for solids and $Z = \sqrt{(D \cdot K)}$ for fluids, E and K being Young's modulus and the bulk modulus respectively, and D being density. Therefore, to maximize the reflection, the difference between $Z1$ and $Z2$ should be as great as possible. Of course, the situation in such a

Table 3.1 Some Physical Properties of Three Bone Tissues

Property	*Antler*	*Femur*	*Bulla*
Work of fracture/J m^{-2}	6,190	1,710	200
Bending strength/MPa	179	247	33
Young's modulus/GPa	7.4	13.5	31.3
Mineral content/% Ash	59	67	86
Density/10^3 kg m^{-3}	1.86	2.06	2.47

complicated structure as the cochlea is not like the planar meeting of two very large volumes of media implied by the formula above. Nevertheless, whatever the particular shape of the interface, the requirement to maximize the difference between $Z1$ and $Z2$ remains.

To function efficiently as a receptor capable of determining accurately the direction of a sound source, the ear should not be excited by vibration reaching it from the bones of the skull. The inner ears of many mammals, though not man, have fibrous pads between the otic bone and the rest of the skull, or are isolated in other ways. This isolation is not complete, and it remains important that bone-borne energy should not pass into the otic bone. The energy transferred can again be minimized by maximizing the impedance mismatch between the otic bone and the rest of the surrounding tissues. Unfortunately for our purposes, auditory ossicles and the otic bone are small and convoluted, so it is difficult to make test specimens from them. However, the tympanic bulla of the fin whale, *Balaenoptera physalus*, is large, and it has a mineral content, density, hardness, and general histology much like that of the other ear bones. It is reasonable to think that its mechanical properties will be similar.

The three types of bone: the antler of a red deer, *Cervus elaphus*, the femur of a cow, and the tympanic bulla of a fin whale were tested for a variety of properties. These were work of fracture (the work needed to drive a crack through the material, which gives a good idea of impact resistance); bending strength; Young's modulus of elasticity, mineral content, and density. The results are shown in table 3.1 and figure 3.1. Later work in my laboratory shows that the femur of the cow probably has a higher work of fracture than it is credited with here; probably 2,800 J m^{-2} would be more typical.

The differences in mechanical properties are very large, and it is likely that they are produced mainly by differences in the amount of mineralization, though there could be some effect of histology; in particular the bulla material was less regularly arranged than the other two bone types.

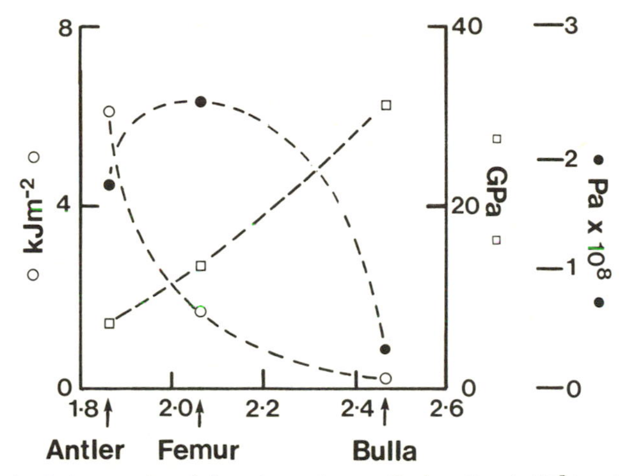

Fig. 3.1. Mechanical properties of three bony tissues. Abscissa: Density/10^3 kg m^{-3}. Ordinate: Three different mechanical properties. Work of fracture shown by open circles, bending strength by solid circles, Young's modulus of elasticity by open squares. The interrupted lines serve merely to connect values of different mechanical properties in different bones; they are not intended to imply any smooth relationship between density and these mechanical properties.

The high mineralization of the bulla is the cause of its high modulus. However, it is just this great mineralization that makes the bulla both stiff and brittle. As we saw in the first two chapters, in "ordinary" bone the mineral is divided into tiny crystallites, and this makes it difficult for cracks to initiate and to travel through the bone. The bulla has so much mineral that it must almost behave as a single block, in which the thermodynamic criteria for crack extension are easily met. Tiny cracks of small length that are stable and do not spread, which seem to be a feature of ordinary bone's "plastic" behavior, cannot develop in the bulla because here cracks are unstable and spread easily. When seen under the scanning electron microscope, the fracture surface of broken bulla is quite smooth, an appearance unlike that of the pitted and ragged surface of a fracture in ordinary bone.

The antler has a rather low mineralization and a very high work of fracture. In fact it is exceedingly difficult to break an antler specimen in impact if it is loaded across the grain; it usually deforms into a U-shape but does not fracture. The antler, therefore, has just the properties required of it. Compared with the "standard" femur it is rather compliant, but any slight disadvantage this may produce in the pushing match is more than overcome by the very high work of fracture. On the other hand, the bulla, and presumably the other ear bones, have a much higher modulus even than standard bone. The question remains whether this is

Table 3.2 Proportion of Sound Energy Reflected between Various Media

	Air	*Water*	*Tendon*	*Femoral Bone*
Water	.999	—	—	—
Tendon	.999	.010	—	—
Femoral Bone	1.000	.307	.231	—
Bulla	1.000	.504	.431	.066

Note: Calculated from Reflection = $((Z1 - Z2)/(Z1 + Z2))^2$.

likely to be significant in hearing. In particular, is the impedance mismatch likely to be significantly enhanced by the high modulus and slightly higher density of auditory bones?

Table 3.2 shows the amount of sound energy reflected at the interface between different media. "Tendon" is as near as we can get at the moment to characterizing the collagenous connective tissue between the skull bones and the bone of the otic capsule. It is clear from the table that the interface between air and any of the other media we are considering is very good at reflecting sound energy. Therefore, it would make effectively no difference whether the bone material was like femur or was more highly mineralized. However, if the otic bone is partially isolated from the rest of the skull by fibrous connective tissue, as is often the case in mammals, the advantage of having high-modulus bone is clear. Between connective tissue and ordinary bone there is 23% reflection; between connective tissue and high-modulus bone the reflection is about 43%. Similarly, inside the cochlea, the amount reflected between fluid (which can be taken as water for our purposes) and the otic bone would be 31% if the bone were like femur, 50% if like tympanic bulla.

Clearly, the high modulus of the bone of the ear structures is important for their various functions. It is bought at great mechanical cost, however. Its work of fracture and bending strength are derisibly low. However, the ear bones are not usually exposed to large forces, and so their inability to resist impact does not matter. The auditory ossicles are so placed that if they were exposed to large loads it would almost certainly be from a blow to the skull that would be mortal. The otic bone in most mammals is almost surrounded by foramina, so that loads passing through the skull are forced to go around the otic bone. This is for good acoustical reasons, as we have seen, but it is very convenient, to say the least, that the otic bone is difficult to load strongly. However, in some mammals, particularly man, the otic bone is an integral part of the floor of the skull. This may be because the skull of man has altered so much in the last few million years that all the changes concomitant with an increase in brain size have not be completed. Whatever the reason, the otic bone is a weak region of the skull and is likely to fracture if the skull is subjected to a hard blow. I said above that the auditory ossicles should, ideally, be stiff yet light. The figures in table 3.1 show that, compared with the femur, a very considerable increase in stiffness (230%) is achieved by a modest (20%) increase in density.

In summary, examination of these three bones shows that bone is capable of exhibiting considerable differences in mechanical properties, and where the reasons for these mechanical differences can be made out, the bone seems to be modified in ways exactly suiting it for its various functions.

3.2 Property Changes during Ontogeny

The example of these three types of bone shows one way in which structure and function vary: between different bones of the same or different animals. We can also consider the changes that take place during development, for a particular bone does not necessarily have the same functions throughout life. Currey and Butler in 1975, and Currey in 1979b, investigated various mechanical properties of the femur of humans aged from 3.5 to 93 years. Here I shall consider only the first 40 years or so, as later changes tend to be senile and not necessarily adaptive. We tested the modulus of elasticity, the bending strength, and the impact strength. The results for impact strength and modulus are shown in figure 3.2. The modulus of elasticity increases markedly with age, particularly in the earliest years, and the impact strength *decreases* even more markedly. The bending strength increases slowly. The changes are caused mainly by changes in the mineralization of the bone: greater mineralization leads to reduced impact strength but to an increase in other values. (Of course, the mineralization in these bones never approaches the very high values found in the ear bones, which are so high that they make the bones weak in bending as well as in impact.)

What, if anything, is the significance of these results? A possible explanation is that the higher impact strength of the young bones is merely a necessary correlate of their lower mineralization. Bone takes time to become fully mineralized. Natural selection produces a bone with a particular stiffness, and this requires some particular amount of mineralization. The immature bone is on its way to the fully mineralized state and meanwhile happens, fortuitously, to be stronger in impact than adult bone material.

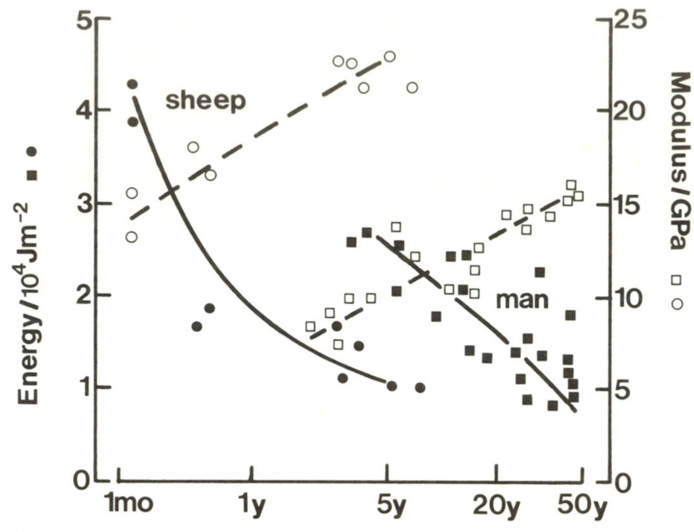

Fig. 3.2. The energy absorption and Young's modulus of human and sheep bone as a function of age. Sheep: circles; man: squares.

Another way of looking at these results is to consider whether bone might at each age be well adapted for the needs of the body. This view has initially two things to commend it: the differences in mechanical properties are so large that it is unreasonable to think that they will not be subjected to strong selection, and it is more profitable to think in terms of features of organisms being adaptive rather than accidental. To think in this way makes one set up testable hypotheses whereas, by its nature, to think that things are accidental prevents further analysis.

If we consider the long time before man became urbanized, it is clear that efficiency in locomotion would have been of great importance. For this, high stiffness would be the feature most highly selected. It is, after all, the prime property of bone to be stiff, and the stiffer the bone material, the lighter the bones can be and still allow efficient locomotion. We know, from our discussion of antlers and ear bones, that bone is capable of having a wide range of values of mineralization. The whale's tympanic bulla achieves in a couple of years a mineralization that the femur of man never achieves. It would be *possible*, therefore, for the femur to be stiffer than it actually is by increasing its mineralization. The fact that the stiffness is not greater must mean that the tendency to increased mineralization is balanced, selectively, by some disadvantage. Since the resistance to impact decreases with increasing mineralization, it is probable that this is the counterbalancing disadvantage.

Children are acted on by rather different selective pressures from those of their parents. They do not trot for miles and miles after wounded prey nor, when really young, walk for miles and miles collecting berries or fetching water. For children, locomotor efficiency is not very important. On the other hand, they are extremely exploratory and have an alarming tendency to fall over, off, and out of things. For children, bones that do not break easily are of great selective importance. The mechanical properties of their bones fit very well with their requirements. Although they are not very stiff, they are remarkably resistant to impact. When children's bones break, they do not break cleanly; often the crack runs out of energy, and a greenstick fracture results which, even in societies devoid of orthopedic surgeons, may heal pretty well.

This explanation of the ontogeny of the mechanical properties of bone seems reasonable, but it is hardly conclusive. However, it is possible to perform a natural experiment, by looking at bones that have a different set of selective pressures during life. Ungulates such as sheep must be able to stand very shortly after birth, and run within an hour or so. There is little time for much of the years of clownish gamboling that so develops the intelligence of the higher primates. Therefore, we would expect the time during which the bones are resistant to impact to be very short. Unfortunately, the femora of sheep are not readily obtainable, and the comparison had to be made on metacarpals. The results, also shown in figure 3.2, are very clear. The sheep's bone attains a modulus of elasticity greater than that of human adult bone within a few months of birth, and pays for it by having an impact strength equivalent to a thirty- to forty-year-old human.

It might still be argued that these apparent differences are the result of the human bone's *actual* age being much less than its *apparent* age, but that the same is not true for sheep. If a long bone is increasing in diameter quickly, and keeps a reasonably constant wall thickness, the bone will be added to on the outside at the same time as it will be eroded on the inside. As a result it will be continually changing its constitution (figure 3.3). Similar things might be true of sheep's metacarpals, but in fact they seem to achieve nearly adult diameter after only a few months' growth. However, studies on the growth of the human femur show that although the external diameter increases rapidly in the earlier years of life, much of the bone in a specimen from the middle of the cortex of a seven-year-old child would have been in existence for several years. Therefore, the marked difference in impact strength between the bones of a child and of a six-month-old sheep in impact cannot be attributed to immaturity of the child's bone.

The low Young's modulus of young bone is carried to an extreme in the fetus. McPherson and Kriewall (1980a,b) show that the bone in the

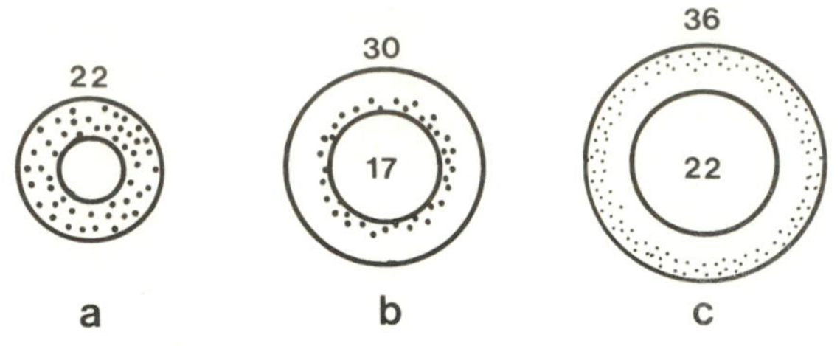

Fig. 3.3. How the age of bone tissue changes with the animal's age. At (*a*) the external diameter of a bone is 22 mm. At (*b*) the external diameter is 30 mm; the internal surface is eroded away, leaving less of the original bone (heavily stippled). At (*c*) none of the original bone is left. New bone laid down between (*b*) and (*c*) is shown lightly stippled. Figures on the sections show the external and internal diameters.

calvarium of the fetal skull has a Young's modulus of about 4 GPa at birth. This value is about one-third of the value for femoral cortical bone of a twelve-year-old. Fetal calvarial bone is unlike adult calvarial bone in that it does not consist of two layers of compact bone separated by cancellous bone; rather, it is uniform, though somewhat porous. Its very low modulus allows the skull to be molded without cracking as it passes down the birth canal. The fetal skull grows very quickly, and so most of the bone in it must be recently formed. One would expect it, therefore, to be rather lightly mineralized and inevitably to have a low Young's modulus. However, Kriewall, McPherson, and Tsai (1981) claim that the mineral content at term is not significantly different from that of an adult. This seems most improbable. Kriewall and his colleagues did not measure the mineral content of adult bone but took my values (Currey, 1969). Since the methods we used were slightly different, it would be best to suspend judgment on this until age comparisons have been carried out using the same specimen size and experimental procedures for all ages. I should be astonished if such a study did not show a clear difference.

Although immature bones may be mechanically different from mature bones, size itself is not an important determinant of mechanical properties. Biewener (1982) shows that the strength of bones of small mammals and birds is about the same as that of the bone of large mammals. I have recently tested the femur of a Galápagos tortoise, *Geochelone elephantopus*. This species of huge land-living tortoises shows marked differences of morphology on different islands of the Galápagos archipelago, and was of great importance in developing Darwin's views on evolution by natural selection. Be that as it may, the bone had a very low modulus of elasticity, about 10 GPa, and a correspondingly low mineralization and high work of fracture. Although the age of the animal was unknown, the size of the femur made it clear that it was many years old. Efficiency of locomotion cannot be very important for tortoises, and they do frequently fall around on their rocky islands, so perhaps impact resistance is important.

Finally, I will mention the humerus of the King Penguin, *Aptenodytes patagonia*, whose bone is the strongest of any ordinary bone that I have tested. The bone has a bending strength of about 290 MPa, a Young's modulus of about 22 GPa (which is, in fact, the highest I know, except for the ear, for Young's modulus tested slowly in bending), and a work of fracture of about 2,000 J m^{-2} (though this last was difficult to test because the specimens had a great tendency to crack catastrophically). Why does this bone have these extreme properties? I think that the prime necessity of this bone material is to be stiff. The penguin's flipper is its means of propulsion and is subjected to large forces. It is important for the flipper to be stiff, so that it does not lose its correct hydrodynamic shape under the influence of these forces. The humerus could be stiff by having a deep section, when seen in cross section, because this would make its second moment of area large. But the bone must be fairly shallow, because it is part of a hydrofoil, which in turn must be slim for hydrodynamic reasons. The stiffness of the humerus can be obtained only by the bone material having a high Young's modulus. The static strength is probably fortuitously rather high, but the toughness is rather low. Obviously, however, more comparisons will have to be made before the significance of such findings as these becomes clear!

The purpose of this short chapter is to show that in healthy bone there is great variation in mechanical properties between animals and between bones, and also that much of this variation is mechanically adaptive. As testing methods become more refined, I would expect more subtle relationships between properties and function to appear. In particular, it would be very interesting to have good data on nonmammalian bone. For example, the selective forces acting on fish skeletons are very different from those acting on land animals, and their adaptive response should be very different.

4. The Shapes of Bones

The bones making up the skeleton of a vertebrate, though forming an infinity of shapes in detail, are nevertheless really of a few types only. This is seen in any typical mammal. There are long bones, such as the humerus, radius, ulna, femur, tibia, fibula, and the metacarpals and metatarsals (figure 4.1a). These are long, thin, fairly straight and, like nearly all bones, hollow. There are bones, such as the carpals and tarsals, which are not much longer in one direction than any other. These tend to be thin-walled and to be filled with cancellous bone (figure 4.1b). There are flat bones, such as the bones making the pelvis, the scapula and, if we can imagine it spread out, the vault of the skull (figure 4.1c). These bones have two fairly thin plates of compact bone with, usually, cancellous bone in between. The vertebrae are in two parts: the centra are like the carpals, but the projections around the spinal cord are like the flat bones (figure 4.1d). The ribs are curved and rather flat, the plane in which they are flattened varying down the thorax. The turbinals in the nose are very thin and scroll-like. There are some other bones, particularly in the skull, that have shapes that do not fit into this classification, but in general most bones are like those listed above.

I shall discuss these types and show what they are useful for and how they must become modified in detail. The fact that bone has as its primary property the necessity to be stiff has been labored, perhaps too much. However, the point must be made again. The long bones of vertebrates act primarily to exert forces on the environment, and in so doing they must withstand large bending moments. These bending moments will, of course, tend to deflect the bones. They must not deflect too much.

4.1 Designing for Minimum Mass

Natural selection will favor animals that can perform particular locomotory functions with the greatest efficiency. By efficiency in this context I mean merely that the function can be performed, and that the cost to the animal in metabolic and other terms is minimized. It is obviously *pos-*

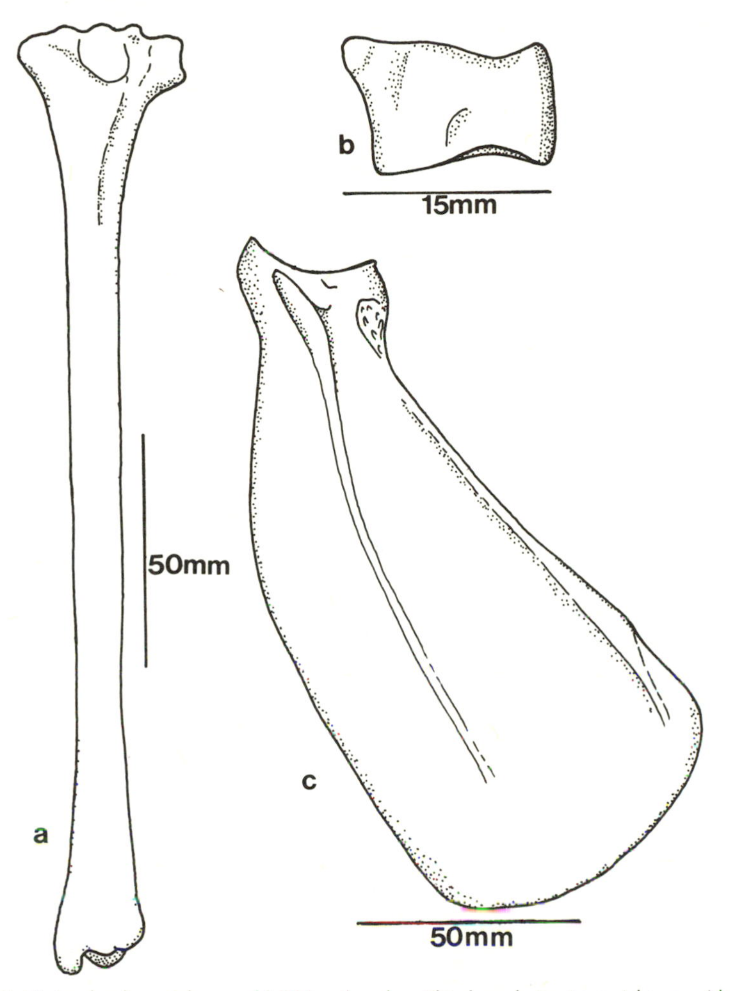

Fig. 4.1. Variously shaped bones. (*a*) Tibia of a dog. This long bone is straight, considerably laterally expanded at the proximal, upper end, and slightly expanded at the lower end. (*b*) A tarsal bone of a dog. This has a complex but basically boxlike shape. It has thin walls and is filled with cancellous bone. (*c*) The scapula of a dog. This is greatly flattened in the plane of the paper. (*d*) A lumbar vertebra of a fallow deer, *Dama dama*. The centrum, *C*, is thin-walled and filled with cancellous bone. The transverse, *T*, and spinous, *S*, processes are greatly flattened.

sible to perform mechanical functions with materials or structures that seem, even to the layman, to be wildly unsuitable. The Incas made fish hooks of gold. Gold is unsuitable, not because it is rare (besides copper it was the only metal the Incas had), but because it has such a low modulus of elasticity and low yield stress.

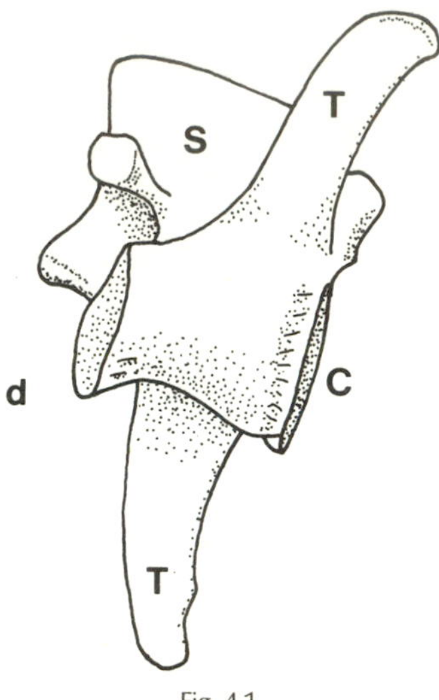

Fig. 4.1.

For locomotory structures, the feature that is likely to be subjected to particularly stringent selection is the mass of the material that will perform a particular function. The reason for this is perhaps obvious, but deserves a little comment. Any animal traveling with some velocity has some kinetic energy. This energy has two components, the "external" and the "internal" kinetic energy (Alexander, 1975a). The external kinetic energy is the mass of the animal times half the square of the velocity of its center of mass. However, individual bits of the animal will have velocities relative to the center of mass, and these will also have kinetic energies associated with them. This kinetic energy, associated with the relative movement of bits of the animal, is called the internal kinetic energy.

A clear example is a flying bird. It has external kinetic energy from its movement through the air, but it also has internal kinetic energy associated with the beating up and down of its wings. All the kinetic energy of the bird must have come from its muscles. If the bird has some mass, and needs to travel at a particular velocity, its external kinetic energy is essentially fixed. However, the energy associated with the beating of the wings will depend to a considerable extent on the mass of the wing bones. Furthermore, unless it has a means of storing the kinetic energy of the downstroke, or making it do work, this energy will be lost, and the muscles must do work to re-create it during each cycle. Therefore, though at any moment the external kinetic energy may be considerably greater than the internal energy, the internal energy is more likely to be wasted. It has been calculated that, in a man running at 6 $m\,s^{-1}$, about 40% of the total power output provides external kinetic energy and 32% internal kinetic energy (Cavagna, Saibene, and Margaria, 1964). Alexander and his co-workers have produced a fine series of papers in which the way in which animals minimize the energy lost in locomotion is discussed (Alexander and Jayes, 1978; Alexander, 1982). In some circumstances, particularly in flying and swimming animals, weight is important as well as mass, but in these cases it is the weight of the whole animal that is important; how this is distributed between the various parts of the body is irrelevant.

In all our discussions of bone there must always be, at the back of our minds, the question: how can the mass of material necessary to do the job be minimized?

4.2 Why Are Long Bones Thick-Walled Hollow Tubes?

Perhaps the most obvious features of long bones are that they are thick-walled hollow tubes, expanded at the ends, and having cancellous bone, rather than compact bone, under these expanded ends. I shall discuss these features in turn, and I hope to show how mechanical analysis of the likely functional requirements of long bones shows that they are well designed.

Finding out how to build a structure that will perform a particular function with minimum mass is called minimum mass (or weight) analysis (Shanley, 1957). Let us take a simple example: a limb bone of uniform cross section loaded as a beam (figure 4.2). It is required to support a pair of bending moments M along a distance L, and in so doing the middle of the length must deform by an amount z only. Note we are saying nothing about how *strong* the bone has to be. Our criterion of failure is excessive deformation, not rupture. This is not because strength is not important but merely that we here assume that if the bone is stiff enough it will be strong enough.

The beam is loaded in so-called pure bending; that is, it is exposed to the same bending moment all along its length. Although limb bones are not often loaded in pure bending, this is handy for present purposes because it means that we can ignore shear stresses. The formula giving the deflection of a beam of uniform shape all along its length in pure bending

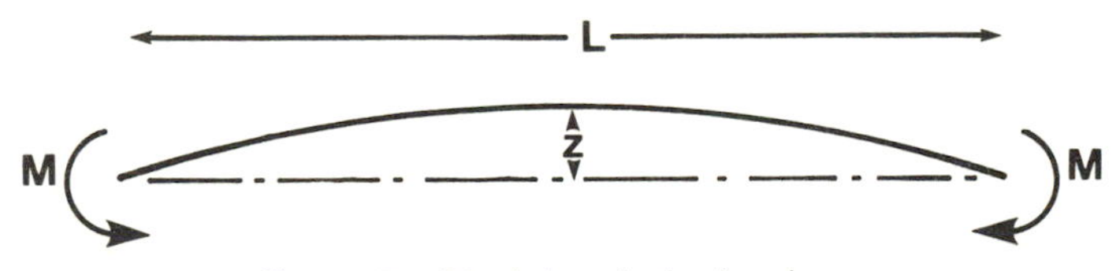

Fig. 4.2. Possible design criterion for a bone.

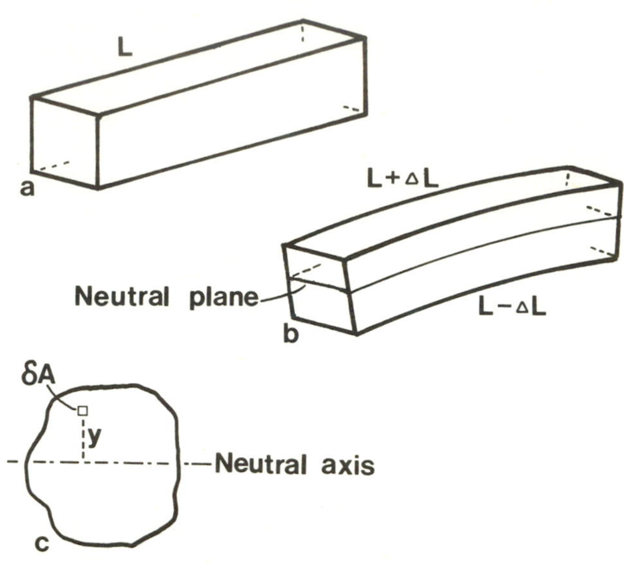

Fig. 4.3. (*a*) A beam of length originally *L* is bent (*b*), changing the length of all fibers except those lying in the neutral plane. (It is usually called a neutral plane, though after bending it is not a plane. "Neutral surface" would be perhaps better.) (*c*) The cross section has a second moment of area $\sum y^2 \delta A$ about the neutral axis. If the bone were loaded about a different axis, the second moment of area would be different.

is $Z = ML^2/8EI$. The terms Z, L, and M are already fixed for us so that we can alter only E, Young's modulus of elasticity, and I, the second moment of area. Let us for the moment assume that E is also fixed. How should the build of the bone be arranged to minimize the mass of the bone? If a beam is bent, as here, it will get shorter on one side and longer on the other (figure 4.3). The amount of deformation will decrease toward the middle of the depth of the beam and there will be a neutral surface, which will remain the same length (though it will become curved) when the bone is bent. Seen in section this neutral surface is a neutral axis. The second moment of area is defined as $I = \sum y^2 \delta A$ (figure 4.3). As shown above, because Z is inversely proportional to I, to minimize Z, I should be as large as possible. What effect will this have on the mass of the beam? If the density of the bone is D, then the mass of the whole bone is: $DLA = DL \sum \delta A$. Therefore, the ratio of mass to second moment of area is $DLA/I = DL \sum \delta A / \sum y^2 \delta A$. This implies that to minimize mass for a fixed value of I the whole of the area should be as far as possible from the neutral axis, because this will minimize the ratio $\sum \delta A / \sum y^2 \delta A$. The larger the second moment of area, the less area, and therefore the less mass necessary to maintain a given deflection.

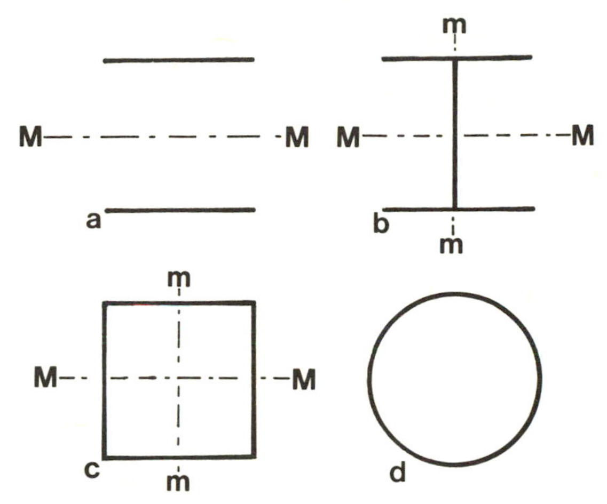

Fig. 4.4. Possible weight-minimizing shapes for cross sections of beams. (*a*) Theoretically best arrangement for neutral axis *M–M*. However, a web is needed (*b*) to keep the flanges apart and to bear the shear stresses. Even so, if the neutral axis were *m–m*, weight would be *concentrated* near the neutral axis, a bad arrangement. (*c*) The best arrangement for two mutually perpendicular neutral axes. (*d*) The hollow cylinder—often the best arrangement.

In fact, the mass can be made as small as one likes by making I sufficiently large. The ideal shape for doing this would be two very thin sheets a long way from the neutral plane. But there would be nothing to stop the two sheets collapsing together, and so a web would have to be inserted to keep the flanges apart (figure 4.4a,b). This arrangement is the theoretical ideal, but would work only if the orientation of the neutral axis were fixed. Nevertheless, this is apparently the ideal solution for the problem as initially set up. It contains a feature, however, that makes it totally impracticable in life, but we shall deal with this after we have made the loading assumptions more lifelike.

If the neutral axis were m–m, rather than M–M (figure 4.4b), then there would be a great deal of bone close to the neutral axis. For bending moments acting in two directions at right angles the box is the best shape (figure 4.4c). The sides act as flanges or webs according to the orientation of the neutral axis, and in either case there is little bone close to the neutral axis. In real long bones the direction of loading is quite likely to come, occasionally, from any direction. In that case the hollow cylinder is the least mass solution (figure 4.4d). This is also the least mass solution for torsional loading, which is frequently of importance in long bones.

So far it would seem that an indefinitely fat cylinder with vanishingly thin walls would be the least mass solution to the problem set by natural selection. In fact, most bones, though hollow, are quite thick-walled. Why is this? The first limitation on the thinness of walls is the possibility of local buckling. Local buckling is the form of buckling seen when a thin-walled plastic drinking cup or an aluminum beer can is crushed. The buckling starts as a little wrinkle, which spreads initially at its two ends and then usually in many directions, leading to a general collapse. Local buckling occurs when the walls are so thin relative to the overall size of the structure that the shape of the structure does not support the wall at some point sufficiently to prevent it from bending in an "easy" direction (figure 4.5). The analysis of local buckling is very complex (Roark, 1965) and for many situations is not properly worked out. Such theoretical results as exist are likely to be considerably in error unless the structures are perfect: any stress concentrations are likely to be very injurious.

For a cylindrical tube loaded in bending or compression the stress in a material that is likely to make it fail by local buckling is Stress $= kEt/D$, where t is the wall thickness, D the overall diameter, and k is a constant which, rather pessimistically, we can take to be 0.5. So, when a tube of bone of Young's modulus of elasticity 14 GPa is loaded, it will collapse by local buckling if the stress somewhere reaches $7 \times 10^9(t/D)$ Pa. It is rather easier to visualize D/t than t/D, and I shall use it from now on. If D/t is small, say 4, the bone will rupture long before it starts to fail by local buckling. We can calculate the critical value for D/t at which it should both rupture and fail by buckling. For a compressive strength of 200 MPa this value is about 35. So, if a bone is designed to achieve some particular *strength* that depends upon bone being loaded to its material strength, it will not be adaptive to have D/t greater than 35 because, at this point, the minimum mass required will start to increase. The reason is that, although the bone is getting relatively thinner-walled, the actual cross-sectional area needed, and therefore the mass of the bone, will start to increase. If, as in the case we are considering, the minimum mass for a particular *stiffness* is being found, we can say nothing a priori because the stresses in the bone may be low, well below the critical stress. However, as will be shown in chapter 7, skeletons are designed so that they are not greatly understressed; that is, they are likely to be designed so that the shape required to fulfill a stiffness criterion is not very different from that required to fulfill a strength criterion. If this is so, even if a structure is designed for a minimum mass for a particular stiffness, it is unadaptive for it to have a very high value for D/t.

The second limitation to the thinness of the walls becomes clearer from a consideration of what is in the lumen of the bone. We have seen that the greater the size of the internal cavity, the less the mass of bone necessary to make a cylinder of the required stiffness. The long bones of vertebrates, except some bones of some birds, and of pterodactyls, are filled with marrow. This is of two kinds: red and yellow. The red marrow is blood forming, and though important in young bones is restricted in adult mam-

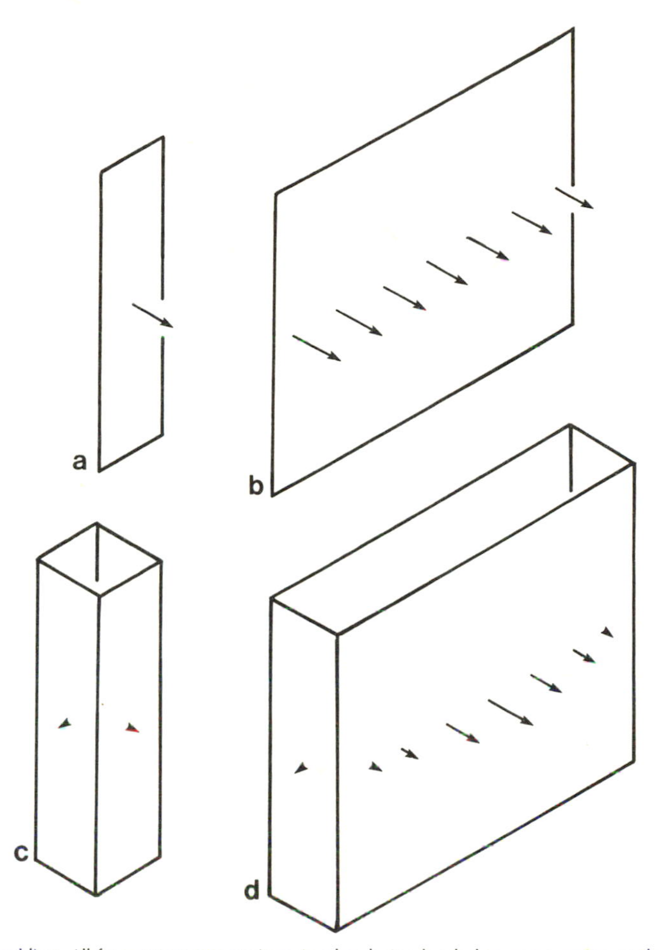

Fig. 4.5. Buckling. All four structures are imagined as being loaded in compression at the top and bottom ends. The resulting displacement of the middle is shown by arrows. (*a*) and (*b*) A narrow and a wide sheet will deflect by the same amount. (*c*) A hollow box, subjected to the same stresses in the walls as (*a*) and (*b*), will deflect little because the walls stabilize each other. (*d*) If the walls of the box are wide, the supporting effect of the neighboring walls is lost, progressively, away from the corners. The complex stress systems set up can produce a local wrinkling.

malian long bones to the very ends, if present at all. In humans, virtually all long-bone marrow is hematopoietically inactive by adulthood (Piney, 1922). In all mammals investigated there is a strong tendency for the concentration of blood-forming marrow to decrease in the proximal-distal direction. Skull bones, and the marrow of the spine, do form blood cells, the limb bones barely do so (Ascenzi, 1976). The rest of the marrow, the yellow marrow, is just fat. This seems to have little physiological function,

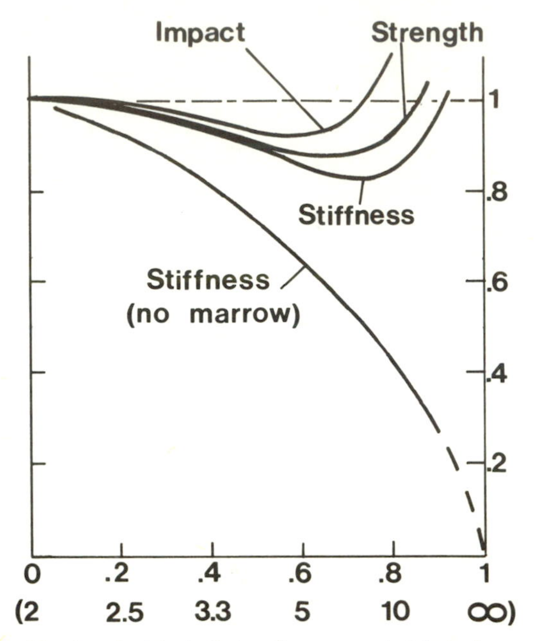

Fig. 4.6. The effect of having physiologically inactive marrow in tubular bones on the optimum shape for weight saving. The abscissa is the ratio of the inner to the outer diameter (the equivalent ratio of outer diameter to wall thickness (D/t) is shown in parentheses). The ordinate is the proportional mass of bone plus marrow needed to produce a particular stiffness, strength, or impact resistance compared with a solid bone.

and has a very low rate of turnover. For instance, it seems not to be mobilized during starvation, in the short term anyhow (Tavassoli, 1974). It appears to be acting merely as a packing material. The calculations made above assume that the bone material is the only part of the anatomical bone to have mass, and that all this mass is contributing to the stiffness. But fat has a minute effect on the stiffness and the strength, though it does contribute to the mass. The density of marrow fat is about 930 kg m^{-3}, that of bone about 2,100 kg m^{-3}. What is the effect of taking marrow into account?

The results of the simple calculations involved are shown in figure 4.6. The lower solid line shows the mass needed to produce a given stiffness at various values of D/t compared with the mass of a solid bone of the same stiffness ($D/t = 2$). The mass declines monotonically as D/t increases. I have dotted the line at the right-hand side to show that such a thin-walled bone would probably buckle at realistic loads. The lowest of the upper thick lines shows the effect of taking the mass of marrow into account. There is now a minimum at an intermediate value of D/t of about

8, and the mass of the bone is now only about 17% less than the mass of a solid bone of the same value of *I*. The curve near the minimum is rather flat.

These results imply that for a hollow cylinder of bone filled with more or less useless fat, the minimum mass solution for stiffness requires fairly thick walls, much too thick to be likely to fail by local buckling. Also, the range of wall thicknesses producing mass values quite close to the minimum is large. Therefore, selection is unlikely to act strongly to produce great uniformity between different bones and between different animals. An important feature of this analysis is that the actual bending moments and the length and deflection of the bone do not figure in the solution; the minimum mass solution gives the best *shape* of the cross section.

Of course, all we have done so far is to set up a situation in which the minimum mass for a given deflection is required. What happens if selection requires that the bone, loaded as before, should not break? The analysis of breaking, in bending, of a real material like bone is difficult because bone undergoes plastic deformation after yielding, and this makes the mathematics difficult (Burstein et al., 1972). However, it is unlikely that bone is often loaded into the yield region; this is presumably a last resort. It is more likely that selection will be for the minimum mass that will support a bending moment *elastically*. This situation is easy to analyze because the bone yields when the outermost part of it (the outermost "fiber") is loaded to the yield stress. For a symmetrical bone the formula for this is: Bending moment $= 2 \cdot I \cdot$ Yield stress$/D$, where D is the diameter of the specimen. Therefore, for a given bending moment, the mass will be minimized by minimizing D/I. Allowing for the mass of the marrow produces a curve of the same general shape as for stiffness, but the minimum mass is at a slightly lower value for D/t, about 6.1, and the thin-walled tube rapidly becomes more massive than a solid bone of the same strength. The mass of such a tubular bone with its cavity is about 13% less than a solid bone.

This analysis shows that for either stiffness or strength, the minimum mass solution with marrow present implies a ratio of diameter-to-wall thickness of about 7. The mechanical advantages of hollow bones become even less clear if we consider impact loading. Suppose that the important mechanical feature of a long bone is that it be able to absorb a certain amount of kinetic energy without breaking in bending. Some such requirement must often be acting in life. Now the strength of a bone in resisting *static* bending is proportional to I/c, where c is half the depth of the cross section. Therefore, if two bones, made of material of the same strength, have the same value for I/c, they will resist the same statically applied load without breaking. But the stiffness of the bones, supposing the bone material to be the same mechanically, is proportional to I. (It is important to be clear that the stiffness of a whole bone—how much it deflects under load—depends upon its build and upon the stiffness of the bone material. I try to distinguish these stiffnesses by talking about the Young's modulus of the bone material.) Therefore, of the two bones, the one with the higher value of c must have a higher value of I, and so be stiffer. But of two structures of the same static strength, the stiffer will absorb less kinetic energy before reaching its failure stress. In fact, one can show (e.g., Singer, 1963) that resistance to impact bending is proportional to $\sqrt{I}/c$. Figure 4.6 shows the curve of the mass of bones of the same value of $\sqrt{I}/c$, assuming the same values of densities of bone and fat as before. The curve is even shallower than the others, the optimum value of D/t being about 4.6, and the saving in weight being only about 8%.

Whatever the loading system that is most important, and is being selected for, the weight saved by having hollow bones with marrow is not spectacular. Indeed, Pauwels (1974, translated 1980) is very unimpressed. "The saving in weight ... is maximum if the diameter is about 65% of the outside diameter. But, even in these optimal conditions, the saving of weight in the diaphysis would attain only about 8%. It would thus be minimal." Pauwels is considering I/c here, and his smaller calculated saving in weight, as opposed to my calculated 13% may be accounted for by the rather lower value he assumes for bone density. Pauwels produces a reason for the presence of bone marrow, based on the remodeling control system, which I find unconvincing, but which the interested reader can refer to. However, the important question here is whether a saving of 10% or so is "minimal." Surely not. In the long bones particularly, weight does not merely have to be carried around, but also has to be accelerated and decelerated during each stride. Consider the implications of this.

In a splendid set of papers Taylor and his co-workers (Fedak, Heglund, and Taylor, 1982; Heglund, Cavagna, and Taylor, 1982a; Heglund et al., 1982b; Taylor, Heglund, and Maloiy, 1982) have analyzed the contribution of internal energy changes and energy changes of the center of mass to the overall power requirements of running terrestrial birds and mammals. These papers are full of good things, but for our present purposes their important equation is

$$\text{Power required/body mass} = (0.478\ \text{velocity}^{1.53}) + (0.685\ \text{velocity} + 0.072)$$

where velocity is in m s^{-1}, and power/mass is in W kg^{-1}. This equation is calculated from observed changes in the kinetic energy of the various body segments, and does not consider the possibility of energy storage in tendons, or inefficiencies in the system. In the equation the first term in parentheses refers to internal energy changes, that is, accelerations of the segments relative to the center of mass. The two other enclosed terms refer to kinetic energy changes of the center of mass.

Consider a horse running at 15 m s^{-1}. The equation implies that 30.1 W kg^{-1} will be required for internal energy changes, and 10.3 W kg^{-1} for center of mass changes. So, about 3/4 of the energy required is for accelerating the limb segments. Other data in these papers show that about 88% of the 30.1 W kg^{-1} is accounted for by the three distal segments of the limbs. About 2/3 of the power required for galloping is caused by fluctuations in internal energy (accelerations of parts of the body relative to the center of mass). In the horse roughly 80% of the mass of these distal segments is bone, and so about 50% of the power required for galloping at 15 m s^{-1} is used to accelerate and decelerate the bones of the distal limb. In this context, therefore, a saving of 10% in the mass of the bone of the distal segments will produce a 5% saving in power required which, from the point of view of natural selection, will be a very important saving. In fact, the analysis of Taylor and his colleagues pays no attention to the energy storage in tendons, so the savings may be less; even so, it is clear that quite small savings in mass in the distal bones will be selected for strongly.

Pauwels produces another argument in favor of the uselessness of hollow bones for weight saving. The argument points out that many long bones are expanded at the ends, and that "the weight of the bone must be much greater than that of a solid structure of the same resistance to bending and of the same length because of the filling and of the considerably denser cancellous bone." This, though true, is irrelevant because, as I shall show below, the problems of the very ends of bones are much less those of resisting bending, and much more those of cushioning impacts and allowing sufficient footing for the soft and weak synovial cartilage. Because of these requirements the ends of bones must be expanded and, if they were of solid bone, they would be disastrously rigid.

Theoretical calculations of the structure producing minimum mass should be matched with reality. Figure 4.7 shows values for D/t of a rather random collection of bones that Neill Alexander of the University of Leeds and I have been able to measure.

The results are rather striking. The values for the nonflying mammals' bones were mostly taken from the literature, so I do not know what kind of marrow they contained. However, it does seem that most adult mammals have yellow marrow in their long bones. If this is so, they have in general ratios of D/t that would be appropriate for minimizing mass while resisting impact loading or being adequately stiff. They are perhaps a little on the low side, the median value for D/t being 4.4, which is slightly less than the theoretical value for minimum mass for energy absorption, 4.6. It should be said that the median value for the femora was 5.4, the other bones—tibiae, humeri, and metapodials—having a correspondingly lower value. Of the birds' bones, most had red marrow

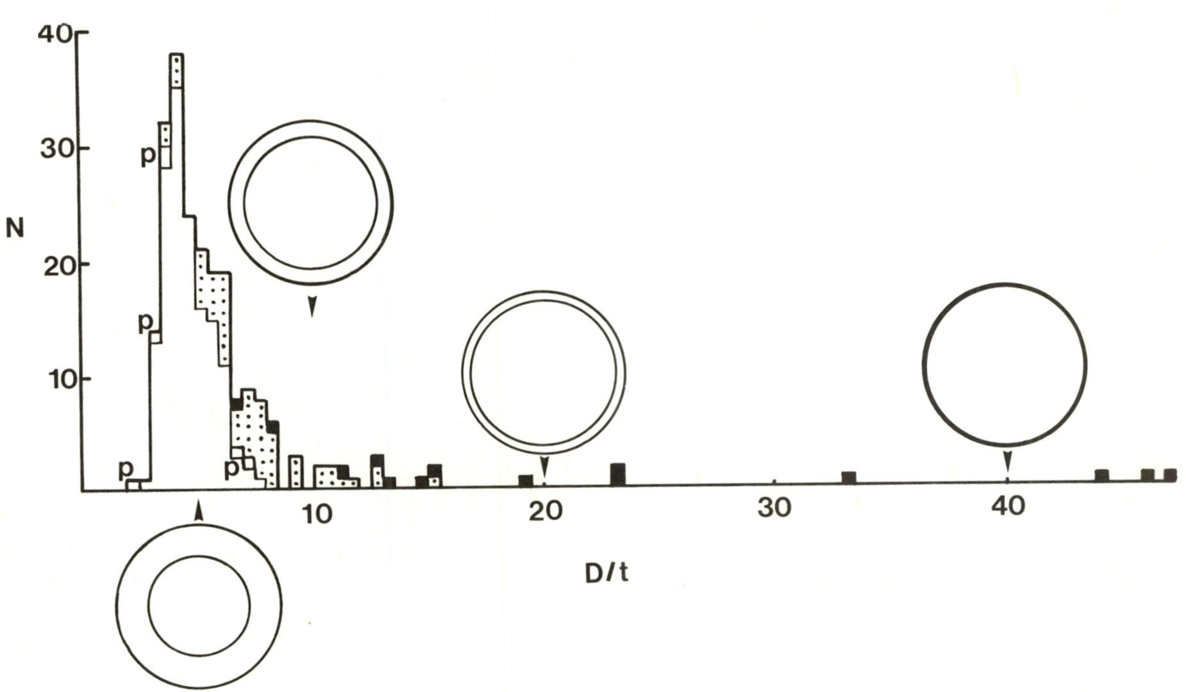

Fig. 4.7. Values of D/t for a variety of vertebrate bones. Abscissa: value of D/t; ordinate: number of bones. The circles represent the cross-sectional appearance of hollow tubes at the values of D/t indicated by the arrowheads. Blank: land mammals. *P*: A pipistrelle bat. Dots: Birds' bones containing marrow. Solid squares: Pneumatized bones—those having a value of D/t greater than 17 are pterodactyls; the rest are birds' bones.

and so the marrow still had a hematopoetic function. These bones have a wide spread, the median value of *D/t* being 7. They are, therefore, roughly minimum mass solutions but, because the marrow is physiologically useful, and therefore has to be put somewhere, one might have expected them to have had a somewhat higher value. The marrow of the gulls was yellower than the others, suggesting a less active physiological role, and the gulls' values for *D/t* are rather lower than those for the other birds, though not as low as for the main mass of mammals. The empty bones did not have startlingly high values of *D/t*, though they were among the highest in the birds, having a median value of 13. This general agreement between the contents of the lumen of these long bones and the relative thinness of the walls is pleasing, but obviously needs to be tested in a much more thorough way.

The bones of a flying mammal, a pipistrelle bat, *Pipistrellus pipistrellus*, are interesting. Two bones, the humerus and the ulna, lie on the right-hand side of the general mammal distribution. The second to fifth phalanges are extraordinarily long and thin, and the marrow cavity is a small proportion of the total cross-sectional area. I suspect that the reason these bones have a meagre marrow cavity (which is filled with yellow marrow) is that there is a limit in the *outer diameter* of the bones because they are part of the generally extremely thin wing. If this is so, then the advantage of a reduction in weight produced by a larger, thinner-walled bone would be outweighed by aerodynamic disadvantages, and the bones are not, therefore, least weight structures. In the previous chapter I mentioned the humerus of the king penguin, whose bone was more mineralized and stiffer than usual. I attributed this to hydrodynamic limitations on the depth of the bone. Although the bone does not have even a roughly circular cross section, so a mean value for *D/t* is not relevant, it is clear that it is rather thick-walled relative to its overall depth. This finding also fits in with the idea of hydrodynamic limitation of depth.

The bones of the pterodactyl are extraordinary. They belong to *Pteranodon*, an advanced pterodactyl, which had a very low flying speed. It is almost certain that the bones were filled with gas. The values of *D/t* are so high that these bones would fail by *local* buckling before they failed by breaking in tension or compression. However, most of them are very slender, so they would be in great danger from Euler buckling if there were a significiant axial compressive component in the way they were loaded. Such thin-walled bones would be very vulnerable to locally applied loads, and the general feeling one has about these pterodactyl bones is that they have been pushed to the limit in the pursuit of lightness. Indeed, most of what we know about pterodactyls shows them to have been animals leading sedate, well-ordered, and precarious lives. The hurly-burly of the life of, say, a crow (*Corvus*) would have been fatal for them (Bramwell and

Whitfield, 1974). The metacarpal and the three phalanges of the fourth digit form the leading edge of the membranous wing. Although they are extremely thin-walled, it is interesting that, as is the case in the bat, the bones forming part of the outboard wing are less relatively thin-walled than the more proximally situated bones. It would be very interesting to know whether these pterodactyl bones were relatively highly mineralized. If so the bone would be stiffer, with a small cost in density. This would have been advantageous, as will be explained in section 4.2.2, because the disadvantage of having highly mineralized bones is a reduction in *strength*, and these bones would fail because they were not stiff enough long before they failed by not being strong enough.

In this discussion of hollow bones I have so far considered the possibility that natural selection is acting to produce a minimum *mass* solution to various mechanical problems. However, other constraints may be important. For instance, the metabolic costs of laying down and maintaining bone and fat may be very different. If bone is more expensive than fat, and is significantly costly, then one would expect the balance to be shifted slightly in the direction of having a larger marrow cavity. However, it is so difficult to obtain reasonable values for the metabolic costs of tissues that I shall not take this line of argument further, except fleetingly in chapter 7.

4.2.1 The Mechanical Importance of Marrow Fat

Marrow fat, at body temperature, is a viscous fluid, and so far I have assumed that, whatever mechanical properties it may have, they will not contribute to the strength or stiffness of the bone. Before leaving the subject of the hollowness of bones, let us see whether marrow could, or even does, have any strengthening function.

First, consider static bending of a long bone. One side is put into tension, and the other into compression. Because a fluid is free to move, and cannot resist shear forces, the marrow will merely move in the bone until it is unstressed. Therefore, marrow will be no use in static bending. Because a fluid cannot resist shear, marrow will also be no good in resisting torsion. Bones are probably not often loaded in tension overall (chapter 6). However, they are, of course, often loaded in compression.

Suppose bone material were isotropic, and had the same bulk modulus as marrow. If a bone consisted of a hollow cylinder of bone tissue completely filled with marrow, then a longitudinal compressive load would be borne by the marrow and by the bone tissue in proportion to their cross-sectional areas. This would, of course reduce the stress in the bone (figure 4.8a–c). Another way the marrow might prevent bone failure would

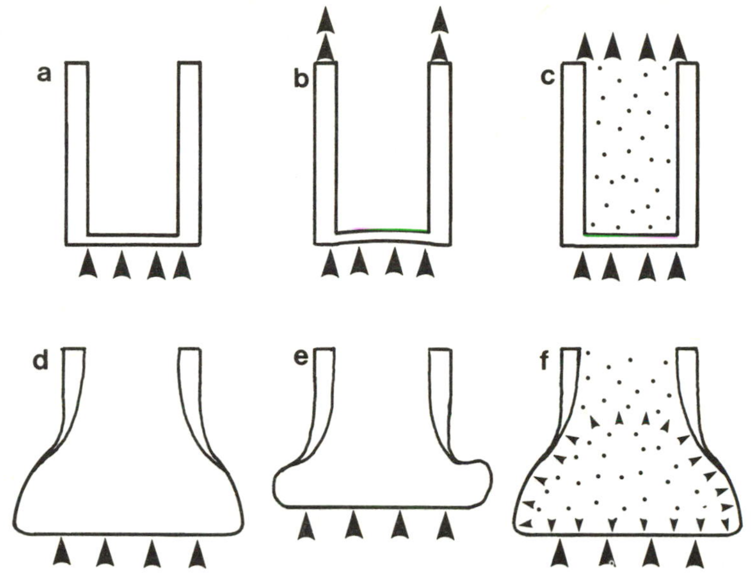

Fig. 4.8 Two possible mechanisms for the bone-protecting effect of marrow. (*a*) A load on the end of the bone can (*b*) be transmitted by bending of the subchondral bone and lead to increased force in the cortex. In a marrow-filled bone (*c*) the load might be partially taken by the marrow. (*d*) Most long bones have thin walls near their ends. (*e*) Loads on the ends could cause the thin walls to buckle. (*f*) Marrow might, by producing hydrostatic pressures, keep the thin walls stable.

be in preventing local distortion of the thin cortex of bone near articulations. It could act by pressing hydrostatically on the walls, preventing them from collapsing (figure 4.8d–e). For either of these strengthening mechanisms to work, however, fairly high hydrostatic pressures must be generated in the marrow. Swanson and Freeman (1966) applied cyclic loads to human femora at a frequency of 2 Hz, and showed that the hydrostatic pressures generated were very small. However, the *rate* of loading produced by cyclic loading at 2 Hz is much less than that imposed during falls.

Recently Bryant has shown that in the sheep's tibia, at loading rates of 500 kN per second, which is roughly that occurring in falls, the greatest hydrostatic pressure in the marrow was only 50 kPa (Bryant, 1983). This was in response to a load on the end of the bone of about 3 kN, which is about what is imposed in a severe fall, and is considerably greater than the loads imposed during locomotion. 50 kPa is very low, and will be insignificant mechanically. Bryant was able to show that if he arranged it that the marrow were compressed by 0.15% of its volume by a small cylinder pushed in by the impacting load, the pressure in impact loading was as high as 500 kPa. The implication of this is that in impact loading the internal volume of the sheep's tibia changes very little. If we may extend these findings to other long bones, we can say that the marrow has little mechanical function to play in long bones, either in static or in impact loading. The question of whether it is important in bones such as the vertebral centra remains open at the moment.

4.2.2 Paying for Strength with Mass

So far in our discussion of long bones we have taken particular values for strength and Young's modulus and have worked out their consequences. We must, before we go further, consider in much more detail the interaction of various properties; we must indulge in some more minimum mass analysis.

In the previous chapter we saw that the mechanical properties of bone change considerably with variations in the amount of mineralization. So, for instance, as bone becomes more mineralized it gets both denser and stiffer. If a bone is being selected for minimum mass that will deflect by some given amount, the question arises: will it be adaptive to increase the mineralization, thereby increasing the density, if the stiffness is thereby also increased? The obvious, and correct, answer is that it depends upon how much stiffer it gets for a particular increase in density. Less obvious, but also true, is that it depends upon how the bone is being loaded. The answers for a bone being loaded as a short column and as a beam are, for instance, different.

1. *Bone loaded as a short column.* Consider a short column (short meaning here that sideways buckling deflections can be ignored). It has some length L and must support a load W and undergo a compressive deformation of Z only. What value of D, the density, and E, the Young's modulus, will produce a bone of minimum mass? We assume that the bone is square in section, of side B. (It can be shown that in the cases we are going to consider the results do not depend on the shape of the section, as long as the shape is constant along the length.)

$$\text{Deformation } Z = \text{Strain}(\varepsilon) \cdot L$$

$$\text{Strain} = \text{Stress}(\sigma)/E$$

$$\text{Stress} = W/B^2$$

$$\text{Therefore } Z = (W/B^2E)L$$

$$\text{Now the mass} = D \cdot L \cdot B^2$$

$$\text{Therefore the mass} = DWL^2/ZE$$

Table 4.1 Expressions That Must Be Minimized to Achieve a Minimum Mass for a Particular Function

	Rupture	*Elastic Energy Absorption*	*Stiffness*	*Euler Buckling*
Short column or tensile member	D/S	ED/S^2	D/E	—
Slender column	—	—	D/E	$D/E^{1/2}$
Beam of constant shape	$D/S^{2/3}$	ED/S^2	$D/E^{1/2}$	—
Beam of fixed width	$D/S^{1/2}$	ED/S^2	$D/E^{1/3}$	—

Note: D: Density; E: Young's modulus of elasticity; S: Yield strength.

But L, W, and Z are all given, so M is proportional to D/E. Therefore, the mass is proportional to the density, and inversely proportional to the Young's modulus.

2. *Bone loaded in bending.* Consider a beam of length L and second moment of area I, loaded as a cantilever with a load at its free end of W, that must deflect by an amount Z only. Again assume that the section is square of side B. Then, from simple beam theory

$$Z = WL^3/3EI$$

$$I = B^4/12, \quad \text{so } Z = 4WL^3/EB^4$$

$$\text{Mass} = LB^2D$$

$$\text{So mass}^2 = L^2B^4D^2$$

$$\text{Mass}^2 = 4WL^5D^2/EZ$$

Again, L, M, and Z are given, so mass$^2 \propto D^2/E$ or, mass $\propto D/\sqrt{E}$.

These results show that if a bone is to be loaded in tension or compression, the effect on the mass of reducing the density of the bone material is the same as increasing the Young's modulus by the same proportion. On the other hand, if the bone is loaded in bending, the Young's modulus appears as a square root term, and so changes in Young's modulus are less important than similar changes in density. Table 4.1 shows the expressions that must be minimized in order to achieve a minimum mass, for bones loaded in various modes, and with different imposed requirements.

Although there are differences between *structures* in the function that must be minimized for minimum mass, the more obvious differences are between the loading systems. In particular, if deformation or buckling is likely to be deleterious, then Young's modulus should be high, but if

Table 4.2 Minimum Mass Expressions for Cancellous Bone

	Rupture	Elastic Energy Absorption	Stiffness	Euler Buckling
Short column or tensile member	$1/D$	1	$1/D^2$	—
Slender column	—	—	$1/D^2$	$1/D^{1/2}$
Beam of constant shape	$1/D^{1/3}$	1	$1/D^{1/2}$	—
Beam of fixed width	1	1	1	—

Note: This is table 4.1 modified, assuming $S \propto D^2$; $E \propto D^3$.

energy absorption is important, then Young's modulus should be low. As was discussed in chapter 3, in compact bone variation in mineral content produces changes in strength and Young's modulus out of proportion to its effect on density. If a bone has merely to achieve a certain stiffness with minimum mass, it should be as fully mineralized as possible, because stiffness will increase more rapidly even than the square of density (a condition that must be fulfilled if increasing density is to be an advantage in the case of a slender column or a beam). But, in life, it is rarely true that only one property is being selected for. The ear bones discussed in the last chapter are exceptional in their single-minded pursuit of stiffness.

Natural selection acting on the mechanical properties of bone material cannot, therefore, produce a material that is best in all circumstances. The compromise that is reached will depend on the relative importance of the different loading modes, and possible modes of failure, to the success of the animal. As we saw in the last chapter the compromise may differ, both between bones, and between different times of life.

Using the minimum mass table we can, for instance, try to make sense of cancellous bone, and to see why it is found where it is. Carter and Hayes (1977a) have shown that bone seems to obey the relationships $E \propto D^3$, where D is the density; compressive and tensile strength $\propto D^2$. Although, as we saw in chapter 3, the bony material itself can show differences in density, the large differences that Carter and Hayes were investigating were caused by the porosity of the bone; in fact density is almost inversely proportional to porosity. Substituting these relationships into table 4.1, we get table 4.2.

None of these functions increases with density, so there will never be an advantage in having porous bone. However, elastic energy absorption does not vary with density nor, interestingly, do any of the functions for a fixed-width beam. A fixed-width beam is not a mere abstraction; many flat plates in the body, such as the bones of the pelvis, are limited in their

Table 4.3–Amount of Energy Absorbed by a Two-tissue Block, Loaded in Impact

		Relative Thickness of Bone						
		0	*1*	*2*	*5*	*10*	*20*	*50*
Young's modulus of bone in GPa	.1	1	2	3	6	11	21	51
	.2	1	1.5	2	3.5	6	11	26
	.5	1	1.2	1.4	2.0	3	5	11
	1	1	1.1	1.2	1.5	2	3	6
	2	1	1.05	1.1	1.25	1.5	2	3.5
	5	1	1.02	1.04	1.10	1.2	1.4	2
	10	1	1.01	1.02	1.05	1.1	1.2	1.5
	20	1	1	1.01	1.02	1.05	1.1	1.25

Note: Unity is the energy that cartilage can absorb on its own. Thickness is relative to thickness of cartilage, whose Young's modulus is taken as 100 MPa = 0.1 GPa.

larger dimensions by anatomical constraints. Nevertheless, there should never be a mass advantage in increasing the porosity of the bone. So the statements, so frequently seen, to the effect that cancellous bone makes the bone structure "lighter" in some way are strictly wrong. Even so, cancellous bone is an important constituent of long bones, as well as wrist and ankle bones and vertebral centra.

The reasons for this are various. An important point to get clear is that these tables refer to structures made of a single type of material. They do not refer to composite structures. Nowhere in the body does cancellous bone exist without at least a thin outer covering of compact bone. (It is a fascinating problem that the echinoderms—the starfish, sea urchins, and so on—have adopted a cancellous structure for effectively all their skeletal elements. The reasons for this are completely obscure.) The combination of cancellous bone with a thin cortex of compact bone can have properties very different from either type on its own. I shall discuss cancellous bone in more detail soon, but here I shall discuss its presence at the end of long bones.

4.3 The Swollen Ends of Long Bones

Returning to long bones, we find that there are considerable geometric constraints imposed on the shapes of the ends. The shafts of long bones have particular sizes, governed by the minimum mass necessary to bear the imposed loads. For reasons that will be developed in the next chapter, synovial joint surfaces have to be large compared with the cross-sectional area of the bone of which they form the end. The forces across the joints are broadly similar to the forces in the shaft. These forces are often *much* larger than the forces the bone seems to be exerting (Hughes, Paul, and Kenedi, 1970). This is because most muscles work at great mechanical disadvantages, for reasons I shall discuss in chapter 6. Figure 4.9 shows this. The top diagram shows a bone, jointed at J, which has to exert a force W at the distal end. The origin of the muscle acting on the bone is at O. The middle diagram shows the forces in the muscle, and across the joint, when the muscle has a fairly large turning moment about the joint, as might be produced by a large flange on the bone. The bottom diagram shows perhaps a more usual situation, in which the muscle passes quite close to the center of rotation of the joint. The forces in the muscle and across the joint are now over fifteen times as great as the force being exerted at the far end of the bone.

Although I shall discuss the articulations between bones more extensively in chapter 5, there is one feature of joints that has such an important effect on the design of the shape of bones as a whole that I must discuss it briefly here. The bearing surfaces of most joints are lined with synovial cartilage. Synovial cartilage is not very strong, its tensile strength (which is probably the relevant strength) being about 20 MPa (Kempson, 1980). Therefore, the loads being transferred from one bone to the next must be spread out over a rather large area of cartilage. The apposed surfaces in synovial joints have to slide past each other. Many vertebrate joints have large angles of excursion, and the requirement for low stress implies large radii of curvature (figure 4.10). These together imply large ends to bones. It is informative to compare the situation in vertebrates with that in arthropods. Arthropods have rather hard bearing surfaces, capable of bearing high stresses. The load can be taken over small areas, the radii of curvature can be small, and as a result arthropod limb elements become pinched in at their ends, instead of being expanded.

The loads in vertebrates must, therefore, be transmitted from the cortex of one bone, with its relatively small cross-sectional area, to the cortex of the next via an interface of large surface area. It is universally found that the very thin subchondral bone lying underneath the cartilage is itself underlain by cancellous bone that leads the loads from the subchondral bone to the dense cortex. The adaptive reason for this is not quite as straightforward as might seem to be the case. Various possibilities for the design of the ends of bones, and their mechanical consequences, are shown in figure 4.11. If the bone underneath the cartilage were an unsupported plate, it would have to be thick if it were not to undergo quite large deformations. Large deformations would not be good for joint functioning. Finite element analysis by J. Bryant shows that there is little advantage,

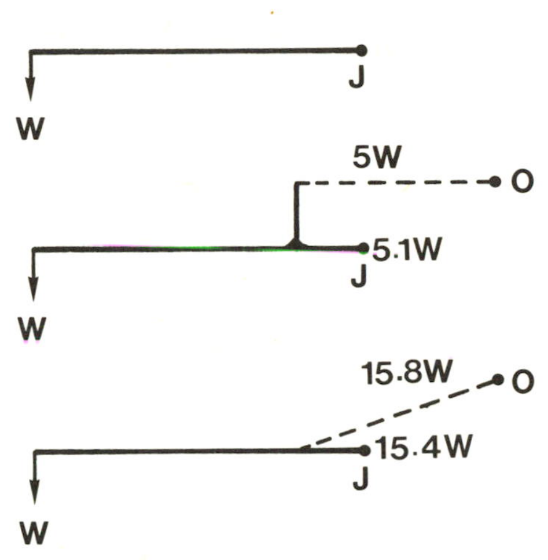

Fig. 4.9. Diagram showing forces in a muscle and across a joint.

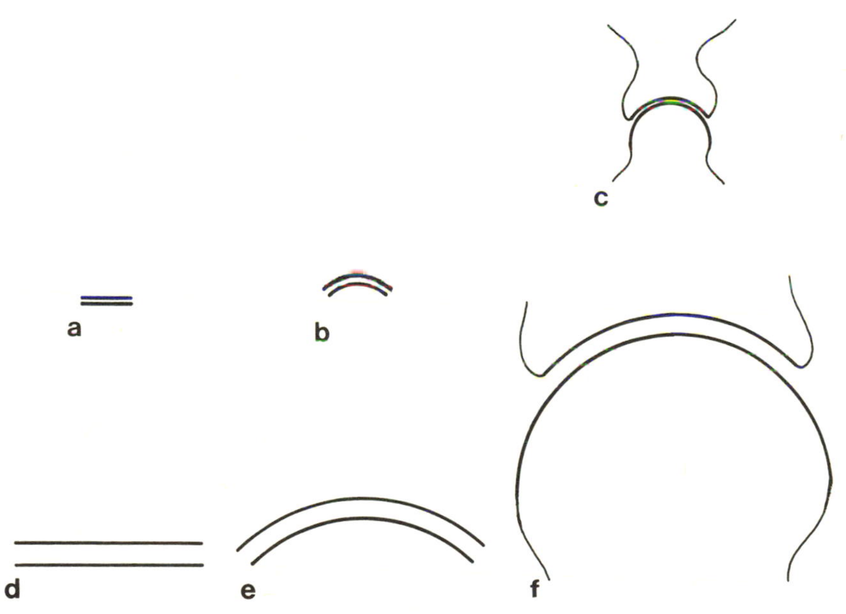

Fig. 4.10. The effect of articular surface strength on the shape of articulations. (*a*) A high-strength surface can bear the loads over a small area, so (*b*) the radius of curvature can be small, allowing (*c*) the size of the articulation to be small relative to the size of the limb elements it joins. This state of affairs is characteristic of arthropods. (*d,e,f*) A weak surface, as is characteristic of the vertebrates, leads to large articulations relative to the size of the limb elements they join.

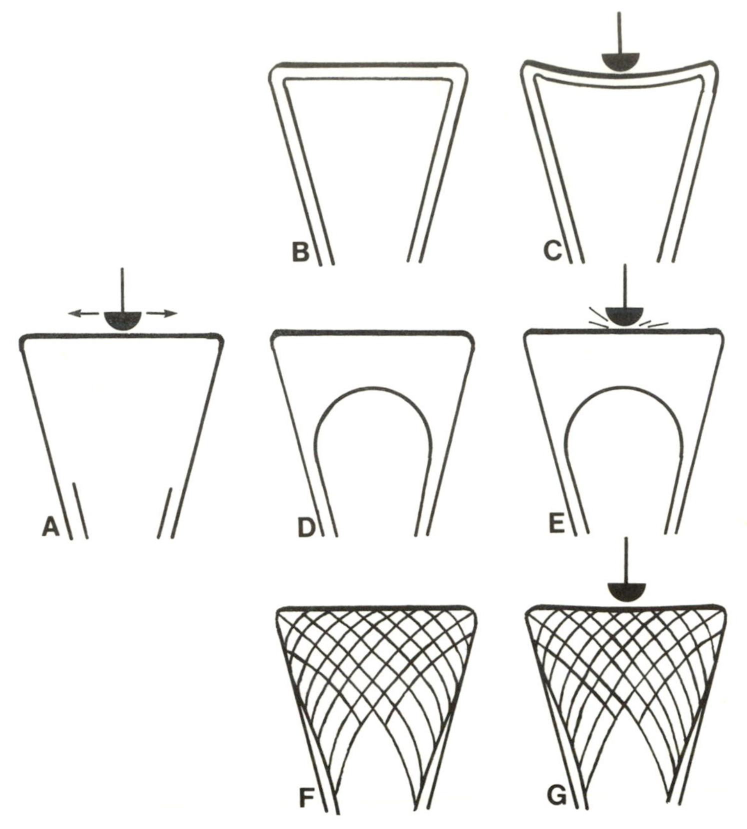

Fig. 4.11. Design for the ends of long bones. (*A*) The basic problem: a load, which may be variable in position, must be transmitted to the cortex away from the joint. (*B*) The thin cortex is continued right around the end of the bone. (*C*) Such a solution would allow large deformations on loading. (*D*) A thickness of bone that is stiff enough to prevent large deformations would (*E*) produce high local stresses in impact loading. (*F*) Cancellous bone will (*G*), when loaded, produce adequate overall deformations to make impact loading innocuous, but will not allow large deformations of the joint surface.

in terms of weight and amount of material used, in having the loads borne by a plate in bending compared with having the bone end made of a solid lump of bone, in which most of the load can be taken as compressive stresses (figure 4.11).

However, there is a grave disadvantage to this solution as compared with the solution of having cancellous bone underneath a very thin shell: solid bone would destroy the cartilage during loading in impact.

Consider two blocks of bone of cross-sectional area A, one lying on top of the other, loaded statically by a load P. The stress in each block will be the same, P/A. The stress in one block of tissue will not be affected by the presence of the other. (I am ignoring the slight complication produced by the possible different sideways strain in the two tissues, which might lead to a complex stress system at their interface.) Probably the only important effect of a more compliant footing for the cartilage in joints would be that the bone's deformation would allow a rather larger area of contact between the two cartilage surfaces.

Now consider how much *energy* has been absorbed by the tissue blocks when being loaded. Each block will have a stiffness, directly proportional to its cross-sectional area and to its Young's modulus, and inversely proportional to its thickness. The work done on the blocks will be stored between them in a ratio inversely proportional to their stiffnesses. Suppose now that the material is loaded in impact and given mechanical energy that must be absorbed. The limit to the process will be when the weaker material ruptures. The stress in this system will be the same everywhere. The system should be so arranged so that the energy absorbed is maximized before this happens. In the case of joints the strength, and Young's modulus, and thickness of the cartilage are more or less fixed by the requirements of lubrication. Suppose we have a layer of cartilage 1 mm thick with a Young's modulus of 100 MPa, which seems a reasonable value for high rates of loading (Unsworth, 1981) and that this is able to absorb a unit amount of energy before rupturing. For example, table 4.3 shows the effect of adding a layer of bone underneath with various values of thickness and Young's modulus. The amount of energy that can be absorbed increases as the thickness of the bony layer increases, but this increase is much more marked when the underlying bone has a low Young's modulus. Table 4.3 shows a reasonable value for the thickness of the underlying bone: twenty times thicker than the cartilage. If the bone has a modulus twice that of cartilage, the system will absorb eleven times more energy than cartilage on its own. However, fully dense bone with a Young's modulus of 20 GPa would allow a total energy absorption only 10% greater than cartilage on its own. The presence of cancellous bone under joint surfaces, therefore, will reduce the total weight of bone needed, compared with a solid block, but will also have the probably much more important effect of allowing the ends of the bones to be loaded in impact without the cartilage being squished.

This discussion of cartilage and bone has introduced, by the back stairs, a topic that is of great importance in structures such as bones that are likely to be severely loaded in impact. These structures should, if possible, be built so that there is a *uniform stress* throughout the tissue. If the stress is uniform, then the amount of energy that can be absorbed will be maximized. Suppose we have a bar, loaded in tension. Its length is L, and it is composed of material of Young's modulus E. If the maximum stress the material can bear is S, and if the bar is of uniform cross-section A, the maximum energy it can absorb before failure is $S^2 \times L \times A/2E = \frac{1}{2}(P^2L/AE)$, where P is the load. Suppose half the length of the bar was doubled in cross-sectional area. (We ignore the stress-concentrating effect of the sharp corner.) The energy that can be absorbed is now:

$$(P/A)^2 \times (L/2) \times A/2E + (P/2A)^2 \times (L/2) \times 2A/2E = P^2L/4AE + P^2L/8AE = \tfrac{3}{8}(P^2L/AE)$$

Adding material to the structure has actually reduced the amount of energy it can absorb. This effect is undoubtedly important in bones. It is difficult, however, to demonstrate that selection acts to bring it about, because selection is also probably acting to produce bones of minimum weight, and this action will produce a similar result.

At the ends of long bones, however, there is a conflict. The cancellous bone is compliant to shield the cartilage. This rather large compliance must not be carried right through the length of the bone, for it is the job of bones to be stiff. Therefore, energy from impact will be absorbed disproportionately at the ends of the bones.

4.4 Euler Buckling

If the bone becomes very thin-walled relative to its diameter, it is likely to fail by local buckling. It is likely to collapse, that is, before the value of load/cross-sectional area reaches the compressive or tensile strength of the bone. There is another situation in which apparently premature collapse is likely to occur: this is when the bone is very slender.

If a reasonably cylindrical and stocky bone is loaded in compression with a force P, the stress at any level will be roughly P/A, where A is the cross-sectional area at that level. If, however, the bone is slender, the situation is more complicated. The bone is unlikely to be completely straight or, if it is, it must be subjected in real life to sideways forces along its length. In the latter case these sideways forces will deform the bone slightly. In both cases, therefore, the straight line joining the two points at each end where the forces are acting will not pass down the middle of the bone (figure 4.12). As a result the initial axial force will have a bending moment about some part of the bone. This bending moment will, of course, tend to deflect the bone still more. This tendency will be resisted by the stiffness of the bone, which will, in general, be a function of both the Young's modulus and of I, the second moment of area about the appropriate axis.

However, if the bone is sufficiently slender, the axial force sufficiently large, and the Young's modulus and the second moment of area suffi-

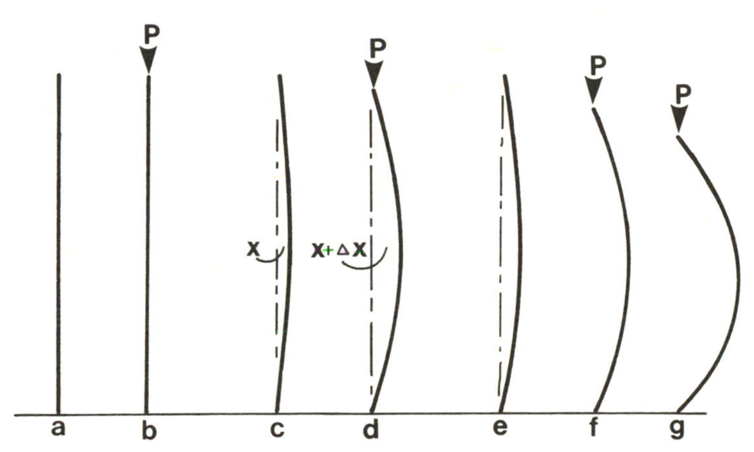

Fig. 4.12. Euler buckling. (*a*) A straight column is (*b*) loaded axially. (*c*) A slightly bent column is (*d*) loaded axially. It deforms more than the straight column because it bows out sideways, but the stiffness of the column counteracts the bending moment $P(x + \Delta x)$, and the column remains stable. (*e*) A bent column is (*f*) loaded axially. The stiffness of the column cannot prevent it from bending more (*g*), and it becomes unstable and collapses.

ciently small, an unstable situation arises. Now the deflection produced by the bending moment is sufficiently large for the bending moment to be increased beyond the ability of the stiffness of the bone to prevent further deflection taking place. The bending moment is thereby increased still more, and so on. In this unstable situation the bone will collapse. What usually happens is that the deformation becomes more and more extreme, resulting in large bending stresses near the middle of the length of the bone. The stress on the convex side of the bone, which started as compressive, becomes zero and then tensile. The stress on the concave side of the bone, which also started as compressive, becomes more and more compressive. Eventually, either the tensile or compressive strength of the bone material is reached and the bone ruptures.

This mode of collapse is called Euler buckling. It is characterized by a deformation of the whole structure rather than the deformation of a small part, which is seen in local buckling. Euler buckling is easy to demonstrate with a long stick or a long piece of steel rod. A small piece of material that seems completely rigid when its length is twice its diameter can usually be broken, or irreversibly deformed, if its length is fifty times its diameter.

Of course, the real loading situations in life are usually more complex than the simple case outlined here, the bone being subjected to bending moments by the action of muscles and by other adventitious forces. Taking the simplest possible case, that of a straight slender column quite

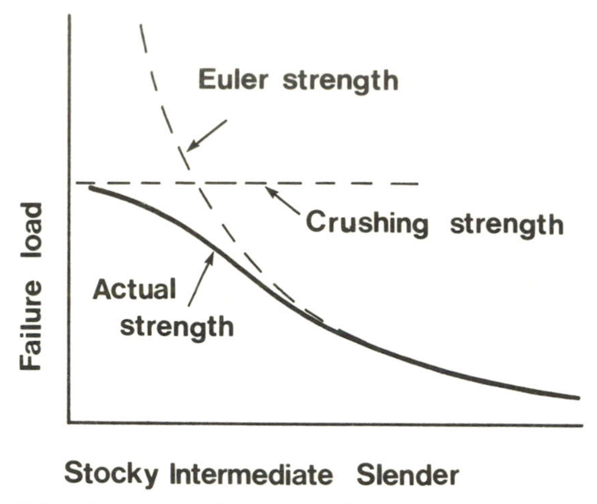

Fig. 4.13. The relationship between the theoretical Euler strength, the theoretical crushing strength, and the actual strength of columns of varying slenderness. Ordinate: load required to make the column fail. Abscissa: slenderness. The actual strength is always less than either of the theoretical strengths, and considerably less in the region where the curves for Euler strength and crushing strength approach and meet each other.

free to rotate about its ends (the pin-jointed condition), it is a standard result that the load Fe that will just cause the column to collapse by Euler buckling is $Fe = \pi^2 EI/L^2$. In this case I is taken in the direction in which it is least. L is the length of the column.

We are interested in whether real bones are ever likely to fail by Euler buckling. Obviously, any bone can be loaded axially in compression until the load equals the critical Euler load Fe. However, if this loading has already made the bone rupture in compression, the value of Fe would be of academic interest only. Bone material has a compressive strength, say S, and so if a bone is loaded in compression it will fail by compressive rupture if Force $= S \cdot A$, where A is the cross-sectional area. It will fail by Euler buckling if Force $= \pi^2 EI/L^2$. From these formulae it might seem that the bone, loaded in compression, would fail first by Euler buckling if $S \cdot A > \pi^2 EI/L^2$. In fact, the force necessary to cause failure in a column of any particular slenderness tends always to be lower than the lower of these two values, unless the column is very slender or very short and stocky (figure 4.13).

Instead of dealing with the factor I/L it is often useful to think in terms of the "slenderness ratio," L/Rg, where Rg is the "radius of gyration." The radius of gyration is the distance that, were the whole area of a section concentrated in a ring that distance from an axis, would give a value of moment of inertia equal to the actual moment of inertia. It is somewhat

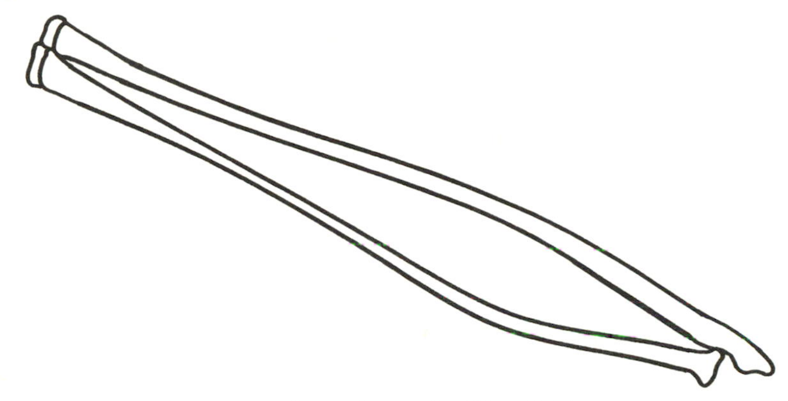

Fig. 4.14. Slender arm bones of a gibbon, *Hylobates lar*.

analogous to the center of gravity of a body—for many purposes the whole mass of the body can be considered to be concentrated there.

$$Rg = \sqrt{I/A}, \quad \text{so} \quad I/A = Rg^2; \qquad I = Rg^2 \cdot A$$

Therefore, the critical Euler force $= \pi^2 \cdot EA/(L/Rg)^2$. We can calculate the slenderness ratio at which bone should, theoretically, fail by both Euler buckling and by yielding in compression. Take $E = 200$ GPa, $S =$ 200 MPa.

$$2 \times 10^8 \times A = \pi^2 \times 10^{10} \times A/(L/Rg)^2; \qquad L/Rg = 32$$

If the bone cross section departs far from being circular, the least value of the second moment of area should be used, not the radius of gyration.

The humerus of an immature gibbon, *Hylobates lar*, had a length of 173 mm (figure 4.14), the midshaft an external diameter of 7.35 mm and an internal diameter of 3.85 mm (Currey, 1967). This produces a value of Rg of 2.07 mm, and $L/Rg = 83.6$. Obviously, this gibbon's bone will be prone to severe weakening because of Euler buckling if loaded in compression. The Euler stress is about 30 MPa, well below 200 MPa, the compressive strength. Similarly (Currey, 1967), a stork (*Ciconia* sp.) tibiotarsus would theoretically undergo Euler buckling at a nominal stress of about 20 MPa, a flamingo (*Phoenicopterus* sp.) tibiotarsus at about 10 MPa, and a willow warbler (*Phylloscopus trochilis*) tibiotarsus at 20 MPa. When we bear in mind that the actual collapse stresses will be somewhat lower than this, these are remarkably low values. The values for stiffness and compressive strength that I have assumed have been taken from other bones. It is likely that, because the bones are so slender and liable to buckling, the compressive and tensile strengths would never be the cause of failure in life. It would be adaptive, if this were the case, and would produce a bone of lower mass to perform the same function adequately, to increase the Young's modulus of the bone, even if it were at some cost to the strength and density. We do not know whether this has happened.

Bones are usually loaded in such complex ways that it is difficult to know, after the event, why they failed. Borden (1974) made some observations on bones that must have started to undergo Euler buckling, yet did not fail. These are the arm bones of children who fell on the outstretched arm. The bones were bent into a bow and then, because they had been loaded into the plastic region of the load-deformation curve, remained bent when the shock was over. Adult bones do not show this behavior because they have much less ability to distort into the plastic region.

In general, however, limb bones that have a value of L/Rg sufficiently high for them to be likely to be severely affected by Euler buckling are found in animals that probably do not load them very severely. Most obvious among these are the long bones of brachiating apes and monkeys (figure 4.14). In brachiators the large loads on the *limb* are tensile loads, produced when the animal is swinging on a bough while suspended by its hand. Although the limb is loaded in tension, it is almost certain (Oxnard, 1971) that the bones themselves are loaded in compression. The muscles surrounding the bones exert a force slightly greater than the force exerted by the body, thereby putting the bones into compression.

Nevertheless, the limb cannot do without the bones, however lightly they are loaded, because they are needed as a rigid strut to poke the hand onto the bough, where it can grasp and call into play the muscles, which do all the work. The adaptive response to these minimal requirements is a set of very slender bones, which are rarely, if ever, liable to fail through Euler buckling. This does not mean that such bones rarely break. In chapter 7 we shall see that such delicate bones are, in fact, quite often broken. The very slender tibiotarsal bones of birds are also probably never loaded very strongly except accidently. In storks and flamingos, the legs are used in rather stately bipedal walking; such birds do not run more than a few paces.

In general, then, bones that would fail, if loaded in compression, by Euler buckling are found in positions where they are rather lightly loaded or, as in the case of ribs, where the loading is not in compression.

4.5 Epiphyseal Plates

In chapter 1 I described briefly the epiphyses of developing bone. In many lower vertebrates the epiphysis is solely cartilaginous. However, in mammals, birds, and some reptiles there is often a bony epiphysis separated from the metaphysis by a cartilaginous plate. The advantage of this

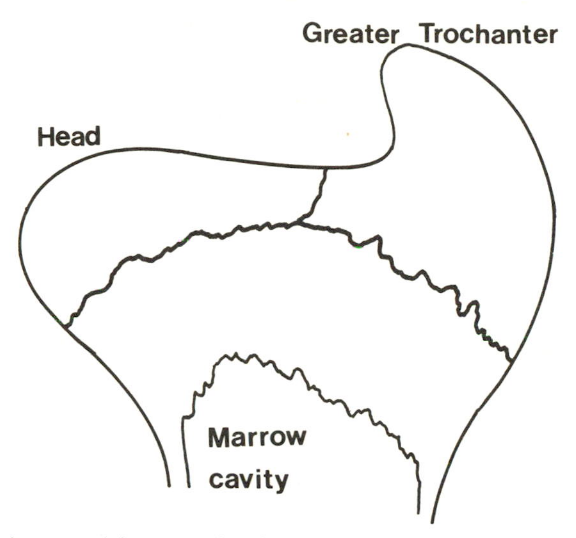

Fig. 4.15. Frontal section of the proximal end of a calf femur. Compact bone is not shown separately from cancellous bone. The epiphyseal cartilage is shown as a thick line.

arrangement is that it allows the epiphysis to be stronger and thicker than would be possible were it made solely of cartilage. However, there inevitably remains, until skeletal maturity, the cartilage of the growth plate, and this tends to be weaker than the bone around it. Bright, Burstein, and Elmore (1974) showed that if the tibia of the rat is subjected to shear, the cartilaginous epiphyseal plate breaks at a rather low force, and even if the epiphysis does not break off, the cartilage plate may show tears in planes where the highest shear strains occur.

Smith (1962a,b), in two undervalued papers, discussed ways in which mammals overcome the weakness of the epiphyseal plate of developing bones. Usually, if the general direction of the plate is straight across the bone, the plate will actually form an extremely crumpled sheet, so much so that the bony ends of the metaphysis and the epiphysis interdigitate (figure 4.15). (This interdigitation is not usually apparent on radiographs, because the X-rays travel the whole breadth of the plate and so the peaks and craters of the interdigitations average out over the image.) As a result of the interdigitation the shear between the epiphysis and the metaphysis is to a great extent borne by the bone. The cartilage is not subjected to much strain, and therefore not much stress (figure 4.16).

Apart from these rather fine-grained irregularities in the line of the epiphyseal plate there are, as Smith shows, larger-scale undulations which also have a mechanical function. Smith demonstrated the stress in various bones, mainly those of man and the cow, by making Plexiglass models

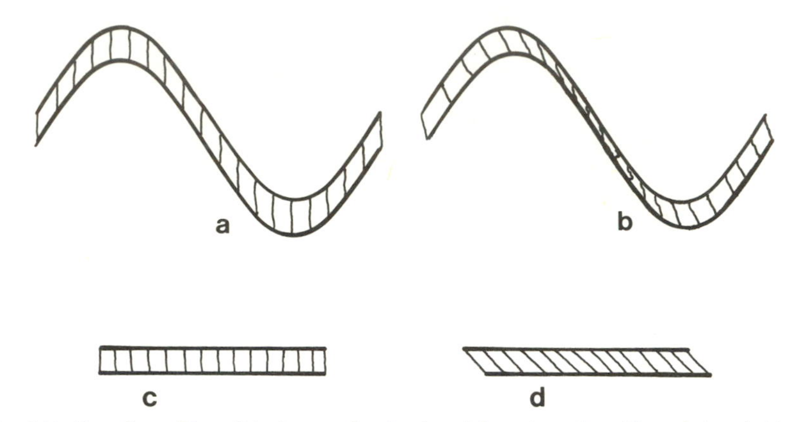

Fig. 4.16. The effect of interdigitation on shearing in epiphyseal cartilage. The subchondral bone is shown as a thick line, cartilage as thin lines. (*a,b*) When a crumpled cartilage plate is sheared, the cartilage is not deformed much before the subchondral bony sheets butt against each other. (*c,d*) A straight plate can only resist shear by the strength of the cartilage.

of them, loading them in appropriate ways, and observing the pattern of strain so produced through crossed polaroids. There are many objections to representing the three-dimensional structure of bone by a Plexiglass sheet. Even so, the bones Smith modeled were probably uniform enough for this technique to be reasonably valid.

The results of the analysis for the human calcaneus are shown in figure 4.17. When the heel is raised, it is compressed by the tibia and put into tension by the plantar ligament and the Achilles tendon. The directions of the principal tensile strains (that is, the directions in which there are no shear strains) are shown by interrupted lines, principal compressive strains by continuous lines. The line of the epiphyseal plate is shown stippled. The greatest shear strains are at 45° to the lines of the principal strains. Along the bottom of the calcaneus the plate is normal to the line of the compressive strains, so it will be compressed. The plate is also in line with the tensile strain. However, the compliant cartilage will be strained in parallel with the much stiffer bone, which will prevent it from being greatly strained in tension.

In the upper part of the calcaneus near the insertion of the Achilles tendon things are not so simple. The epiphyseal plate passes at about 45° to the direction of both the principal strains. This is the worst possible direction for shear. However, the danger of shear failure is mitigated by the plate being stepped: lying alternately parallel to the principal compressive and tensile strains.

A similar analysis of the distal end of the femur and the proximal end of the tibia with load across the joint and tension in the cruciate ligaments

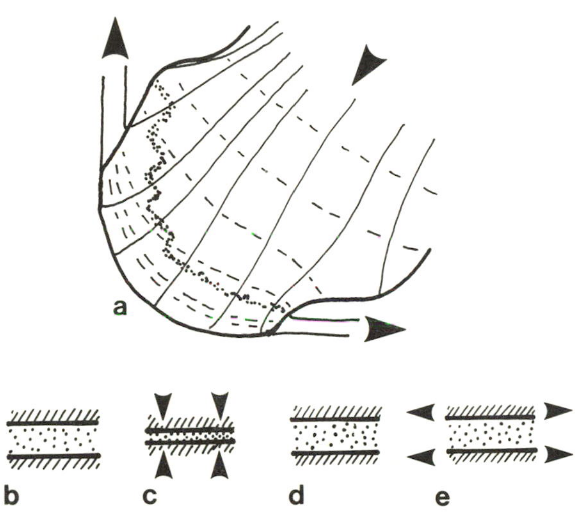

Fig. 4.17. (*a*) Diagram of a Plexiglass sheet modeling the human calcaneus. The arrowheads show the directions of the important forces. Principal compressive strains: thin solid lines; principal tensile strains: thin interrupted lines. The epiphyseal plate is shown dotted. (*b,c*) The plate may be loaded safely in compression. (*d,e*) If loaded in tension; the load is borne almost entirely by the bone acting in parallel with the cartilage.

(a situation occurring often during locomotion) shows that the plates are arranged everywhere nearly normal to the principal compressive strains, a most adaptive arrangement.

In general then, epiphyseal plates seem to be arranged so that the cartilage is in compression. If it is in tension, the epiphyseal and metaphyseal bone take most of the load, and so the cartilage undergoes little strain. The geometry of the plate is such that it is rarely subjected to much shear strain. Occasionally, however, it is not possible to prevent at least part of the plate being loaded in tension. Here are two examples. When a calf is stationary, each femur bears about one-fifth of the animal's weight. The reaction to the weight is upward through the acetabulum (figure 4.18). Equilibrium is achieved by the downward weight of the body and the force produced by the hamstring muscles on the ischium (Smith, 1962b). The forces load the pelvis in what is effectively three-point bending. The ilium and the ischium are attached to each other by an epiphyseal plate, which runs dorsally from the acetabulum. The ventral part of the plate will be in compression and the dorsal part in tension. No artful deviation of the path of the epiphyseal plate from the direct one can really prevent the cartilage from being loaded in tension. Similarly, in the tibia of the cow the action of the patellar ligament, which is inserted onto the epiphysis, makes it inevitable that the plate be loaded in tension at its

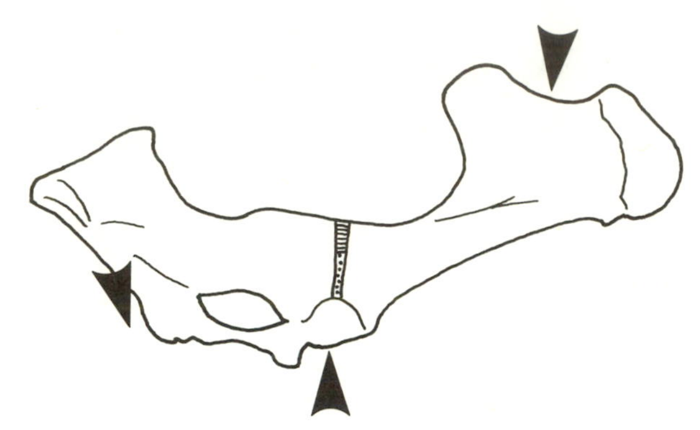

Fig. 4.18. Side view of the pelvis of a calf. The three major forces acting on it are shown. The epiphyseal plate passes dorsally from the acetabulum. The lower part is hyaline cartilage (dots), the dorsal part is fibrocartilage (lines).

anterior end, although the reaction ot the femur will load the posterior region in compressıon (figure 4.19). In these places, where tensile loading of the epiphyseal plate is inevitable, the plate is quite different in structure from where it is loaded in compression: it is composed almost entirely of collagen fibers, which bridge the gap between the epiphysis and the metaphysis and are orientated in the direction of the tensile forces. Smith describes several plates in which the cartilage changes from being of typical, hyaline form with much matrix to being fibrous, as the loading changes from being compressive to being tensile.

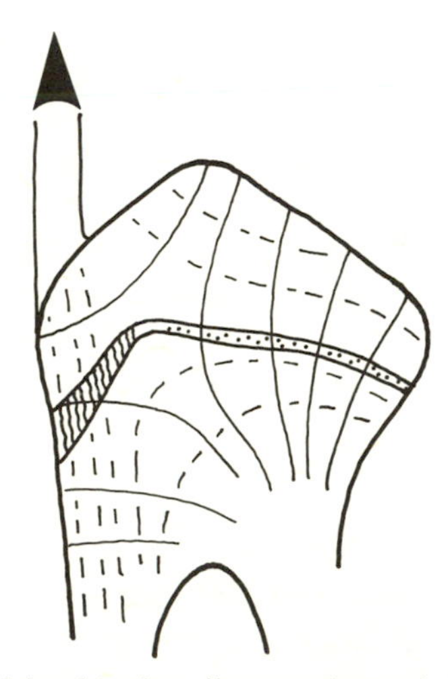

Fig. 4.19. Plexiglass model of the tibia of a calf, conventions as in figure 4.18. The action of the patellar tendon inevitably loads the anterior part of the epiphyseal plate in tension.

4.6 Methods of Analyzing Stresses in Bones

Bones have complicated shapes, and they are acted on by complex loads. It is usually very difficult to discover the stresses and deformations that are likely to occur in bone by using, for instance, the structural analysis applied to beams. Beam formulae will work well enough on long bones of fairly uniform cross section, though even here the analysis tends to break down near the ends (Huiskes, Janssen, and Sloof, 1981). In complex structures, such as the ends of long bones, vertebrae, or the skull, analytical methods are almost useless. The situation is made worse by the presence of cancellous bone, so that the structures of complicated shape are also made of materials of different mechanical properties.

There are other methods of analysis possible. One is to attach strain gages to the bone or to a model of it. This method is direct, allowing one to measure the actual strains on the surface, but it is expensive in time and allows surface strains only to be measured. The important study by Huiskes, Janssen, and Sloof, mentioned above, involved attaching more than 100 rosette strain gages to a femur, which ended up looking like a fully decorated Christmas tree.

Another method is to use photoelasticity. Strained models of bones can be viewed or photographed after being placed between crossed polaroids. The strain affects the plane of polarization of light, and a colored pattern, indicative of the strain in the model, appears. Although this method produces visually convincing patterns, it is quite difficult to derive the actual strains in the model from the pattern. The method is readily applicable to plane models, but if it is extended to three-dimensional analyses a feasible, but complicated, process must be carried out. This is the "frozen stress" technique, in which the model is loaded while hot, and is then allowed to cool under load. Some strains remain in the cooled model even after it is unloaded, which can then be sliced into planar shapes. The main disadvantages of this technique are the difficulties of making the models, and the fact that materials of different modulus cannot be modeled in the same structure. However, it is certainly capable of producing useful insights.

Another method is finite element analysis (FEA). This entirely mathematical approach depends on the computer, because the calculations involved are unbelievably long-winded. I shall not discuss the theory of the method in any detail. Essentially, the structure is divided into a set of small elements of relatively simple shape whose mechanical behavior is known analytically (figure 4.20). The program then has to calculate deflections in the structure as a whole in response to applied loads, making use of boundary requirements such as continuity of strain. The stresses are calculated from the strains. A short description of FEA is in

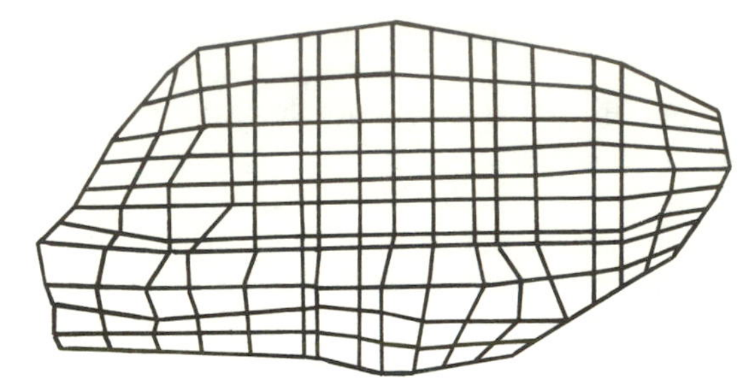

Fig. 4.20. An example of a finite element grid. This one models the human patella and is a slice taken from the midline. Articular surface at the bottom. There are 168 elements, and the material in each has one of 23 different concatenations of mechanical properties. This allows variation in cancellous bone density and anisotropy to be accommodated. (Modified from Hayes and Snyder, 1981.)

Rockey et al. (1975), a more detailed description in Zienkiewicz (1977).

The advantages of FEA are many. It can be used to investigate structures of any shape and complexity. There are no requirements that the Young's modulus of different elements should be the same. It is easy to alter the shape of the model, or of the mechanical properties of its constituent materials, by altering a few input variables. Graphical output is easily obtained, and stresses deep in the structure can be examined. The disadvantages of FEA are that it is an approximate method, extremely expensive in computer time, and it is not easy to check that the answers are even roughly correct.

Most people who use FEA use ready-made packages, which to a large extent they have to treat as black boxes. There are methods for checking the accuracy of the results, but they are not completely satisfactory. A result of the large amount of expensive computer time required is a fatal tendency of people to do an insufficiently fine-grained analysis. In the early days of the use of FEA on bone, the late sixties and early seventies, there were some ludicrous cases of people applying two-dimensional analyses to trabecular architecture. This is not always invalid, but in these cases it was like analyzing the stiffness of an I beam while ignoring the effect of the central web. A recent symposium (Simon, 1980) shows the stage of sophistication that has been reached. There is still a great tendency to take the results as they come, without independent checks. FEA has undoubtedly a very important role to play in the analysis of bone, particularly as the programs are becoming able to cope with features such as anisotropy. Nevertheless, it is not rash to predict that a large number of bad and misleading papers using it will appear in the next ten years. Readers intending to use FEA are urged to read Huiskes and Chao (1983).

4.7 Cancellous Bone

Cancellous bone is found in many places in bone. I think it useful to distinguish four locations:

1. At the ends of long bones, under synovial joints.
2. Where it completely fills short bones.
3. Where it acts as a filling in flattened bones.
4. Under protuberances to which tendons attach.

I have considered the first already. I shall consider the rest in turn. However, we must now deal further with the general properties of cancellous bone.

Cancellous bone may in some ways be thought of as compact bone with interconnecting holes in it, yet the ideas about stiffness and strength applicable to compact bone have to be severely modified for cancellous bone. This is because the arrangement of the struts and little beams of cancellous bones in space mean that it is behaving as a *structure* as well as a *material*. The loads in cancellous bone can be transferred from place to place by bending moments, and compressive loads may cause individual trabeculae to buckle. Therefore an analytical account of cancellous bone would have to explain not only the reaction of very small volumes of bone to force, just as in compact bone, but also how it was that those particular small volumes came to be subjected to those forces. This latter is not a problem with carefully machined test pieces of cortical bone.

Nevertheless, in three important papers (Carter and Hayes, 1976b, 1977a; Carter, Schwab, and Spengler, 1980) Carter and his co-workers have put forward the idea that "all bone can be mechanically viewed as a single material" (Carter and Hayes, 1976b). These workers have isolated specimens of human and cattle cancellous bone and determined their elastic and strength properties in relation to their apparent density. By apparent density I mean the mass of the bone, with marrow removed, divided by the volume of the specimen determined from its outside dimensions. They found that both the strength and the modulus fitted power-law equations very well, and that compact bone fell on a continuation of the curve (figure 4.21). The relationships were, as has been mentioned above, Young's modulus = kD^3; compressive strength = $k'D^2$, where D is the apparent density. This is only an empirical observation, and does not explain *why* the strength varies as the square of the density, and the modulus with the cube. Nevertheless, the observations are very useful as a guide in trying to work out why cancellous bone is found in some situations and not in others, as I shall show below.

Recently Carter, Schwab, and Spengler (1980) have shown that the *tensile* strength and Young's modulus obey the same power law, and

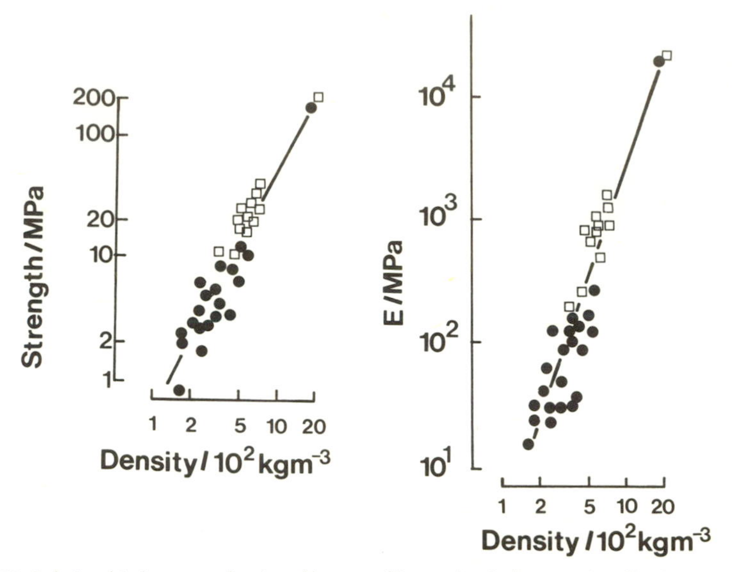

Fig. 4.21. Relationship between density of bone and its mechanical properties. Abscissa: apparent density. Note logarithmic scales. Open squares: cows; closed circles: humans. (Derived from Carter and Hayes, 1975.) The symbols at the top have been confirmed by many investigators, and so should be heavily weighted in the mind's eye.

indeed the constants k and k' were nearly the same. This is to be expected of Young's modulus; one would not expect the modulus measured in tension and in compression to be different. However, it is interesting that the tensile and compressive *strengths* are the same. This is not true for fully compact bone, of course.

The theory of the relationship between porosity and mechanical properties is not well worked out. Ceramicists are mainly interested in noninterconnecting cavities. For instance, spherical cavities in a material of Poisson's ratio 0.3 have an effect on Young's modulus described by $E = E_0(1 - 1.9p + 0.9p^2)$, where E is the modulus of the material with pores, E_0 is the modulus of a fully dense material, and p the porosity (Davidge, 1979). Not surprisingly, this does not fit the bone data nearly as well as the power-law curve. Gibson and Ashby (1982) have made a theoretical analysis of foams with interconnecting and noninterconnecting cavities and have compared their results with experimentally derived data. They found that both the elastic collapse stress *and* Young's modulus should be, and were, proportional to the *square* of the density and, in the case of Young's modulus, not proportional to the cube. Although the relationships derived by Carter and his colleagues are derived from tests on bone material itself, it may be borne in mind that the results of

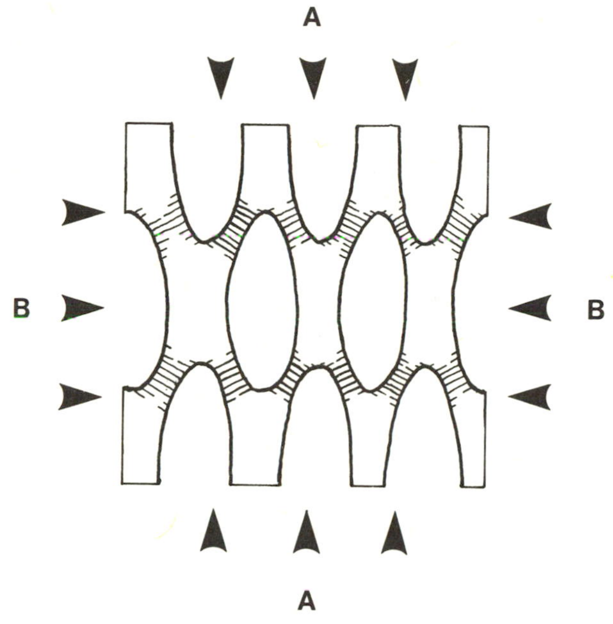

Fig. 4.22. Diagram showing anisotropy of cancellous bone. The finely hatched regions are subjected to bending.

Gibson and Ashby are not consonant with those of Carter for Young's modulus, though they are for strength.

The grain of the individual lamellae in cancellous bone seems to lie along the length of the struts, and therefore, because the struts are loaded only from their ends (they cannot be loaded via the marrow), the bone material in the struts will be loaded fairly well in the direction of its grain. If, as is likely, the bone material in cancellous bone is like that in compact bone, it will obviously be advantageous that the material should be loaded in the direction of its grain, because it is strongest in that direction. Anisotropy in cancellous bone will be produced, not by its fine-scale structure, but by the arrangement of the trabeculae themselves. Figure 4.22 shows diagrammatically a small block of rather anisotropic cancellous bone. Obviously, it will be much stiffer when loaded in the *A–A* direction than when loaded in the *B–B* direction. When the block is loaded in the latter direction, the finely hatched regions will be subjected to much larger bending moments than when the block is loaded in the *A–A* direction. They will therefore deform much more, and the block as a whole will be more compliant. There are few studies of the anisotropic mechanical behavior of cancellous bone, but it is clear that it can be quite anisotropic. Townsend et al. (1975) examined the anisotropic

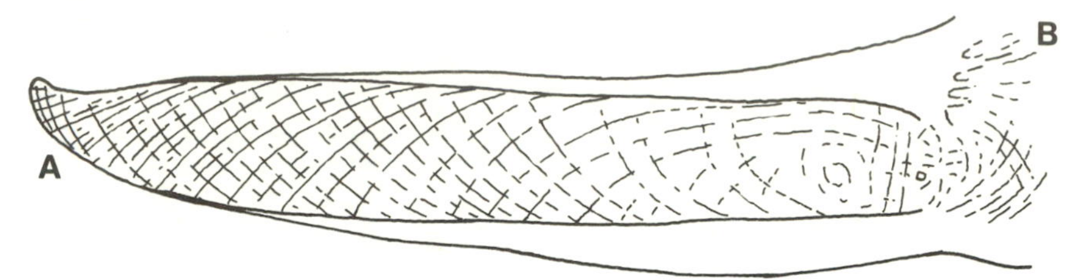

Fig. 4.23. Cross section of the ischium of a horse. This is a fairly flat sheet loaded by adductor muscles at its free end, *A*. The predominant orientation of the trabeculae is sketched in a rather idealized way. In region *B* the loading system is quite different.

elasticity of 18 cubes of cancellous bone from the human patella. The median of the ratio of the greatest to the least value for modulus in each section (they tested in three orthogonal directions) was 1.7. The greatest ratio was 4.8.

A glance at any section of cancellous bone shows it to be far from randomly organized. Indeed, it was the apparent relationship between the cancellous bone in the head of a femur, and a crane-shaped bar, that led Wolff to produce his famous (though unhelpful) "law of bone transforma-

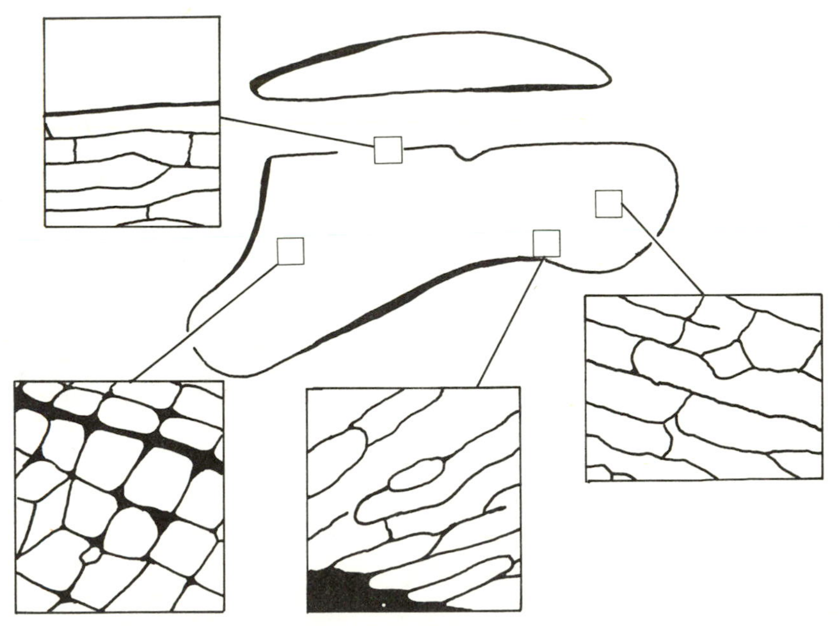

Fig. 4.24. Sagittal section of a cervical vertebra of a horse. The boxes show regions enlarged. Note that the trabeculae are in general arranged at right angles to each other.

tion" (Wolff, 1869). The subject of the intellectual origins of Wolff's law is dealt with most interestingly by Roesler (1981).

Figures 4.23 and 4.24 show two bones in which cancellous bone occurs densely. Two things are at once apparent. (1) There is a fairly clear distinction between compact and cancellous bone, there being little bone that it is difficult to classify as one or the other. (2) The trabeculae, as they fan out from the cortical bone, appear to cross each other roughly at right angles. This latter feature is not universal in cancellous bone, but does occur often enough to merit explanation.

4.7.1 Principal Stresses

Think of a piece of bone with, at some moment, a complicated set of forces acting on it. Inside the bone we can imagine a very small cube, sufficiently small that the forces in it do not change through the thickness of the cube. This cube can have an arbitrarily complex set of normal and shear stresses acting on its faces. It can be shown that it is always possible to rotate the cube in such a way that there are no shear stresses acting on its faces. There will be three normal stresses (any of which can be zero, of course) acting on each of the opposite pairs of faces, and therefore acting mutually at right angles. These stresses are called the **principal** stresses. To simplify the discussion, suppose that we are considering stresses only in a plane, ignoring the third dimension. The two-dimensional sheet we now have has very many small areas, each of which could, in the imagination, be rotated so that only principal stresses act on it. It is possible to follow stresses through the sheet; they will change direction in various ways. However, these stresses do have the important characteristic that at any point there will be two at right angles.

If the bony sheet is solid, then the stresses are not constrained to go in any particular place by the bone itself, but only by the forces acting on the bone. However, suppose that the bone were filled with a finely spaced set of small holes, set sufficiently close that the general direction the lines of stress follow are effectively unaltered from the situation in the solid sheet. Are there any particularly advantageous or disadvantageous ways the holes could be placed? The holes should, in fact, be set so that the solid bone remaining lies in the direction of the principal stresses.

As an example, figure 4.25 shows a small portion of a cancellous network consisting of struts left between holes. Each strut joins at its end to other struts at a node. Suppose the length of each strut is ten units and that they have a square section of side one unit. Suppose the nodes

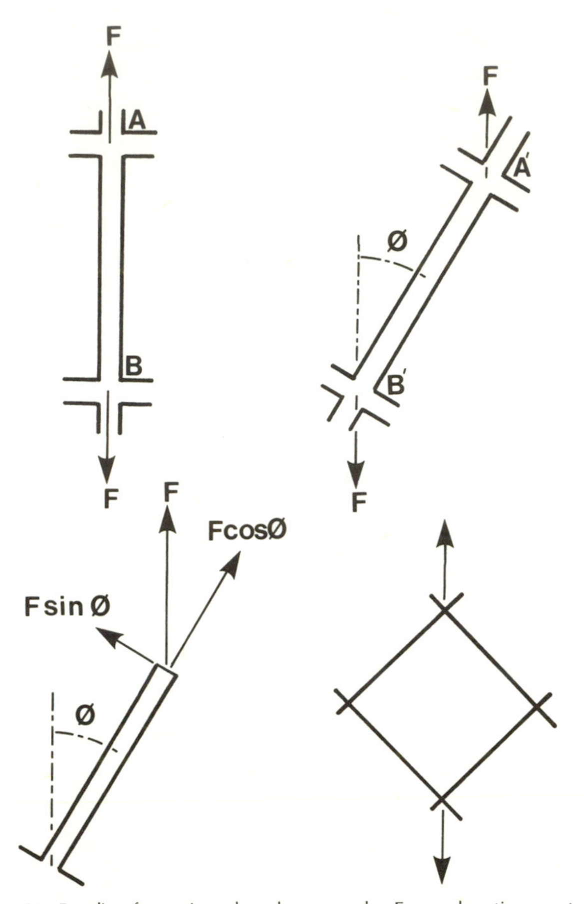

Fig. 4.25. Bending forces in trabecular networks. For explanation see text.

A, *B*, and *A*′, *B*′ have to carry a load *F* between them. What will be the maximum stress in the struts connecting the two nodes? In the case of *A*, *B* it will be *F*. In the case of *A*′, *B*′ it will depend on the value of ϕ. Take the node *B*′ as being held rigid. The force *F* can then be decomposed into two forces: $F \cdot \cos \phi$ acting along the strut, and $F \cdot \sin \phi$ acting to produce a bending moment of $10 \cdot F \sin \phi$ at the other end of the strut, at the node *B*′. The maximum stress caused by a bending moment *M* in a beam is Mc/I, where *c* is half the depth of the section, and *I* is the second moment of area of the cross section. These have values of 0.5 and 1/12 respectively. Therefore, the greatest tensile stress caused by the off-axis component of *F* is $60 \cdot F \sin \phi$. This will be added to the tensile force to give a total of $F \cdot \cos \phi + 60 \cdot F \cdot \sin \phi$. When the strut is at right angles to the load's line of action it will, notionally, be bearing a maximum stress of $60 \cdot F$, which is 60 times greater than the maximum stress when it is aligned with the load. However, the situation is not nearly as bad as this because, so far, I have ignored all the other struts in the meshwork. If ϕ were 90°, another strut would be in line with F and would be bearing the load. In fact, the worst orientation is at $\phi = 45°$, when two struts are sharing the load borne by one strut when ϕ is zero. Even so, given the quite arbitrary ratio of length to breadth we have set up, the tensile stress will be 22 times as great as in the strut oriented in line with the force. In fact, things rapidly become quite complex as we consider the other principal stress, which may be the same as, or different from, the first one in both magnitude and sign. However, despite these complications, it remains true that if the struts are oriented in the direction of the principal stresses, they will experience no bending moments, and this will minimize the maximum stress they experience. It is easy to show, by extension, for the three-dimensional case, that all these struts ought to be mutually perpendicular.

4.7.2 The Arrangement of Trabeculae in Bones

The question of whether the trabeculae in cancellous bone are, or are not, arranged in relation to the principal stresses has been the subject of much rather desultory argument (Murray, 1936; Roesler, 1981). One problem is that it is usually very difficult to determine what forces are acting on a bone, so it is not clear what stresses will be acting inside it. Another related problem is that bone with cancellous bone inside it is no longer a homogeneous body, and the distribution of stress is itself constrained by the presence or absence of bony tissue at particular points. Nevertheless, in some bones the distribution of stress is intuitively clear, and we can see whether the cancellous struts are well arranged.

The ischium of the horse has a posterior flange which is loaded in bending by the action of adductor muscles. Its mechanical situation is like that of a cantilevered beam loaded at its free end. The principal stresses acting in a cantilever are shown in figure 4.26, the trabecular arrangement in the ischium in figure 4.23.

The spinous process of a thoracic vertebra of a horse is shown in figure 4.27. The spine measures 185 mm dorsoventrally, 12 mm from side to side, and 33 mm anteroposteriorly. The main function of the spine is to bear muscle pulls in the anteroposterior direction. However, because it is about three times longer in this direction than it is laterally, it is likely, quite often, to bend in the direction shown by the small arrows. The trabeculae seem to be arranged to resist such bending.

Vertebral centra are, because of the stress-distributing property of the intervertebral disc, usually loaded almost axially, even if the spine is flexed.

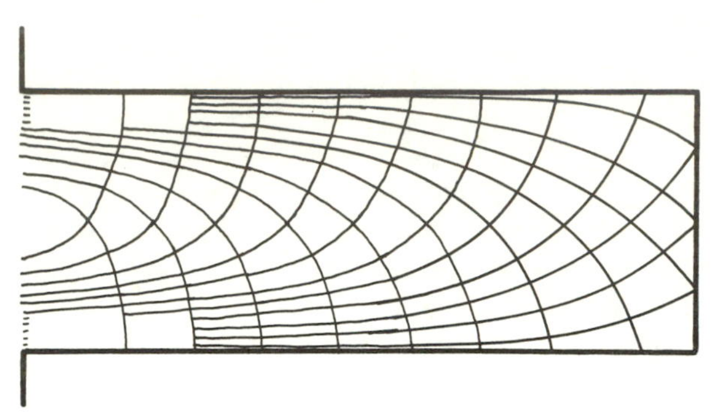

Fig. 4.26. Principal stress trajectories in a cantilever loaded at its free end. The more crowded the trajectories, the greater the stress. They have been omitted for clarity near the root of the cantilever.

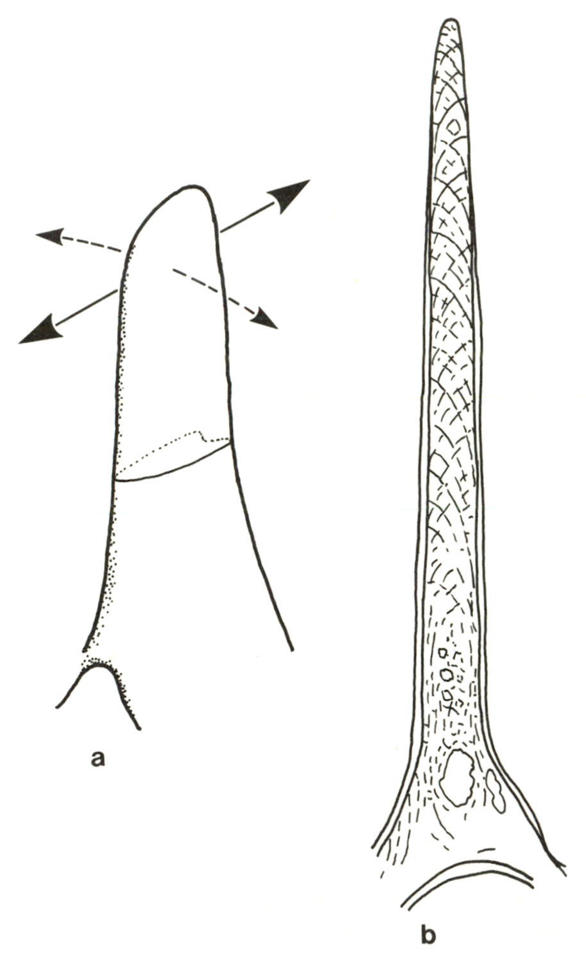

Fig. 4.27. Spinous process of a thoracic vertebra of a horse. (*a*) The direction of loading. (*b*) Transverse section showing trabeculae.

A principal compressive stress will run, therefore, from one end of the centrum to the other. Possibly (Frost, 1964), because the body of the centrum is concave along its length, compressive forces between the two ends will cause the walls of the centrum to bow inward, causing radially directed compressive stresses. This effect may not be very marked but, however large it is, it will be less than the effect of the end loading in producing axial compressive stresses, and so the expected direction of the principal stresses will be longitudinal and radial. This is the orientation of the trabeculae. Galante, Rostoker, and Ray (1970) showed that trabecular plugs from the human centrum were about 2.5 times stronger when loaded in the longitudinal direction than when loaded dorsoventrally or laterally, so the radial fibers certainly have significant strength.

In all these bones the trabeculae seem to be oriented in the direction one would expect if they were aligned with the direction of the stresses in a solid body. They seem to be a bony model of the principal stresses, arranged so that they will not be subjected to significant bending moments. However, the trabeculae are not merely embodiments of mathematical abstractions; they have their own lives to lead. If there were no radial stresses in the centrum, and it is not clear that there are any, it would still be bad design to have only longitudinal struts. These longitudinal trabeculae would have a very high ratio of length to second moment of area, and therefore would buckle in an Euler fashion at a very low load. The radial trabeculae may be resisting radial stresses, but they also are pinning the longitudinal trabeculae and preventing buckling. Unfortunately, therefore, the distribution of trabeculae in the vertebral centra has two rather different types of explanation.

Recently, Hayes and Snyder (1981) have adopted an apparently more objective approach to the explanation of trabecular orientation. They have examined a rather detailed finite element model of the patella (figure 4.20). They quantified the density and geometrical anisotropy in some detail, and then used them in the finite element model along with information about the relationship of mechanical properties to the density and anisotropy. Several of their findings are of great interest. First, they found that the orientation of the trabeculae corresponded rather well to the direction of the principal stresses. Second, they found that the trabecular density was greatest in those regions where the *difference* in the values of the principal stresses was greatest. Hayes and Snyder found a better correlation between the trabecular density and this difference in stress (the so-called Tresca stress criterion *pace* Hayes and Snyder, who call it the von Mises criterion) than they did with maximum tensile stress or maximum compressive stress. They write, "This tends to refute the hypothesis that either tensile or compressive stresses govern trabecular architecture." While this is true as stated, examination of their computer graphics output shows that there are large tensile stresses along the anterior border of the patella, and fairly large compressive stresses in the middle of the posterior border. In both these regions the other principal stress is rather small in magnitude, and both have dense cancellous bone. It would also be true, therefore, that the density of cancellous bone could be correlated with the presence of large principal stresses, without respect to sign. The question remains open.

The correlation between the directions of the principal stresses and the trabecular orientation is very strong. However, since anisotropy of Young's modulus was fed into the finite element model, and this was highly dependent on trabecular orientation, it is not clear whether, to some extent, the correlation was imposed by the answer. Still, this is an interesting study, and certainly corroborates the hypothesis that trabeculae are oriented in relation to principal stresses.

Hayes and his associates have not tried to model the stresses in trabecular bone in terms of its actual geometric shape, but considered only its bulk properties. Indeed, there have been very few attempts to model trabecular architecture because a set of interconnecting struts and sheets is very difficult to analyze. Finite element analysis seems, at the moment, the only way forward.

Lanyon (1974) has shown a clear similarity between the principal strains, as measured with strain gages in vivo, on the cortex of a bone and the arrangement of the trabeculae in the cancellous bone beneath the surface. The bone was the calcaneus of the sheep, which is a suitable bone for this experiment because it is loaded symmetrically about its midline, so the flat lateral cortices, to which the strain gages are stuck, are likely to have the same direction of strain as the trabeculae. Although the directions of the principal strains vary somewhat during the locomotory cycle, the direction of the trabeculae conforms quite closely to that of the principal strains when these are greatest.

One of the great difficulties in considering cancellous bone is that of determining to what extent the sheets are orthogonal. In some cases the pattern is clear, in others much less so. Oxnard and Yang have been developing an optical method for coping with this. Essentially, an X-radiograph, or some other suitable pictorial representation, is scanned by a laser beam and a Fourier transform of the resulting diffraction pattern made. The result is a rather easily comprehended contour map showing the density of elements lying in particular directions (Oxnard and Yang, 1981). The information they obtain shows, for instance, that in the human fourth lumbar vertebra there does seem to be a great preponderance of orthogonally directed struts, lying longitudinally and transversely. In the second lumbar vertebra there are, in addition to similarly arranged struts, a set lying at about 30° to the long axis of the centrum. Oxnard and Yang suggest that these elements are concerned with stresses lying at an angle to the long axis, and that these stresses come because the second lumbar vertebra is subjected to much more bending than the fourth lumbar. This is so because it lies farther away from the load line of the body. Although I do not find this explanation wholly convincing, it is clear that their technique enables one fairly easily to determine the anisotropy of the geometry of the trabeculae. Another example Oxnard and Yang produce—a rather clear orthogonal arrangement in the lumbar vertebrae of the gorilla, while the orangutan shows complete isotropy—is perhaps more convincing. So is their functional explanation: the orangutan imposes a more complex set of loads on its vertebrae than does the gorilla, because it has a much more varied locomotory repertoire, and as a result no particular set of orthogonally arranged struts would be appropriate for the orangutan.

So far I have discussed the apparent orthogonality of trabeculae as being related to the directions of the principal strains in a solid bone of which, as it were, the cancellous bone is a skeleton. However, a superficially completely different approach also suggests why the observed structure may be adaptive. This is the consideration of cancellous bone structures as minimum weight braced frameworks.

Consider the structure shown in figure 4.28a. It is pin-jointed at *p*, *q*, *r*, and *s*; that is, it is free to rotate at these points. A sideways load (*b*) will obviously make it collapse. This can be prevented if one or more of the pin joints is replaced by a rigid joint (*c*). The sideways load is now resisted by *bending* moments set up in the structure (*d*). An alternative way of making the structure stable would be to add an extra member (*e*). The structure now bears the loads as *axial* stresses, and deforms little under load (*f*). A structure in which load is resisted by changes in length of its members, rather than by bending, is known as a braced framework. A clear introduction to the subject of braced frameworks is by Parkes (1974). The fact that braced frameworks do not bear bending loads shows the probable relationship they will have with cancellous bone, in which I have shown it will be advantageous not to have bending moments, even though the struts might be capable of bearing them.

The method of deriving braced frameworks of least weight is not simple. It is based on two mechanical theorems: Maxwell's lemma and Michell's theorem, which I shall not discuss. They are reviewed in Parkes' book. What is interesting is the *type* of structure that results. They are orthogonal nets. Depending on the loading system they may be rectangular or, in systems with concentrated loads, they are often developments of equiangular spirals.

Figure 4.29 shows the minimum weight framework for a cantilever, pinned at *A* and *B*, and loaded at *C*. The elements that carry tension are

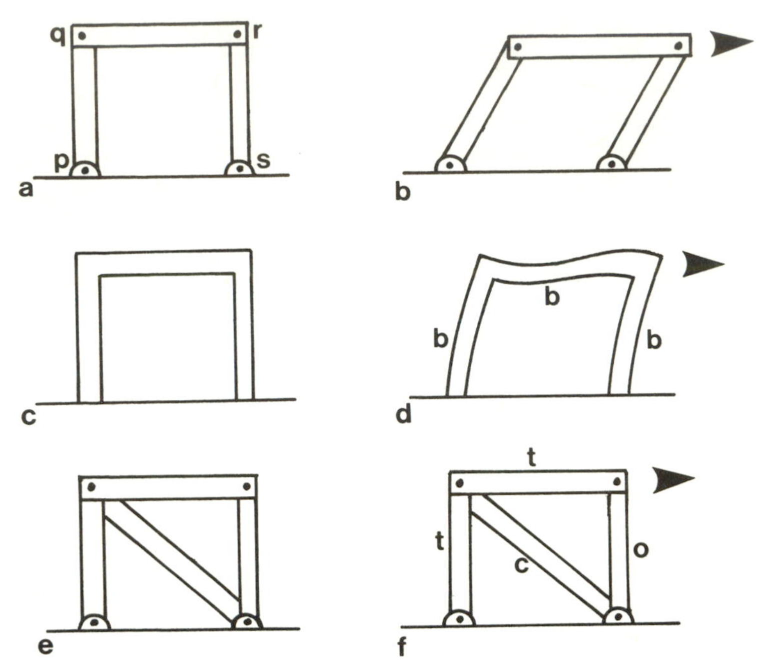

Fig. 4.28. Braced frameworks. The structure in (*a*) has no sideways rigidity (*b*). If one or more joints are made rigid (*c*), the structure resists bending (*d*) by means of bending stresses set up in the members *b*. If one extra diagonal element is put in (*e*), the structure is rigid (*f*), resisting the load by tension *t* or compression *c* stresses. The member marked *o* is, in fact, unloaded in this situation.

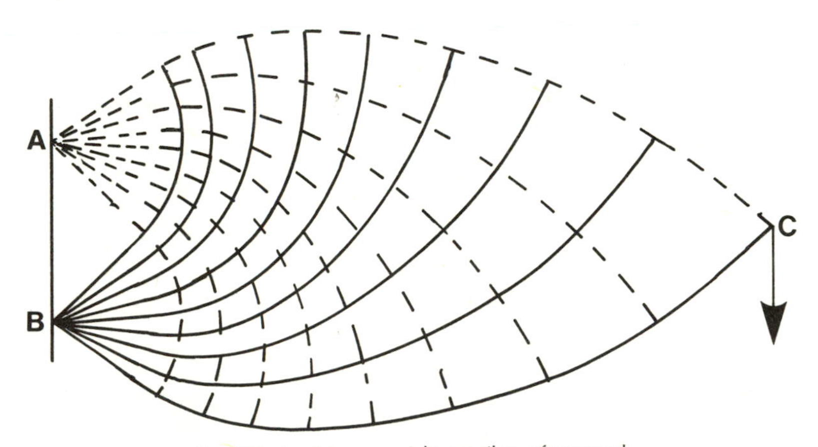

Fig. 4.29. A minimum weight cantilever framework.

shown interrupted; the compression members are shown continuous. The resemblance of this general pattern to, for instance, the pattern of trabeculae in the ischium of the horse is striking (figure 4.23). Of course, just because it is striking we should remember that there are great differences in the two situations. The cancellous bone is enclosed, as always, between two sheets of compact bone, and the loading system is more complicated in the bone. However, the resemblance is such that it seems very likely that cancellous bone is often a distorted version of a minimum weight braced framework. Unfortunately, deriving minimum weight frameworks for complex loading systems is so difficult that it will be some time before this type of analysis will be really useful in analyzing cancellous networks.

4.7.3 The Energy Absorption of Cancellous Bone

The load-deformation curve of cancellous bone has a characteristic shape (Carter, Schwab, and Spengler, 1980). In compressive loading there is an initial linear region, then there is a slight drop in the curve, and then a long plateau region (figure 4.30). When the curve dips, trabeculae have started to buckle. The plateau region is where more and more of

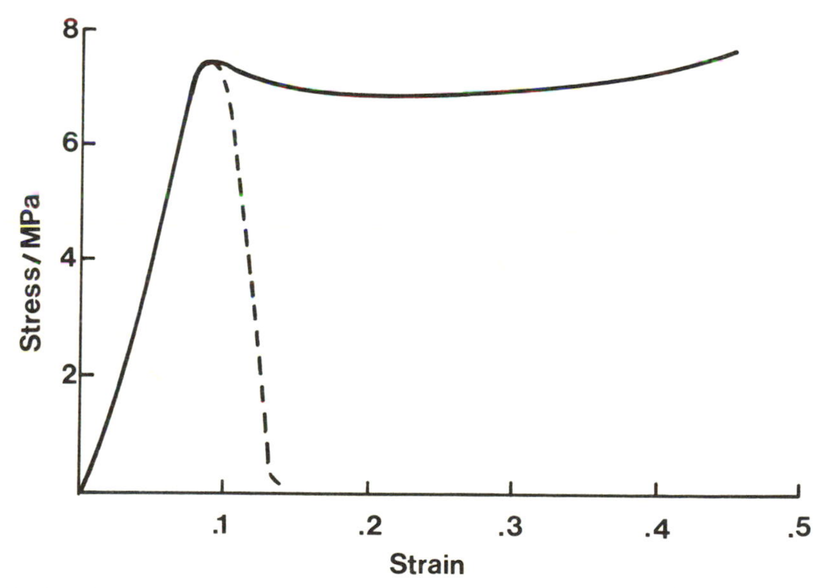

Fig. 4.30. Stress-strain curves for cancellous bone in both tension (interrupted line) and compression (solid line).

the length of the specimen collapses. When trabeculae collapse, the bone becomes compacted in that region, and so resists further collapse. In tensile loading, on the other hand, the linear region is much the same as in compression, but once the specimen starts to fail, the trabeculae tear across at one level, and the load-deformation curve is roughly triangular. The energy, shown by the area under the load-deformation curve, is much greater in compression than in tension.

It seems often to be assumed that cancellous bone is good at absorbing energy. The findings of Carter and Hayes and other workers concerning the relationship between density and Young's modulus and strength show that this is not true if we concern ourselves with elastic, preyield behavior. Consider a piece of elastic material loaded to its yield stress S. If it has a Young's modulus E, then the strain at fracture, ε, will be S/E. The area under the stress-strain curve, which is equivalent to the work done on the specimen, or the energy it absorbs, is, for a linearly elastic material (stress $\times$ strain)/2 $= S^2/2E$. Since, as Carter and Hayes found that for bone in general, fracture stress $\propto D^2$, and Young's modulus $\propto D^3$ so, $S^2/2E \propto (D^2)^2/D^3 \propto D$. The area under the stress-strain curve will be proportional, then, to the apparent density, and dense bone absorbs more energy per unit volume. If selection were favoring a material of minimum mass to absorb energy, it would select for a material of minimum value of $D \cdot E/S^2$ (table 4.1).

$$\propto D \cdot D^3/D^4 = 1$$

In other words, for the criterion of minimum mass all densities of bone are equal. Of course, if mass of fat is taken into account, then cancellous bone is a disadvantage, because it will add to the mass of the bone without improving its mechanical properties.

These calculations apply only if elastic deformations are being considered. If the trabeculae buckle in compression, then the cancellous load-deformation curve will enter the long plastic zone, and will absorb considerably more energy than would compact bone. Furthermore, cancellous bone often has hematopoietic red marrow, in which case the marrow is useful, not mere packing, and should not be considered in the mass of the bone.

4.7.4 Cancellous Bone in Short Bones

Long bones usually have cancellous bone only near their ends. Many shorter bones, however, are essentially thin shells of compact bone enclosing a core of cancellous bone. Examples of bones showing such a

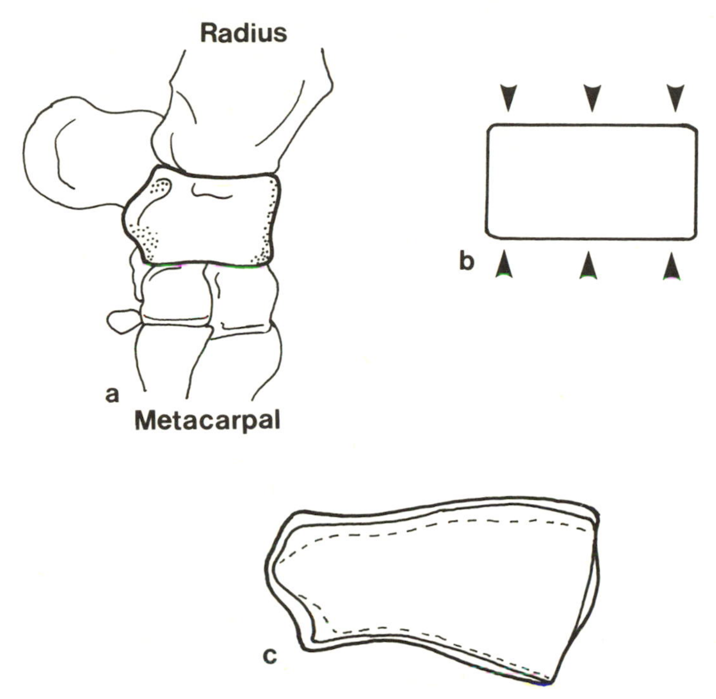

Fig. 4.31. (*a*) Left carpus of a horse, medial view. (*b*) Design requirements of the scaphoid: it must take a variable compressive load over a short distance. (*c*) Cross section of the scaphoid. There is compact bone around the outside. Just inside this is dense cancellous bone. Inside this is more tenuous cancellous bone. The dotted line shows the approximate position of the boundary between the two types.

structure are wrist and ankle bones and the centra of vertebrae. I shall discuss a bone of the wrist of the horse to show that structure is related to function in this bone whose lumen has much cancellous bone. The discussion will have ramifications well beyond the bone in question and will produce results of general interest.

The scaphoid is in the wrist. It lies between the radius and three more distal carpal bones (figure 4.31a). There are complexly but not strongly curved articular surfaces on top and bottom. The bone is about as deep as it is broad in one view, and half as deep as it is broad in another. The scaphoid obviously cannot be considered to be a long bone. The articular surfaces allow some relative movement of the radius and the wrist, and through this of the greatly elongated single metacarpal. During the stepping cycle different parts of the articular surfaces will become more or less heavily loaded. The design requirements are shown very diagrammatically in figure 4.31b. The solution, as shown by the horse, again drawn very diagrammatically, is shown in figure 4.31c. The less dense cancellous

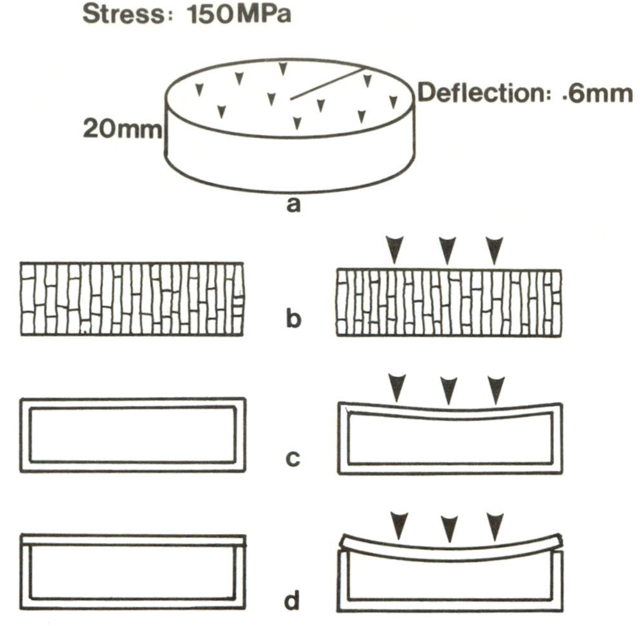

Fig. 4.32. The design of a short bone. (*a*) The specification. (*b*) The cancellous bone solution. (*c*) The box with an integral top. (*d*) The box with a separate plate on top.

bone consists of thin-walled tubes all oriented the same way—from top to bottom of the bone.

A possible design for the scaphoid would be a box shape with thick walls but with no cancellous bone in the lumen, only marrow fat. It is interesting to compare the mass required to produce a given stiffness using either cancellous bone or using a boxlike bone. The analysis below is rather crude, but it will bring out some points of interest. I have used values near those found in bone, for verisimilitude, but the approach could be much more general.

Suppose we have a bone of circular cross section whose function is to take a distributed load from one face to the other. It has a depth of 40 mm, and a radius a, which we shall vary. The force is distributed evenly all over the surface, producing a stress 15 MPa. Suppose Young's modulus of compact bone is 20 GPa, and its density 2,000 kg m^{-3}. The density of the fat filling the space inside the bone is 1,000 kg m^{-3}. Cancellous bone has a bulk density Dc, which can vary, and Young's modulus of elasticity is $20 \times (Dc/2{,}000)^3$ GPa. This is making use of the findings of Carter and Hayes (1977a). The design criterion we now impose is that the deflection produced by the load should not exceed 1.2 mm, taking both ends of the bone into account (figure 4.32). This is a rather arbitrary choice of criterion. However, it is probable that the ability to resist excessive distortion is important in bones in joints, as elsewhere. The value of the distortion chosen would not, in fact, affect the anlaysis; it is chosen merely so that the stresses in the bone do not become unrealistically high. We shall consider the bone to be symmetrical about the section at mid-length, and so from now on will consider only half the length.

As figure 4.32 shows, the mode of deformation is different in the two cases. Cancellous bone will deform uniformly, the hollow compact bone will deflect more in the middle than near the edges.

Take cancellous bone first. Allowable strain = 0.03. Because $E = S/\varepsilon$, the value of E must be $1.5 \times 10^7/0.003 = 500$ MPa

Now $Ec = E\text{comp}(Dc/D\text{comp})^3$; $5 \times 10^8 = 2 \times 10^{10} \times (Dc/2 \times 10^3)^3$

$Dc/D\text{comp} = (5 \times 10^8/2 \times 10^{10})^{1/3} = 0.292$

Porosity = 0.708

Bulk density = $0.292 \times 2 \times 10^3 = 584\ \text{kg}\,\text{m}^{-3}$

Fat occupies 0.708 of the volume, so its density = $708\ \text{kg}\,\text{m}^{-3}$

Total density of cancellous bone = $1{,}292\ \text{kg}\,\text{m}^{-3}$

Mass of bone = $a^2 \times \pi \times 0.02 \times 1{,}292$ kg = $81.2a^2$ kg.

For the hollow bone the situation is more complex. I shall assume first that the bone on the top surface, "the plate," is fixed rigidly to sidewalls whose mass can be ignored and which are themselves completely rigid. (The assumptions made in this part of the analysis favor the boxlike bone over the cancellous bone.) Now it is clear that as the value of a, the radius, increases, there will be a greater distance for the plate to span, and its deflection, if its thickness remains the same, will become greater. So, as the radius increases, the plate must become thicker if the deflection is to remain constant.

The formula for a flat circular plate, loaded uniformly, held rigid at the edges is:

$$\text{Deformation} = 3W(m^2 - 1)a^2/16\pi Em^2t^3$$

(Roark, 1965) where a is the radius, W is the total load (= stress $\times\ \pi a^2$), and m is the reciprocal of Poisson's ratio. I take $m = 3$. The precise value assumed for Poisson's ratio can be seen, from inspection of the formula, not to be very critical. Given the above values $t = (0.208a^4)^{1/3}$, and so the mass of the bone is $\pi a^2 t \times 2{,}000$ kg. The mass of the fat in the lumen is $\pi a^2(0.02 - t) \times 1{,}000$ kg.

The results are shown in figure 4.33. The box solution is lighter up to a value of radius to depth of 1.5:1, after that it becomes heavier. We could, alternatively, assume that the plate of bone is not rigidly fixed to

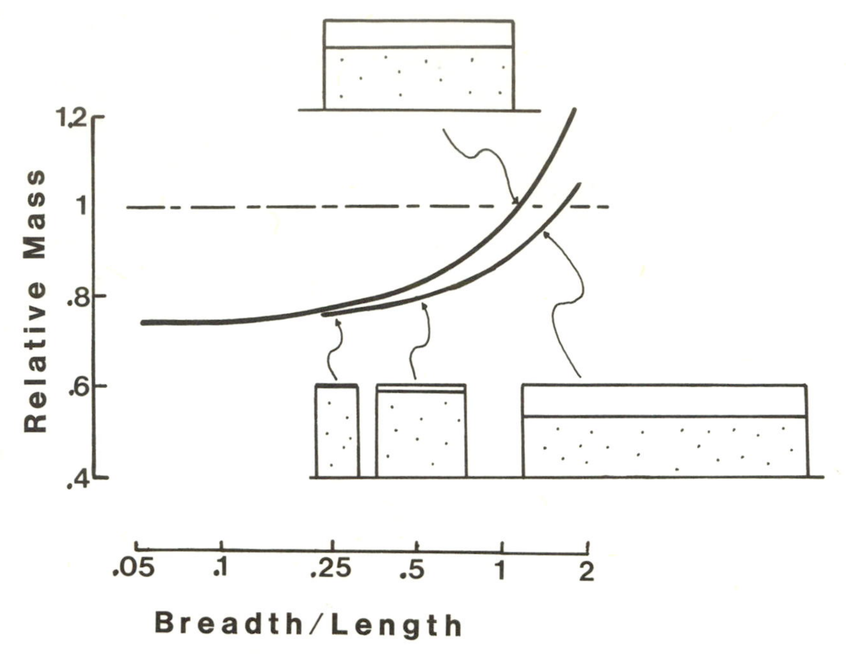

Fig. 4.33. Curves showing the weight of boxlike bones of length 20 mm compared with that of a bone of the same compliance made entirely of cancellous bone. The weight of the marrow is included. Ordinate: weight of box bone relative to weight of cancellous bone. Abscissa: ratio of breadth of bone to its length. Note log scale. Lower curve: end assumed to be rigidly clamped to rigid, weightless sidewalls. Upper curve: end assumed resting free on sidewalls. The little pictures represent the half-length bones. Dots: marrow.

the stiff sidewalls, but is instead simply supported—that is, it can rotate freely at its support (figure 4.32). The result of such an analysis is also shown in figure 4.33. This time, because the plate is less constrained in its movements it deforms more easily. Consequently, the changeover takes place at a smaller radius: when the ratio of the radius to depth is about 1.1:1 (figure 4.33).

The design criterion adopted here is probably not particularly realistic, but the analysis shows that the least mass solution changes from being a box shape to cancellous bone in a region of breadth/width not far removed from that in which bones do, in fact, change over from a box shape with walls of compact bone and with no cancellous bone in the lumen (although there will be some cancellous bone at the ends for reasons we have discussed) to a shape in which the loads are taken by cancellous bone from one end to the other, with compact bone forming only a shell around the outside. The changeover would seem to occur when the box is too flat compared with what one sees in real bones. However, the assumption made in the model was that the sidewalls were massless and rigid. Making them compliant and with mass would cause the curves in figure 4.33 to move upward, and make the changeover occur in a less flat bone.

4.7.5 Sandwich Bone

It is important to be clear that the mechanical factors making it advantageous to have cancellous, rather than compact, bone in a stubby bone like the scaphoid are different from those making it advantageous in the middle of sandwich bone. In sandwich construction the bone is subjected to bending, and the stresses in the compact bone are greater than those borne by the cancellous material. The cancellous bone acts merely to keep the outer sheets of compact bone apart, and to resist shear stresses. In the short bone discussed above the loads are mainly compressive and are taken through the bone by cancellous material, which therefore bears a major part of the load.

Cancellous bone is found also in sandwiches, that is, in bones that have two of their dimensions much greater than the third. Examples are the bones of the vault of the skull, scapulae, parts of the pelvis, the pectoral process of birds, and the carapace of turtles. They are characterized by a layer of cancellous bone that is sandwiched between two thin layers of compact cortical bone. The cortical bone gives way quite abruptly to cancellous bone; there is no gradual increase in porosity as is seen in the cancellous-cortical transition in the ends of long bones (compare figures 4.23 and 4.34). These flat bones are, in fact, almost classical sandwich constructions, such as are used by engineers in making lightweight panels.

The analysis of such panels can be made in various ways. I shall use a rather simple one which, nevertheless, brings out the main properties of sandwiches. Assume the flat plate is a beam loaded in bending. For purposes of calculating the stiffness of the plate we can assume that, instead of being made of two materials of different Young's modulus, it is made of a single material, but that the thickness (not the depth) of the cancellous filling of the sandwich is reduced in proportion to its modulus (figure 4.35). There are therefore two variables in a sandwich of bone that has a particular depth: the ratio of the thickness of the cancellous bone to the overall thickness, and the ratio of the Young's modulus of the cancellous bone to that of the cortical bone. Fortunately, it seems, from measurements I have made, that usually the thickness of the compact bone is very similar on the two sides of the sandwich. This means that the neutral plane is in the middle of the section, which markedly simplifies the sums.

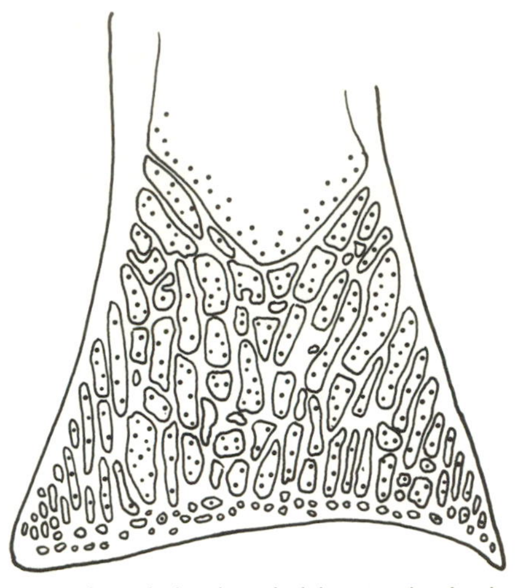

Fig. 4.34. Longitudinal section through the glenoid of the scapula of a sheep. Marrow shown dotted. Note that the cancellous bone does not start abruptly underneath the joint surface; there is a zone of transition. The cancellous struts lead from the joint surface to the cortex.

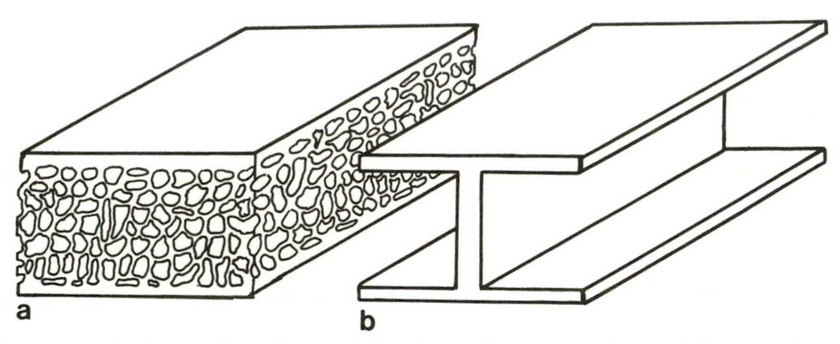

Fig. 4.35. For bending, a sheet of sandwich bone (*a*) can be considered as consisting of two flanges with a web connecting them (*b*). The ratio of the thickness of the web and the width of the flanges should be equal to the ratio of the modulus of the cancellous bone to the modulus of the cortical bone.

The next problem in the analysis is to find the relationship between porosity and Young's modulus of elasticity. We assume that the plate is subjected to "pure" bending, that is, it is bent by a couple at each end, and that therefore the shear stresses can be ignored. As mentioned before, Carter and Hayes (1977a) have produced quite good evidence that the Young's modulus of bone is proportional to its overall density cubed. There is no particular reason for supposing that the bone *material* in can-

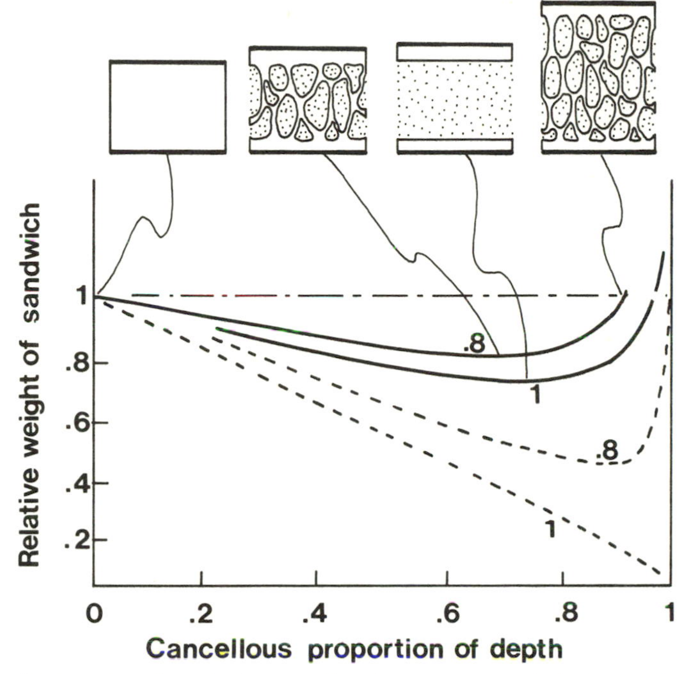

Fig. 4.36. Weight-saving effect of using cancellous bone in a sandwich. Explanation in text. The figures on the curves refer to the porosity of the cancellous bone. The pictures at the top correspond to various points on the lines. Bone: white; marrow: dotted.

cellous bone has a different density from that of compact bone, so we can take it that $E = 2 \times 10^{10}$ (1 − porosity)^3 Pa. Finally, as in section 4.2.2, we find the mass necessary to give a particular bending stiffness to the plate.

Figure 4.36 shows the results making two different assumptions. The hatched lines show the case when the bone is considered to be empty; the solid lines show the case when bone is considered to be filled with marrow whose density is 0.45 that of bone and which does not contribute to the stiffness of the plate.

Take first the case when the bone is considered to be empty. If the cancellous bone were completely porous (that is, it does not exist), then the case reduces to two plates separated by space. The relative mass decreases steadily to the limiting point where the bone consists of two indefinitely thin sheets infinitely far apart. Taking the more meaningful case where the cancellous bone does actually exist (porosity < 1), a minimum always appears. The reason for this is as follows. Table 4.2 shows that the mass of a beam of bone of fixed width of given stiffness is independent of porosity, or density. Therefore, a completely solid beam has the same mass as a completely cancellous one. Where there are two sheets of compact bone sandwiching cancellous bone, the mass will be less than solid bone of the same stiffness. This must, therefore, also be lighter than the completely cancellous bone.

If we make the much more realistic assumption that the cancellous bone has marrow in it, then the situation is more clear-cut, because we can determine the maximum weight saving possible. This is shown by the solid lines in figure 4.36. With completely porous (nonexistent) cancellous bone the maximum saving of mass is about 30%, which occurs when about 75% of the total thickness of the bone is occupied by marrow. For more realistic situations, in which the porosity is between 0.7 and 0.9, the mass saving is about 18%–25% compared with solid bone, and the cancellous bone occupies 55%–75% of the total depth. With porosities of 0.5 or less the mass saving becomes small, 10% or less.

There are implications of the calculations whose results are illustrated in figure 4.36. If we suppose that minimum mass for a given stiffness is being selected, and that any spaces are filled with marrow, then solid bone is not the best. The theoretical best solution is two layers of cortical bone with nothing in between, and this will produce a saving in weight of about 30% compared with solid bone. However, the structure would be unstable in practice because the two layers of compact bone would have nothing to keep them apart. Furthermore, there are some shear loads, and these must be borne through the whole depth of the section. So some internal cancellous stiffening is needed. The more the stiffening, the less the weight saving, but a porosity of 80%, which is about that often found in cancellous bone, produces a weight saving of 20%, with about 70% of the total depth of the section being occupied with cancellous bone. With this arrangement the section would be 15% deeper than an equally stiff solid bone.

It is difficult to compare such theoretical calculations with reality, because real bones do not have regular shapes or loading systems. It should be remembered that these calculations are appropriate for flat sheets loaded in bending. Beams that are deep relative to their length will have shear forces as an important component of the loading system, and this analysis will not be correct. Measurements of representative parts of the horse's thoracic spine and its ischium (figures 4.27, 4.23) give median values of 0.72 and 0.66 for the proportion of the total depth occupied by cancellous bone. These values are at least consonant with the calculations.

If the function of the cancellous bone is to keep the shape of the bone constant and to resist such shear stresses as exist, another general feature of cancellous bone in this kind of position is explained: the rapid changeover from completely solid bone to completely cancellous bone whose porosity does not change much through the depth of the section. There

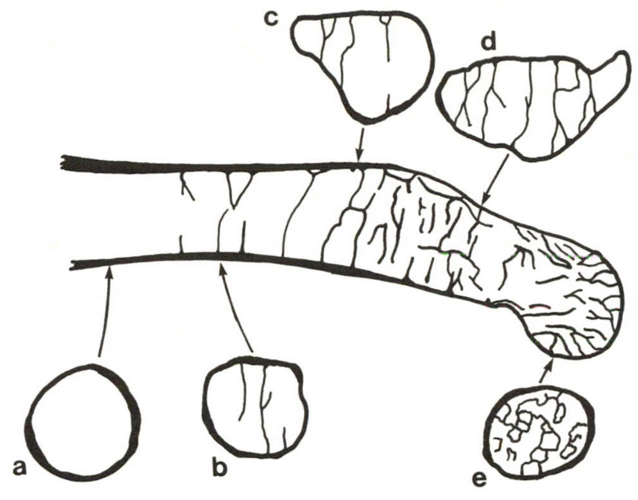

Fig. 4.37. Sagittal and transverse sections of a herring gull's humerus.

would be no advantage, in the performance of either of the functions of the cancellous bone (keeping the cortices apart and resisting shear), in having the Young's modulus of the material vary through the section. So the porosity should be constant. Things are not nearly so straightforward in the cancellous bone of joints, where impact loading and complicated geometrical shapes are encountered. The porosity in these regions is much more variable.

The pattern for a constant value for I/depth, which is a measure of bending strength, is similar to the curves shown in figure 4.36, though the curves are flatter, with their minima nearer to 1, and the optimum ratio of cancellous to cortical bone shifted slightly to the left. (Although the *strength* of bone is a function of density, as is its stiffness, we can ignore this. This is because the fracture should always occur in the compact bone; since the stiffness is proportional to Density^3 and strength to Density^2, the cancellous bone will not be loaded to its breaking stress before the compact bone is loaded to its breaking stress.)

A good example of the way cancellous bone is nicely suited to the build of a bone is shown in the humerus of a large bird such as the herring gull, *Larus argentatus* (figure 4.37). In the middle of the length of the bone the shaft is a hollow cylinder, with rather thin walls. Near the proximal end the bone becomes flattened on what we may call its dorsal side. The bone then dips down to the articular surface. Where the bone is circular in section (*a*), with a fairly small radius of curvature in relation to the wall thickness, there is no cancellous bone. There is no need for it because the forces acting on the bone are carried in the shaft walls, and there is no danger of buckling. In region (*b*), however, the bone becomes flattened quite suddenly, and just where this happens, trabecular struts appear, running mainly from the flattened dorsal surface to the opposite side of the cavity. Notice, too, that these struts first appear in the middle of the bone, farthest away from the sidewalls. This again is what we should expect. When (*d*) is reached, the bone is fairly flat on both sides and is really just a conventional sandwich structure. Most of the trabeculae run dorsoventrally. At (*e*), just below the joint surface, the cancellous bone has a rather different function: to lead the loads from the articular surface to the cortex. Therefore, the trabeculae form tubes and are oriented in a different direction. This appearance of trabeculae underneath places where the cortex becomes flattened is very characteristic of bones in general.

4.7.6 Cancellous Bone in Tuberosities

The cancellous bone in protuberances and tuberosities onto which muscles and tendons attach does not have a shock-absorbing function. These bumps and lumps are not loaded in impact by muscles; this is ensured by the compliance of the tendons of the muscles themselves. The cancellous bone merely has the job of leading the forces from the area under the tendon insertion into the nearby compact bone. The struts of

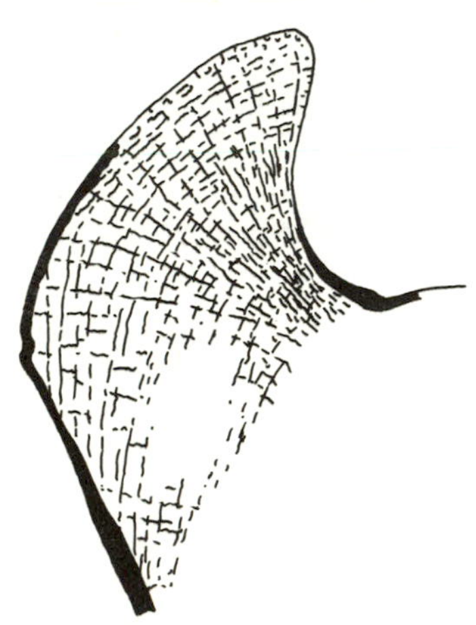

Fig. 4.38. Stylized drawing of a frontal section of the greater trochanter of the femur of a calf, to indicate the main directions and density of the cancellous struts. The blank region has very few trabeculae. Where the adductor tendons insert at the top, the cortex is very thin.

the cancellous bone very often show a clear orthogonal arrangement, shown in figure 4.38, a cross section of the greater trochanter of the femur of a cow. The similarity of this arrangement to that of minimum weight frameworks is again striking. As the line of action of the tendons is often easily determined, and often does not vary much through the cycle of movement, the cancellous bone lying under muscle insertions would seem to be useful for testing the idea that cancellous bone is often arranged optimally for weight saving.

5. Articulations

Because bones are rigid, design problems are inevitable when it is necessary to move one in space relative to another. The joints between bones have a most important influence on the functioning of the skeleton, and we shall consider them at some length.

The relative motions of two rigid bodies in space can be described by means of six independent modes of motion (figure 5.1). There are three of translation and three of rotation. These independent modes are called degrees of freedom. If the two elements are independent, all six modes must be used to describe their relative positions, and they are said to have six degrees of freedom. However, if the elements are connected in some way, then probably the number of degrees of freedom will be reduced. In the jaw joint of the badger the condyle of the mandible fits very snugly into the mandibular fossa on the upper jaw. The lower jaw can be rotated from being tightly shut to being about 45° open. This is one rotational degree of freedom. There is also the possibility of the jaw moving as a whole laterally (figure 5.2). The actual amount of movement in this mode is small. This is a translational degree of freedom. Apart from these two modes, effectively no other movement is possible. The position of the badger's jaw can, therefore, be uniquely defined by specifying two values only. The jaw of the cow, on the other hand, has six degrees of freedom (figure 5.2).

It is important to be clear about two things. One is that the possession of a degree of freedom does not imply any particular value for the angular rotation or the translational movement possible (except that it is greater than zero). The other is that movement in respect of one mode may restrict or even completely abolish movement in respect of another. An example of this is the human knee. This has one degree of freedom in flexion-extension, and another of rotation. When the knee is flexed at about 90°, the lower leg can be rotated about its long axis. This second degree of freedom is abolished when the knee is completely straightened. Of course, the foot can still rotate, but this is achieved by rotation at the hip, not at the knee.

The number of degrees of freedom of a joint must be distinguished from the number of modes of movement possible. For instance, the acromion

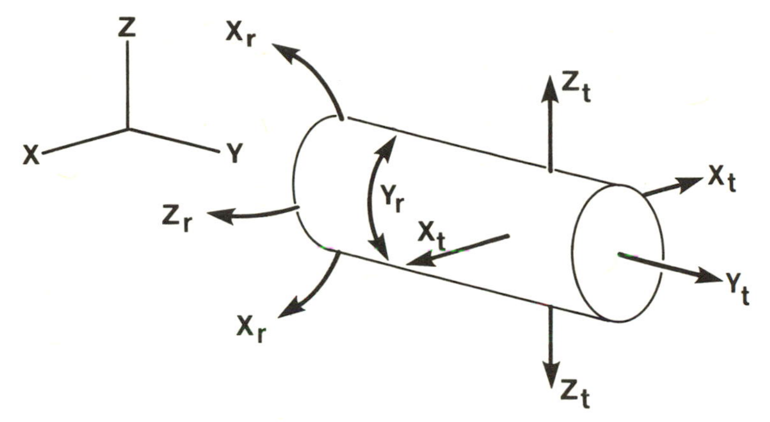

Fig. 5.1. The six possible degrees of freedom of a body with respect to fixed axes in space. Subscript *r* refers to rotation about an axis. Subscript *t* refers to translation along an axis.

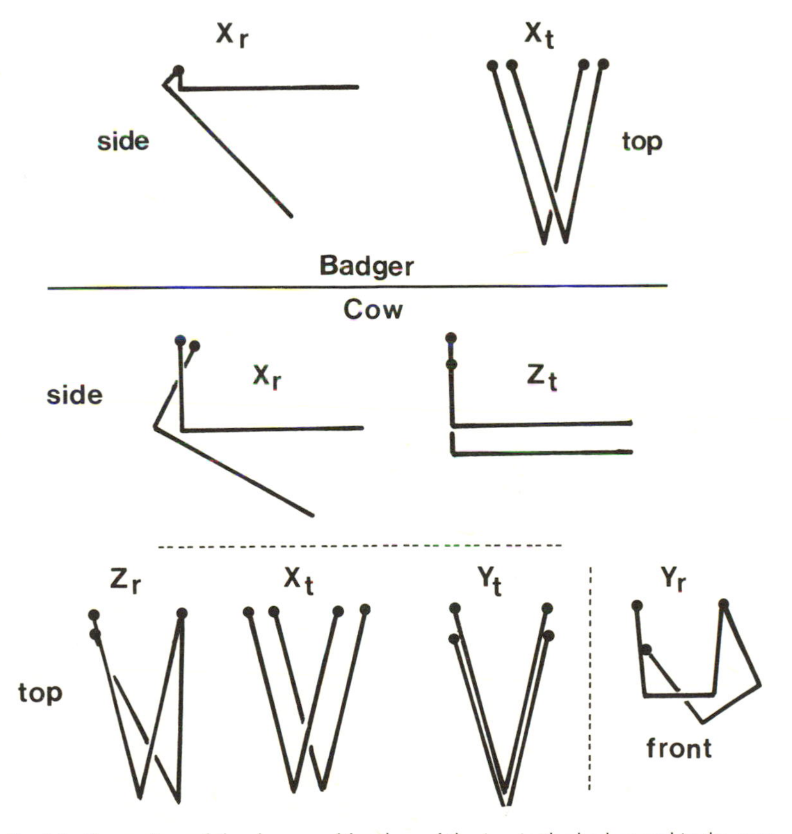

Fig. 5.2. Comparison of the degrees of freedom of the jaw in the badger and in the cow.

process on the scapula of the cat has three translational modes of movement relative to the manubrium of the breast bone, being attached to it only by a ligament whose length can be altered to some extent. This joint has, therefore, three translational degrees of freedom. (I am ignoring rotational freedom here.) In man, however, the same two anatomical features are connected by the bony clavicle, whose length is effectively constant. This connection fixes the distance between the two. Therefore, although the acromion can move, relative to the manubrium, in three directions (anteroposteriorly, dorsoventrally, and lateromedially), it does not have three independent translational degrees of freedom. This is because if the distance between two points in space is fixed, a specification of their relative positions along two perpendicular axes immediately fixes it along the third.

5.1 Some Human Synovial Joints

I shall now write about a few joints in the human skeleton, to give some idea of the range of joints that exists. There is nothing remarkable about the human skeleton, except that it is well known, and the reader can examine him or herself more easily than, say, a lion or a cobra.

5.1.1 The Elbow

Three bones meet in the elbow: the humerus, the ulna, and the radius. The joint has two main functions: to allow the forearm to flex and extend relative to the upper arm, and to allow the wrist to pronate and supinate. The ulna has only one degree of freedom with respect to the humerus. The distal part of the humerus has an articulating surface, the trochlea, which is shaped like a pulley with a central groove. The trochlea articulates with the trochlear notch on the ulna (figure 5.3), The trochlear notch has a ridge which corresponds to the groove on the trochlea. The most proximal part of the ulna is the olecranon process to which various extensor muscles are attached. If the humerus had the same conventional, tubular shape at its distal end as it has at the proximal end, the olecranon process would bump against it during extension. To allow for full extension the humerus has a hollow, the olecranon fossa, into which the olecranon process can fit (figure 5.3). In full flexion the other end of the ulnar notch has the coronoid process, which would similarly bump into the wall of a conventional tubular bone. So, there is a coronoid fossa to accommodate the process. These two fossae deeply indent the humerus, and indeed in some mammals, for instance the dog, there is a hole right through the humerus. The humerus is correspondingly thickened in two great but-

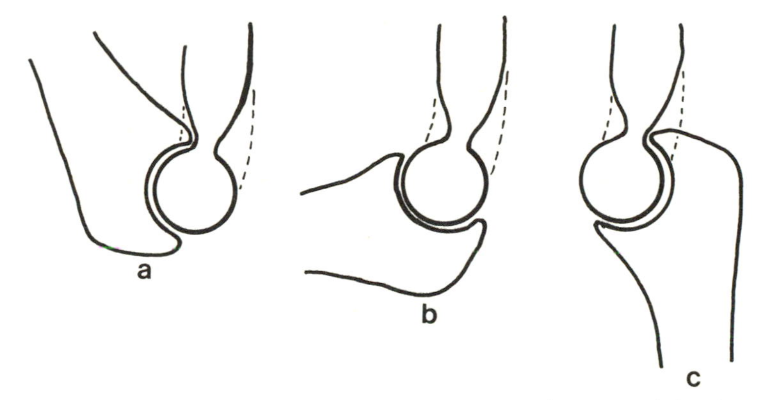

Fig. 5.3. The relationship between the distal end of the human humerus and the ulna; these bones are shown in sagittal section. The dotted lines show the thickness of the humerus out of the plane of the section. (*a*) Elbow fully flexed. The coronoid fossa of the humerus receives the coronoid process of the ulna. (*b*) Elbow at right angles. (*c*) Elbow fully extended. The olecranon fossa of the humerus receives the olecranon process of the ulna.

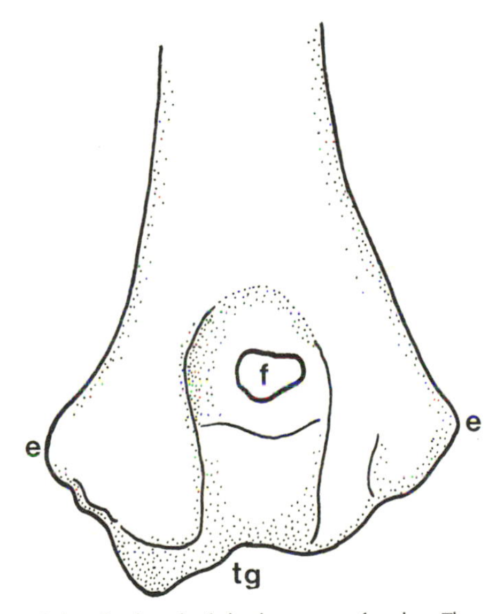

Fig. 5.4. Posterior view of the distal end of the humerus of a dog. The main shaft divides into two, and the bone is actually pierced by the coronoid fossa *f*. The ulna rotates in the trochlear groove *tg*. The epicondyles e are the sites of the origins of ligaments that stabilize the elbow in the lateral direction.

tresses, which diverge sideways to the epicondyles (figure 5.4). The articular surface of the trochlear notch occupies nearly a semicircle. As a result the humerus and ulna will be prevented from disarticulating, whatever the degree of flexion or extension, by bearing surfaces which are being pressed against each other (figure 5.3). This is, as we shall see, quite different from what happens in the knee. The penalty that must be paid for this stability, this lack of tendency to disarticulate, is that the lower end of the humerus has a near-hole in it, just proximal to the articular surface. The sideways-flaring buttresses must be rather expensive in terms of mass compared with a simple tubular end to the bone. The ulna can rotate relative to the humerus, but about one axis only; it has only one degree of freedom.

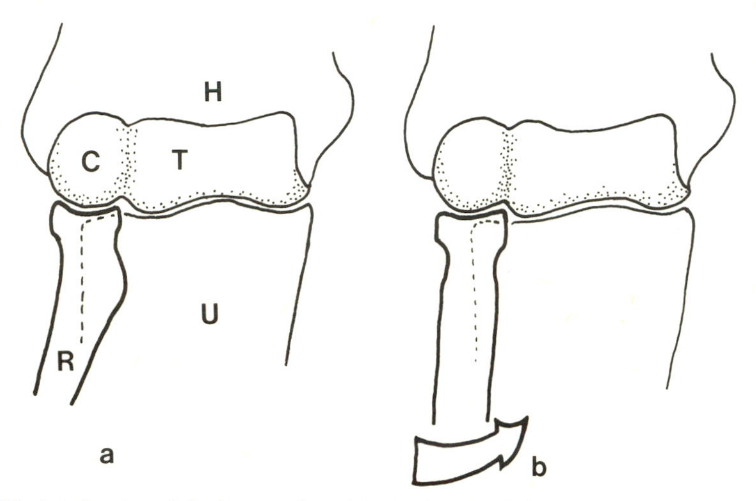

Fig. 5.5. Anterior view of the human elbow joint in the extended position. (*a*) The wrist is supinated. The ulna *U* engages on the trochlea *T* at the end of the humerus *H*. The dished head of the radius *R* engages on the hemispherical capitulum *C*. (*b*) The wrist is pronated. The head of the radius has rotated on the capitulum, producing the movement shown by the arrow. The ulna can move with one degree of freedom only, but the radius can rotate on the capitulum, and so can pronate or supinate the wrist at all angles of extension of the elbow.

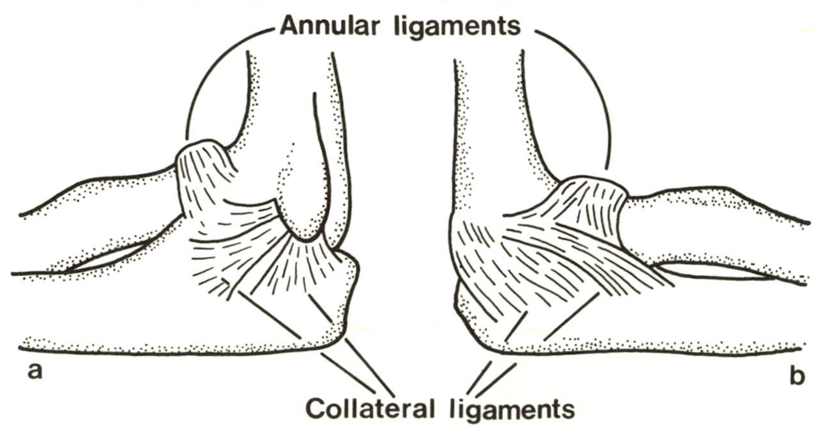

Fig. 5.6. The ligaments of the human elbow joint. (*a*) Medial view; (*b*) lateral view.

The radius has two degrees of freedom. These allow it not only to rotate *with* the ulna during elbow flexion or extension, but also to twist *around* the ulna; this movement allows the wrist, which is attached to both the radius and the ulna, to pronate and supinate. In order to allow this to happen the proximal end of the radius is hollowed out into part of the inner surface of a sphere, and this articulates with the capitulum on the humerus, a spherical surface. Although the joint looks like a ball and socket, it does not have the three degrees of freedom usually associated with a ball and socket. The ulna has one degree of freedom only, and the radius is bound to the ulna at top and bottom. It can rotate with the ulna, and if this were its only possible movement, it would be better to have some form of cylindrical, rather than spherical, bearing surfaces. However, because the radius must rotate around the ulna, it needs another degree of freedom at the elbow. This is allowed by the spherical bearing surfaces (figure 5.5).

The bearing surfaces are kept in contact by a whole network of ligaments (figure 5.6). An annular ligament wraps around the head of the radius and keeps it from separating from the ulna. Two groups of lateral collateral ligaments stretch from the epicondyles to the top of the ulna. Their function is to prevent the ulna from rotating laterally relative to the humerus, and also to keep the humeral, radial, and ulnar joint surfaces in contact. These potential lateral rotations of the ulna cannot readily be prevented by the action of muscles, and so the ligaments are very important. The joint can also potentially undergo hyperflexion or hyperextension. These movements can to a large extent be prevented by muscular action. There is also an anterior ligament, running from the top of the coronoid fossa to the tip of the coronoid process, which helps to prevent hyperextension.

5.1.2. The Knee

Two bones meet in the human knee: the femur and the tibia, though the head of the fibula is not far away. When the knee is fully extended, it is very stable, and the femur and the tibia are effectively locked solid. However, the knee is capable of flexion, and when this has started, the second degree of freedom becomes available; the tibia can be rotated about its own long axis relative to the femur. The flexion-extension of the knee is over about 160°. The rotation of the tibia is about 75°. The rotation of the tibia allows the foot to be pointed laterally or medially. The foot can also rotate when the knee is fully extended, but now the rotation is produced by the femur rotating about its own long axis, from the hip joint.

There is, therefore, an important difference in the functioning of the arm and the leg. The foot can rotate either by rotation of the tibia relative to the femur, *or* by rotation of the whole leg relative to the hip. The former allows about 75°, the latter 90°, but they are not additive. In the arm, on the other hand, the humerus can rotate about 170° and also, the pronation/supination allows a further 170°. Therefore, when the elbow is extended, the hand can be rotated through nearly full circle. This degree of mobility of the hand is unusual among tetrapods, and few animals outside the primates show it.

The top end of the tibia has two facets, lying side by side. The medial facet is biconcave, that is, it is saucer shaped. The lateral facet is convexo-concave. It is concave from side to side, but gently convex from front to back. On this somewhat insecure basis the femur rolls and slides. The lower end of the femur is divided into two condyles, both biconvex. Neither condyle is even remotely congruent with its tibial facet. The lateral condyle, being biconvex and articulating against a convexo-concave surface, cannot, in theory, have contact along more than a line. The medial condyle also, in fact, has a very poor fit. It has a radius of curvature in the anteroposterior direction which is variable but never, in an average-sized adult, greater than 40 mm, whereas the tibial facet has a nearly constant radius of curvature of about 80 mm. So here, also, the contact must be along a line rather than a surface. (This statement is not quite true because the cartilage on each surface will deform, increasing the area of contact. Also, the situation is altered by the presence of the semilunar cartilages, described below.)

In the humeroulnar articulation there is good congruity of the joint surfaces, and so the type of relative motion of the two articular surfaces is determined by their shape. This is not the case in the knee, which could flex by the femur either *rolling* or *sliding* over the tibia (figure 5.7). Although there is some uncertainty about this, it seems that the first 15° or so of flexion involve pure rolling, then a mixture of both movements occurs, and finally there is pure sliding.

The stability of the knee is mostly ensured by four ligaments: the medial and lateral collateral ligaments, and the two cruciate ligaments. The collateral ligaments run as shown in figure 5.8a. Notice particularly that they are not parallel to each other. These collateral ligaments have some effect on the lateral stability of the knee but, compared with the elbow, muscles are far more important in ensuring the stability of the joint. The ligaments are arranged in such a way that they become taut when the knee is fully extended, and they are, therefore, important in preventing hyperextension of the knee. The joint capsule at the back of the knee is greatly thickened and stiffened by collagen fibers, and this also helps to prevent hyperextension. When the knee is flexed, all these ligaments slacken.

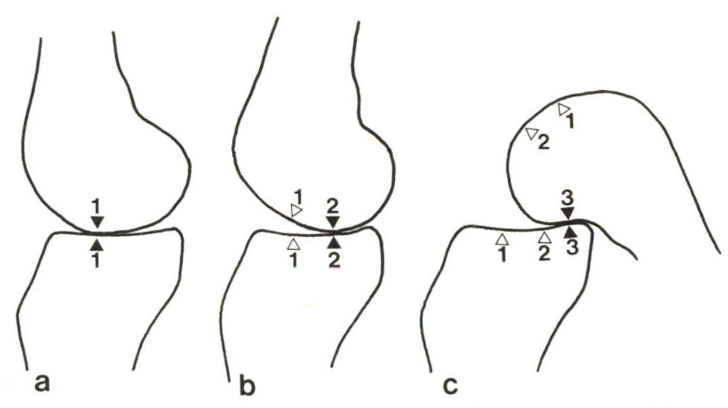

Fig. 5.7. Diagram of flexion of the human knee joint. In each drawing the solid triangle shows the position of closest contact, the open triangles the position of previous closest contact. (*a*) Full extension. (*b*) The first 15° or so of flexion are mostly rolling. If the rolling continued, the two bones would separate from each other. (*c*) Full flexion. The points 2 and 3 on the femur are much farther apart than the equivalent points on the tibia. This shows that there must have been considerable sliding of the two bones against each other.

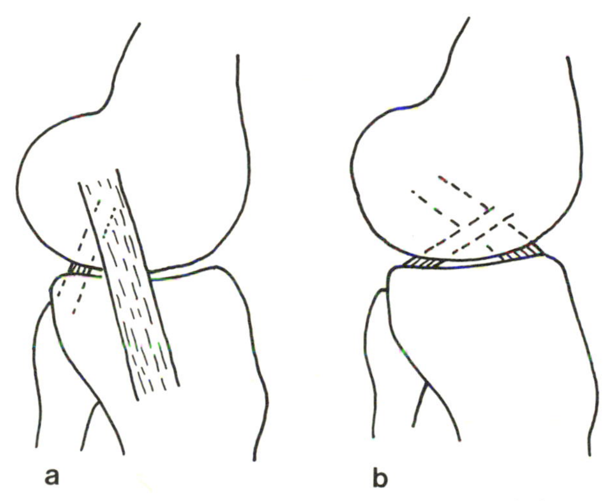

Fig. 5.8. The human left knee seen medially. (*a*) The collateral ligaments; (*b*) the cruciate ligaments. Note that the two sets of ligaments cross with opposite senses, so the two sets resist opposite types of rotation.

The cruciate ligaments run across each other between the two femoral condyles (figure 5.8b). Their particular lengths and points of origin and insertion are such that they almost completely determine the nature of the relative movement of the femur and the tibia during flexion and extension. When the knee is fully extended, each ligament resists further extension. As the knee is flexed, the anterior cruciate becomes slack; the other remains taut, however.

The cruciates and the collateral ligaments account for the inability of the knee to twist when it is fully extended. Suppose one tries to rotate the tibia medially relative to the femur. This will tend to unwind the cruciates, and so they will not resist this movement. The collateral ligaments, on the other hand, because they too lie at an angle, will be tightened by medial rotation, and will tend to screw the femur down onto the tibia and resist the motion. If the tibia is rotated laterally relative to the femur, the opposite will happen: the collateral ligaments will relax, the cruciates will resist and screw the articular surfaces together. When the knee is flexed, the collaterals and the cruciates becomes relaxed, the collaterals more than the cruciates, and rotation of the tibia becomes possible.

Two important structures remain to be described in this extraordinarily complex joint: the patella and the semilunar cartilages. The patella is a sesamoid bone lying at the distal end of the quadriceps tendon just before it inserts onto the anterior crest of the tibia. The part of the tendon distal to the patella is called the tibial tendon or ligament. A sesamoid bone is one that develops within a tendon. The patella is probably the most important sesamoid in the human body. It articulates with the femoral condyles. Its function is to keep the line of action of the quadriceps tendon away from the axis of rotation of the knee joint. In this way the quadriceps is able to exert a large turning moment about the knee joint at all degrees of flexion.

The semilunar cartilages, or menisci, are two C-shaped pads of cartilage lying between the femoral condyles and the tibia. In essence they function to fill the space left by the lack of congruity between the two bones. They are tethered to the tibia at the ends of the *C*, and at some other points, but are able to shuffle around to accommodate the movements of the condyles (figure 5.9). They have a marked and beneficial effect on the distribution of load, making it much more even, and so lessening the stresses. Until recently, orthopedic surgeons would happily chop out patients' cartilages if they were slightly torn. The almost invariably resulting osteoarthritis of the knee has caused this to be a much less frequently performed operation.

Being a zoologist schooled in the belief that animals are optimally designed by natural selection, I must admit to regarding the knee with some disquiet: it seems marvelously designed to make the best of a bad job. The elbow, with its snug-fitting trochlear notch seems so stable compared with the knee, with its more or less planoconvex articulations held rather desperately together by ligaments; yet the hand is able to rotate much more than the foot. It cannot be argued simply that the knee is weight-bearing while the elbow is not, because in many mammals the fore-limbs bear more static load than the hind limbs, and yet the knees of nearly all land mammals are rather similar, as are the elbows. The main

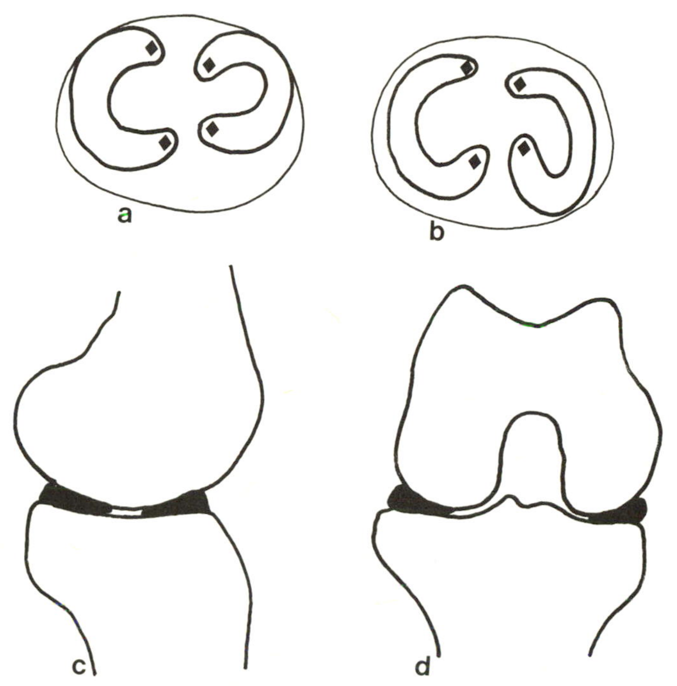

Fig. 5.9. The semilunar cartilages of the human knee. (*a*) Top view of the tibia with the cartilages (shown by thick lines) as they are when the knee is fully extended. The cartilages are free to deform except that they are anchored to the tibia at the points marked by solid lozenges. (*b*) Similar view, showing the shape adopted by the cartilages when the knee is fully flexed. (*c*) Side view of the fully extended knee. The cartilages, shown solid black, greatly increase the area over which the load is transmitted between femur and tibia. (*d*) Frontal view of fully flexed knee.

difference between many land mammals and man is that the radius becomes relatively more important as an articulating and load-bearing bone. Where this happens, it is usually not able to rotate about its own length, and pronation and supination of the forefeet are much reduced.

I give a possible reason for the difference here, but I have little faith in its truth. The ankle joints of nearly all mammals are simple hinges. The tibia and fibula are bound together, and only one degree of freedom is permitted. The mechanical simplicity of the joint is made necessary by the large loads it has to exert during acceleration and running. Although the forelimbs bear more of the *static* load of the body, the hind limbs can thrust through the center of gravity of the body without producing large upsetting turning moments; this the forelimbs cannot do. Nevertheless, the hind foot must be capable of accommodating to irregularities in the ground and to changes in direction. The tibia must, therefore, be capable of rotating about its own long axis. This requires two degrees of freedom.

Fig. 5.10. The problem of stability in a ball-and-socket joint. Each of these joints has 90° of excursion in the plane of the paper. The interrupted line passes through the center of rotation of each. As the female portion (lightly hatched) surrounds the joint more completely, and so prevents disarticulation more or less successfully, so the neck of the male portion must get narrower relative to the head.

But it is not possible to have a joint with two degrees of freedom and with large angular excursions and at the same time have the stability of a shaft rotating in a trough, which is essentially what the elbow joint is. A ball-and-socket joint, such as in the hip, has various other problems. The main difficulty with a ball-and-socket joint is that, if it is to have a reasonable range of excursion, there is a geometrically imposed conflict between stability of the joint and the strength of the neck of its male member (figure 5.10). Some of the worst effects of this can be overcome by making the greatest rotation to be a spinning of the neck about its own long axis. The main part of the bone can then be angled to the neck, and great rotatory freedom is possible. This is what happens in the human femur, which has about 160° of excursion in flexion-extension. Unfortunately, this solution has its own problem: the shaft is offset from the head, so that longitudinal forces on the shaft have to be led into the joint via bending stresses in the neck. The incidence of fractures of the femoral neck in elderly people shows that this is mechanically a very stressful design.

The forelimb is either less weight bearing, as in most primates, or less mobile, as in most cursorial animals, than the hind limb. It can dispense with two degrees of freedom at the elbow, and therefore can have a more stable joint.

5.1.3 The Carpometacarpal Joint of the Thumb

This is not an important weight-bearing joint, but it has a degree of flexibility important to the human hand. The two bones involved are the trapezius and the first metacarpus. The joint has a *saddle* articulation. Both articular surfaces are concave in one direction and convex in the direction normal to the first (figure 5.11). The two surfaces fit snugly

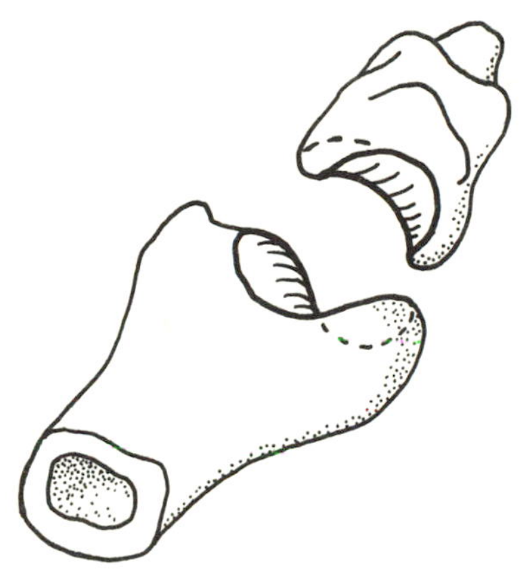

Fig. 5.11. The carpometacarpal joint of the human thumb, exploded to show the saddle joint.

together. This allows two degrees of freedom with considerable stability. However, the amount of rotation allowed by this kind of joint is not very large, about 60° in each plane. Consideration of the geometry of a saddle joint shows that, if the two surfaces fit snugly together, very little movement is possible. However, there is considerable possible movement; the metacarpus can rotate about its own long axis to some extent, which is a third degree of freedom. This movement is made possible because the ligaments holding the articular surfaces together are fairly slack. Indeed, the ligaments stabilizing a saddle joint would need to be very stiff in order to prevent significant rotation about the third axis.

5.2 The Swelling of Bones under Synovial Joints

It is very characteristic of long bones that they swell at their ends, this swelling being capped by synovial cartilage. Beneath the thin layer of cartilage is a thin layer of subchondral cortical bone. Beneath this, in turn, is cancellous bone, which extends for some distance from the cartilage before merging with the compact cortical bone of the shaft.

The synovial cartilage has an overwhelmingly important function: to lubricate the relative motion of the two bones it separates. This it does remarkably well. Measures of the coefficient of friction are difficult to obtain, and the results in the literature are rather varied. Somewhere between 0.002 and 0.02 seems reasonable. Such values are not as good as for a ball race (0.001), but the lower values are rather better than those of a bearing hydrodynamically lubricated with oil. The mode of lubrication is a subject of considerable argument, and is discussed briefly later. The important feature of cartilage in the present context is that, in order to carry out its lubricating function it has to have a rather watery consistency. This makes it both extremely compliant and rather weak. The properties of articular cartilage are dealt with comprehensively in Freeman (1979).

Being watery, cartilage is very viscoelastic, but a reasonable value for the "instantaneous" Young's modulus in tension is 100 MPa (Unsworth, 1981). This is about two orders of magnitude less than compact bone. The tensile strength is in the region of 20 MPa, which is about 1/8 the value for compact bone. Although it is difficult to obtain a very meaningful value for the compressive strength of synovial cartilage, it is likely that the tensile strength is of critical importance in determining whether the cartilage splits under load.

Suppose the cartilage were absent; what would be the likely compressive stresses in the bones at the joints? (This analysis will ignore the hideous wear and pain that result when bony surfaces rub together.) The following analysis is from Weightman and Kempson (1979). If the joint surfaces were completely congruent, the stresses across the joint would be less than in the cortex of the bones, because the area taking the load would be greater. However if, as is the case, the radii of curvature of the joint surfaces are different, say $R1$ and $R2$, with the male surface, of course, having the smaller radius of curvature, then the stress is given by:

$$\text{Stress} = 0.62[PE^2((R1 - R2)/(R1 \cdot R2))^2]^{1/3}$$

where P is the load across the joint and E is Young's modulus of the material of the apposed surfaces. The greater the difference in the radii of curvature, the more the load will be concentrated near one point, and the greater will be the stress there.

Day, Swanson, and Freeman (1975) have investigated the human hip. On applying a load of three times body weight they obtained stresses in various regions of the acetabula of cadavers, the higher values being about 3 MPa. Weightman and Kempson (1979) then calculate from this value what would have been the stresses had the subchondral bone layers been in contact. They assume radii of curvature of 25 and 22.5 mm, and E of cancellous bone of 300 MPa. This produces a contact stress of 5.7 MPa. However, the Young's modulus cannot really be as low as they assume, as there must be a thin layer of *compact* subchondral bone. Had they taken, as the the other extreme, the Young's modulus of compact bone, say 15 GPa, the calculated stress would be about 80 MPa. The real value must lie between these values. The synovial cartilage will, in effect, make the radii of curvature of the male and female parts of the joint more similar. Synovial cartilage certainly acts, therefore, to reduce local stresses by increasing the area of contact, though its importance is still a matter of argument. The cartilage might also act as a cushion for damping dynamic loads. This has been discussed in chapter 4. It has been investigated by Radin and Paul (1971).

If the greatest stress in the cartilage is to be small, the cartilage should be compliant, loaded over a large area, and thick. In fact it is compliant, but it is also very thin, particularly compared with the thickness of the subchondral cancellous bone.

In their experiments Radin and Paul (1971) found that the subchondral bone was only about ten times stiffer than the cartilage capping it. Therefore, as the area loaded was effectively the same in the two tissues, we should expect that ten millimeters of bone would have the same impact stress-attenuating effect as one millimeter of cartilage. In fact, it turned out that the cartilage was rather ineffective. Radin and Paul dropped weights onto the phalanges of cows and measured the peak load. Removing the cartilage increased the peak force at the far end of the phalanx by about 9%, whereas removing about half the length of the phalanx (this would include compact as well as cancellous bone) increased the peak force by a further 31%. Therefore, the cartilage, though significant, clearly does not have a very important function in the attenuation of impact forces.

The swelling of bones at their ends, as we have seen, to a large extent allows a sufficient area for cartilage. However, there is another function that this swelling will serve in many joints. This is to increase their lateral stability (Pauwels, 1948, 1950). Synovial joints are almost without exception stabilized by ligaments, which prevent movement in inappropriate directions. For instance, the tibial and fibular collateral ligaments run down the sides of the human knee joint and prevent the leg from bowing inward or outward. If the knee is hit, or otherwise subjected to a bending moment in the lateral direction, it is the ligament on the tensile side that is primarily responsible for preventing disarticulation. (If the adventitious loading is in the direction of usual movement, then the muscles will also be concerned with resisting it, as well as ligaments such as the cruciates.)

For a given bending moment M, which will not depend upon the morphology of the joint, the tensile force in the ligament will be M/D, where D is the distance from the ligament to the farthest point in the joint that is bearing load; in effect the hinge point. So, the greater the value of D, the more resistant the joint will be to movement in the wrong direction (figure 5.12). A survey I made of a good number of mammalian hinge joints showed that the widest point of the bone, which usually lies near the axis of rotation of the bone with respect to its neighbor, is about twice as wide as the bone in its mid-shaft region, measured in the same direction. If the bones did not expand at the ends, the collateral ligaments would be exposed to about twice the stress for any given bending moment.

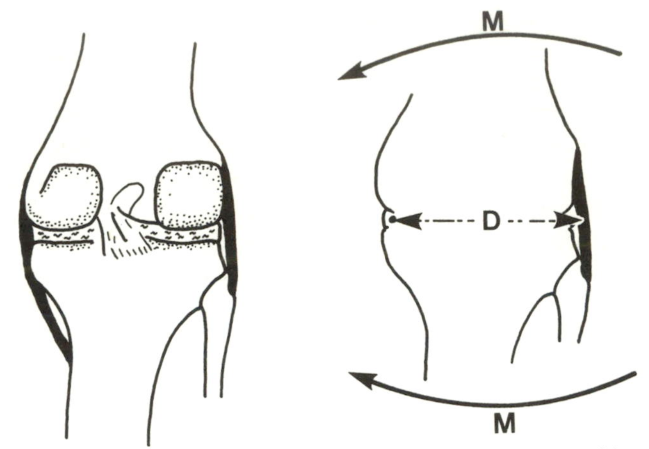

Fig. 5.12. (*Left*) Posterior view of the human knee joint, showing the collateral ligaments (solid black), and semilunar cartilages (squiggles). The femur and tibia are both expanded laterally near the joint. (*Right*) Diagram showing how the expansion reduces the tensile load in the collateral ligament when the joint is subjected to a bending moment M, because the force in the ligament is M/D.

An instructive comparison can be made with the arthropods, to show that joints do not all *have* to be like vertebrate synovial joints. The arthropods have exoskeletons, with the living tissues lying inside the practically nonliving cuticle. The sliding joints have an almost unmodified cuticle at the bearing surface, and this is strong and stiff. There is usually only one degree of freedom, most joints being hinge joints, and simple rotation is all that they are capable of. So the joint surfaces can deal with high contact stresses, and there is no need to spread the load out. As a result, the joint surfaces have a relatively *small* cross-sectional area compared with that of the middle of the element. These differences are shown in figure 5.13. Indeed, as the length of the hinge joints is restricted because the living tissues have to pass through from one element to the next, the loads on the structure near the joint are higher than elsewhere, and the cuticle is stiffened. This is quite the reverse of the situation found in synovial joints. One point of similarity in the two types of joints is that in both there is a tendency for the elements to broaden out in the line of the hinge, to reduce the disarticulating moment.

Unfortunately, we are not out of the wood yet in our discussion of the consequences of the weakness of articular cartilage. The expanded ends of the bone are themselves weaker than the main part of the shaft, and this has bad effects on the ability of some bones to absorb kinetic energy without breaking, such as during falls.

In my laboratory we have studied the load/strain behavior of the human radius when it is loaded in compression (Horsman and Currey, 1983). Two levels, "lower" and "upper," were chosen. At the lower level

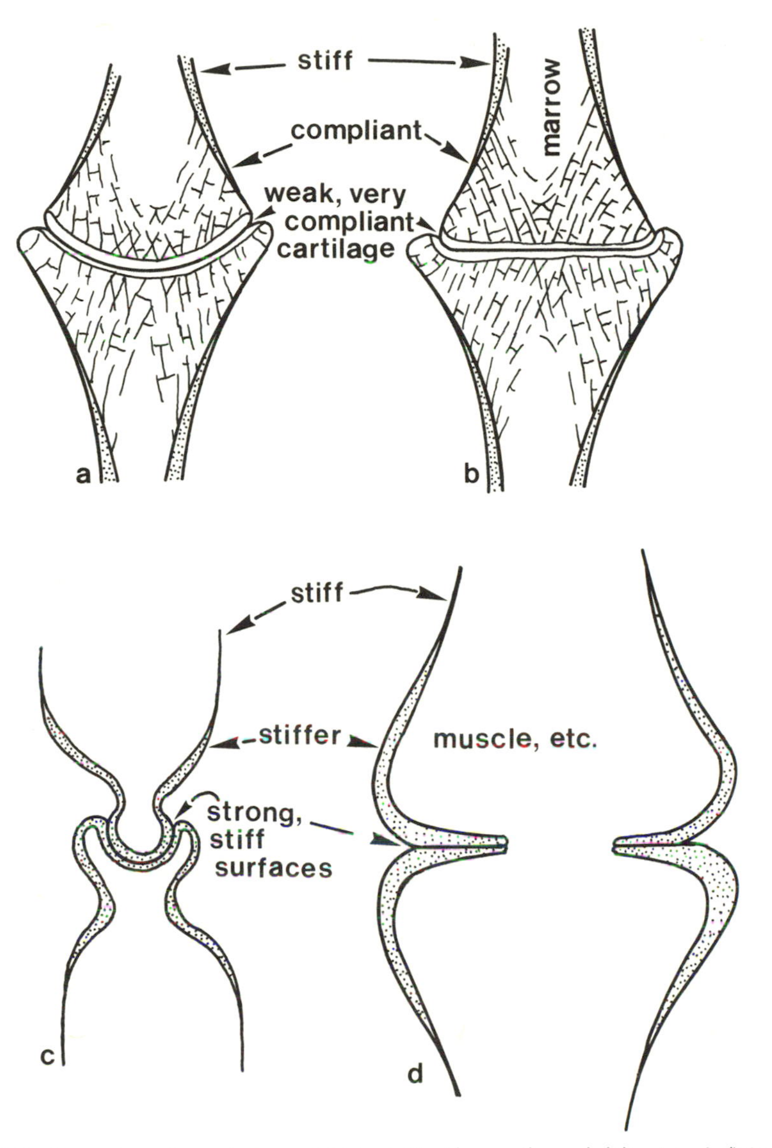

Fig. 5.13. Comparison of a typical synovial joint (*a,b*) with an arthropod sliding joint (*c,d*). In (*a*) and (*c*) the view is along the axis of rotation. In (*b*) and (*d*) the view is normal to the axis of rotation.

there was much cancellous bone, with a thin sheath of compact bone around the outside. At the higher level the bone was completely compact. The amount of bone material at the two levels (determined by photon absorptiometry) did not differ significantly. However, the stiffness, as measured by strain gages, was about 1.7 times greater at the higher than at the lower level. The bones were loaded until they broke; this always happened near the lower level. Even though the strength varied greatly according to the amount of bone material, which itself varied greatly according to the age of the subject, the *strain* at fracture was almost constant. The bone structure fails when the strain in the cortex exceeds some particular value.

The significance of these results is this. Imagine a radius, inside an arm, that in a fall has to absorb a certain amount of energy. What is the best disposition of the bone material so that it nowhere reaches a strain at which it will break? We can consider the bone as consisting of a number of segments of length L. If a segment has a cross-sectional area of bone material (excluding marrow) A, behaves linearly, and is loaded elastically to a load P, then the energy stored is $P/2 \times \Delta L$, where ΔL is the change in length

$$
\begin{aligned}
&= (P/2)\varepsilon L, \quad \text{where } \varepsilon \text{ is the strain} \\
&= (P/2)(PL)/AE, \quad \text{where } E \text{ is Young's modulus} \\
&= P^2L/2AE.
\end{aligned}
$$

For equilibrium, the force all along the bone must be the same. There will be some value of strain, P/AE, which will be greater than the material can bear. Since P and E are, by definition, the same at all levels, so too should A be, because then the strain will be the same everywhere and the whole bone will, in the best situation, be loaded to just under the failure strain. If one part of the bone has twice the cross-sectional area of another, the strain and, therefore, the total energy stored will be half that stored in the more slender part. However, things are more complicated if the Young's modulus varies as well as the area. If the value of failure strain is fixed, but the value of E in the fatter segment is half that in the slender segment, then the energy stored in each segment will be the same.

Now the radius is more compliant near its end than higher up. The adaptive reason for this—mainly to reduce the peak stresses in the cartilage—has already been discussed. The amount of bone mineral per unit length is about the same near the end as it is in the mid-shaft, however. Because the end of the radius consists of a thin sheath of compact bone with cancellous bone inside, the stiffness will, to a large extent, be determined by the stiffness of the compliant cancellous bone. At the end of the radius the compact sheath and the cancellous core are acting almost in parallel; that is, the strains in each will be the same. Stress = strain × Young's modulus. Because the Young's modulus of the cancellous core is less than that of the sheath, the core will be at a lower stress than the sheath. It will not break first, even though it has a lower compressive strength than the sheath because, compared with compact bone, its strength is relatively greater than its Young's modulus. In effect, the compact sheath has to bear a disproportionate share of the load. The compact sheath is breaking at a load that is determined largely by its own meagre cross-sectional area; this results in a breaking strain being reached at a fairly low load. It is a matter of common observation that the radius, when loaded longitudinally, tends to break at its end. The end of the bone has to be compliant in order that the cartilage be spared. Yet the whole bone cannot be compliant, because it is the function of bones to be stiff. As a result, bones like the radius are not designed particularly well from the point of view of energy absorption.

There is a final twist to the story in these particular experiments. The impact, and static, strength of the radius declines with age, concomitantly with a reduction in the amount of bone mineral. However, the stiffness at the lower level declines more rapidly with age than the stiffness at the higher level. Because the deformation at fracture is effectively constant, and because the bone fails when any part of it is loaded to a strain greater than it can bear, the disparity in the amount of energy absorbed by the end, and by the shaft, increases with age.

5.3 Synovial Cartilage

In several places we have seen that the weakness and compliance of synovial cartilage has profound effects on the design of bones. I shall here briefly discuss the lubrication of synovial joints, to show why cartilage has these properties. There is an enormous literature on synovial joints. Four fairly recent articles or books on the subject are Ghista (1982), Sokoloff (1978, 1980), and Hasselbacher (1981).

Synovial cartilage is so different from bone because its mechanical properties are adapted to allow lubricated relative motion of apposed surfaces with a very low coefficient of friction, and therefore very low wear, for many years if need be. There is still much controversy about how the lubrication works. Fortunately, knowledge of the exact mechanism does not really hinder us in understanding cartilage, because the various proposed mechanisms require the same properties for cartilage. However, we must say a little about the kinds of lubrication that are in general possible.

5.3.1 Friction and Wear

When two hard surfaces rub together, their respective high points touch with very high local stresses. In metals the pressure welds the surfaces together, and the welds, or the metal beneath, have to be broken in shear, to allow relative motion to continue. Similar, but poorly understood, processes occur when nonmetallic materials rub together. Suppose the material of the two surfaces is the same, and it has a shear strength of S_s and a compressive yield strength of S_y. Then if a normal force P_n acts across the surfaces, there will be a contact area of $A_c = P_n/S_y$. It turns out that for a stiff material with a sharp yield point in its load-deformation curve the area of contact is proportional to the load. The tangential force P_s necessary to shear the surfaces past each other will be $P_s = A_c \cdot S_s$, where S_s is the tangential shear strength of the material. Therefore, the coefficient of friction (μ) will be $P_s/P_n = S_s/S_y$. Most stiff materials have roughly the same relationship between compressive yield strength and shear strength, and so they tend to have roughly the same value of the coefficient of friction. One way to reduce friction is to reduce the area of contact while keeping the shear strength low. This can be done, for instance, by having a surface film of low shear strength underlain by a stronger material. The coefficient of friction then approaches:

(Shear strength of surface film/Yield strength of bulk material).

With more compliant materials, like cartilage, the yield stress is not reached. Instead, the cartilage will deform elastically until the area of contact is great enough to bear the load. This area is still quite small, however, compared with the area apparently in contact. The area of contact increases with load, but not proportionally to it: $A_c \cdot P_n^x$, where x is less than unity. Therefore, for unlubricated compliant materials the coefficient of friction is greater at low loads than at high ones.

Wear, of course, is the actual plucking of material from one surface and occurs mainly by the mechanism discussed above. Friction is caused by the force necessary to produce shear fractures, but it is also caused by elastic deformations. If two asperities rub past each other, they will be deformed elastically, and work must be done in producing the deformation. When the asperities pass, they will regain their undeformed shape, but the elastic recoil does not help to continue the relative motion, and the stored energy will be lost as heat. Elastic loss also accounts for rolling friction, in which there may be no relative sliding of two surfaces. Nevertheless, the two touching surfaces will distort each other, and although the elastic recoil will in this case *aid* the motion, there will inevitably be viscous losses that will prevent all the stored strain energy being released usefully.

The coefficient of friction of synovial joints is extremely low. Coefficients for various engineering materials are: Steel on steel, 0.6; brass on steel, 0.3; *PTFE* on steel, 0.04–0.2; *PTFE* on *PTFE*, 0.04–0.2. In contrast, Collins and Kingsbury (1982) quote various published values for synovial joint coefficients. These vary from 0.002 to 0.02. Clearly, efficient lubrication must be taking place.

5.3.2 Modes of Lubrication

The two main modes of lubrication used in engineering practice are hydrodynamic, or aerodynamic, lubrication and boundary lubrication. In hydrodynamic lubrication the surfaces are prevented from touching each other at all by a layer of liquid between them. This layer can be induced to form in various ways. One is by active pressurized pumping (this is technically hydrostatic lubrication). The other ways all make use of the fact that if the two bearing surfaces are closer in some regions than in others, then viscous fluid, in being forced through a narrowing gap, will exert a pressure (figure 5.14). Such a narrowing gap will appear, for instance, if a rotating shaft is not quite coaxial with the journal in which it is rotating. In hydrodynamic lubrication the coefficient of friction may be very low, well below 0.01.

In boundary lubrication the lubricant is physically attached to the surfaces, and in effect acts like a material of very low shear stress lying on a stiff bulk material, as I described above (figure 5.14c). This bound layer is

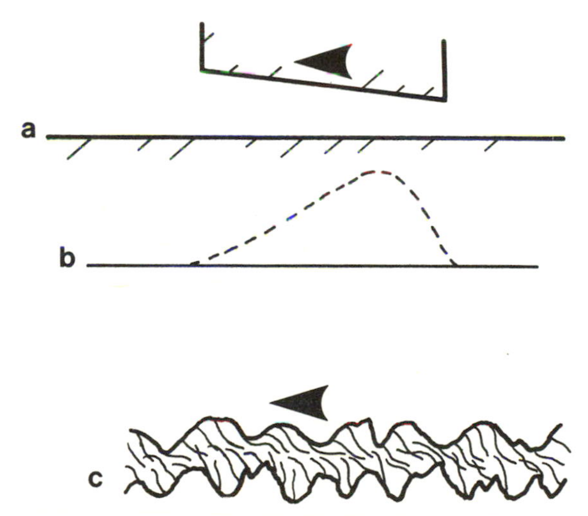

Fig. 5.14. Hydrodynamic and boundary lubrication. (*a*) Two surfaces move relative to each other. They are angled to each other. If the fluid between them is adequately viscous, hydrostatic pressure develops sufficient to keep the surfaces apart. (*b*) The magnitude of the hydrostatic pressure in relation to the nearness of approach of the two surfaces. (*c*) Boundary lubrication. High spots on the surfaces are prevented from touching by the adsorbed boundary lubricant.

very thin and may allow asperities of the more solid underlying material to touch in places. In elastohydrodynamic lubrication the asperities which would be expected to touch do not, because they deform when the stresses become high, and hydrodynamic lubrication occurs between the geometrically altered surfaces. The coefficient of friction for boundary layer lubrication is usually greater than about 0.1, an order of magnitude greater than the coefficient of hydrodynamic lubricants. Against this disadvantage of boundary lubrication is the advantage that it functions well whatever the relative velocity of the two surfaces. Hydrodynamic lubrication, on the other hand, will not work at all at very low shear rates, because a flow of fluid is required to build up sufficient pressure differences to keep the surfaces apart.

5.3.3 Lubrication of Synovial Cartilage

People who have studied synovial cartilage lubrication find values for the coefficient of friction from 0.002 to 0.02 (Collins and Kingsbury, 1982). This coefficient, however, is averaged over many cycles of joint movement. In each reciprocating cycle the joint surfaces must twice be brought to rest with respect to each other. When this happens hydrodynamic lubrication cannot be occurring. As McCutchen (1981) says, "Reluctantly, or otherwise, one concludes that cartilage is unlike any bearing in machinery."

Although there is no consensus about the mechanism of lubrication, certain things are agreed. One is that some form of elastohydrodynamic lubrication is taking place when the surfaces are moving rapidly past each other. This form of lubrication is helped if the bearing surfaces have a fairly low modulus. The other thing that is agreed is that the lubrication taking place when the relative movement is very slow depends on the flow of fluid *through* the cartilage. McCutchen (1981) believes that the fluid is expressed *from* the cartilage in places where the pressure becomes high; this he calls "weeping lubrication." On the other hand Maroudas (1969) and Walker et al. (1968) suppose that fluid moves into the cartilage where pressure is high, leaving behind those long-chain molecules of the synovial fluid unable to pass into the cartilage. These molecules act as very efficient boundary lubricants. Collins and Kingsbury (1982) to some extent effect a synthesis of these views.

Whatever turns out to be the answer, it does seem that two features of synovial cartilage are necessary for its proper functioning. One is that it should have a low modulus of elasticity. The other is that it should be very permeable to low-molecular-weight molecules such as water. To me it is impossible to imagine a biological material having these features and also having a high tensile and shear strength. From the two requirements of compliance and permeability follow many of the most obvious design features of bone. The tail wags the dog indeed.

5.4 Intervertebral Discs

So far, I have discussed synovial joints only. Before we see what generalities about joints can be discovered, I shall discuss the joints between vertebral centra. These are quite different from synovial joints because the relative motion of the bones is allowed by a flexible connection, rather than by the sliding of two cartilage surfaces. The flexibility is produced by the intervertebral discs. Each of these consists of a central nucleus pulposus surrounded by an anulus fibrosus (figure 5.15). The nucleus is a ball of very hydrophilic jelly, bounded peripherally by collagen fibers. The anulus consists of a series of not very clearly distinguished laminae. Each lamina has collagen fibers oriented almost in the same direction. This direction is nearly longitudinal in the peripheral laminae, becoming more oblique to the long axis in the laminae nearer the nucleus. Each lamina has its fibers at an angle to its neighbors, usually at about the same angle from the long axis of the spine, but in opposite directions.

The intervertebral disc is a device that makes use of the tensile strength and stiffness of collagen, even though the disc is loaded as a whole in compression. This seemingly paradoxical state of affairs is brought about by the nucleus pulposus. The loads applied to the discs are sufficiently small for the watery nucleus to be considered incompressible, that is, its volume cannot change, although its shape may, and indeed does, alter.

The hygroscopic gel of the nucleus draws water in, and does so until the tendency to increase in volume is matched by the resisting tension

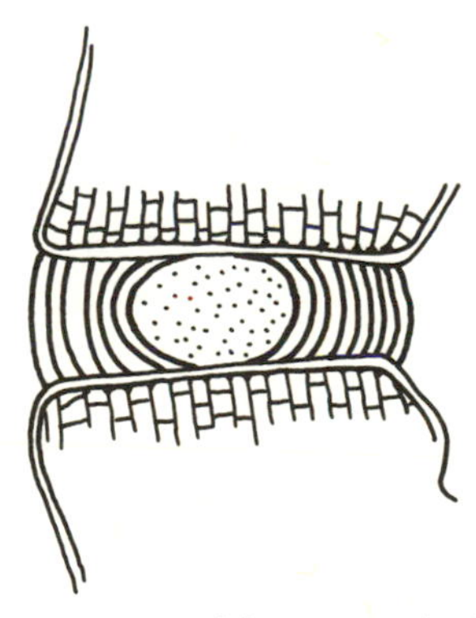

Fig. 5.15. Highly diagrammatic cross section of the intervertebral disc between a sheep's lumbar vertebrae. The nucleus pulposus is shown dotted and is surrounded by the laminae of the anulus fibrosus. Beneath the thin cortical bone at the ends of the centra lies the cancellous bone.

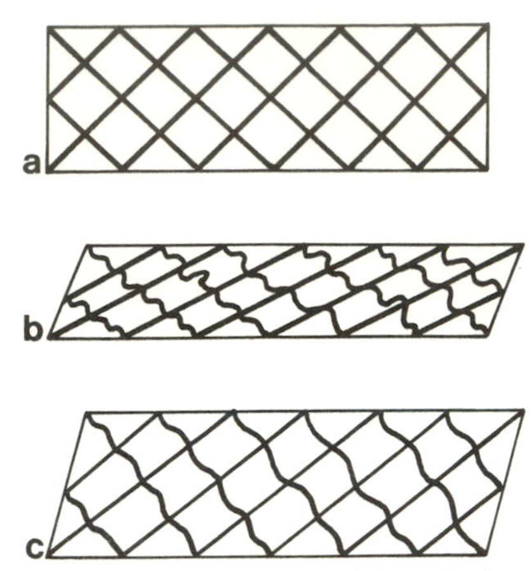

Fig. 5.16. How the intervertebral disc resists torsion and shear. (*a*) The undistorted shape. (*b*) The disc has sheared, but the south–west, north–east collagen fibers are the same length, and so are not stressed. This requires a reduction in volume, which is particularly resisted by the nucleus pulposus. (*c*) There is no change in volume, but now the south–west, north–east fibrils are stretched. Their stiffness will limit the amount of shear possible.

in the fibers of the anulus and the tone of the muscles of the vertebral column. A hydrostatic pressure of about 1 MPa develops in the nucleus and, of course, there will be a tensile stress in the collagen fibers of the anulus to produce equilibrium. The fibers of the anulus are, therefore, slightly prestressed in tension before any compressive load is applied to the disc. When load is applied, the incompressible nucleus is squashed into an ellipsoid that protrudes laterally. This tends to increase the tension in the fibers of the anulus. Furthermore, when there is bending between two vertebral centra, the nucleus distorts so that extra tension is placed on the fibers on the tensile side. If there is shear between adjacent centra, the obliquity of one set of the fibers in the anulus will tend to make the distance between the centra decrease, and this will be resisted by the nucleus (figure 5.16). This is rather like the situation in the knee, where *either* the collaterals *or* the cruciate ligaments will tend to screw the joint surfaces closer together, and in doing so prevent further rotational movement.

It is clear, therefore, that the two-component structure of the disc acts so as to put the collagen fibers of the anulus under tension, for all normal loading situations. If the spine is loaded in tension, the nucleus plays no part, but in compression it is important in allowing the collagen fibers to be stretched.

Unlike the situation in bones with synovial joints, the presence of a disc imposes no particular shape on the ends of the centra.

Table 5.1 Kinematic Properties of Some Joints

Joint	*Degree of Freedom*	*Type*	*Rotation Possible*
Elbow	1	Hinge	145°
Thumb	1	Saddle	70°
	2		50°
	3		60°
Knee	1	Plano-convex	140°
	2		70°
Thoracic centra	1	Flexible	A few degrees
	2		
	3		

5.5 Joints in General

I have discussed three synovial joints and one flexible joint. Each has a somewhat different structure. Table 5.1 shows the number of degrees of freedom of the joints and their range of excursion. Because the type of joint at the end of a bone has such an effect on the shape of the bone, we need to consider what kinds of movement between adjacent bones impose what kinds of joint geometries.

First, in the vertebrates at least, flexible joints, in which there is no relative sliding of joint surfaces, seem confined to three situations. The bones may show little relative movement, or the bones may show a great deal of relative motion, with all six degrees of freedom, or the loads across the joint are small. The hyoid bones show the second state, with adjacent bones often being connected by quite long strips of fibrocartilage. In such joints the form of the end of the bone is not constrained by any requirements of the joint material. Flexible joints that bear little load are, for instance, seen in some rib-sternum joints. Again, the bones are unaffected by the fact of being part of a joint. Flexible joints that carry large loads are rather uncommon in mammals. In fact, in the vertebrates such joints are restricted almost entirely to the vertebral centra, to some of the joints of the pelvis, and to the joint between the two rami of the lower jaw. The amount of relative movement allowed in the lower jaw, for instance, may be small but it is essential. Scapino (1981) shows how various degrees of flexibility of the carnivore's mandibular symphysis are correlated with different masticatory actions. Ride (1959) showed how the flexible symphysis in kangaroos is crucial in allowing the two procumbent first lower incisors

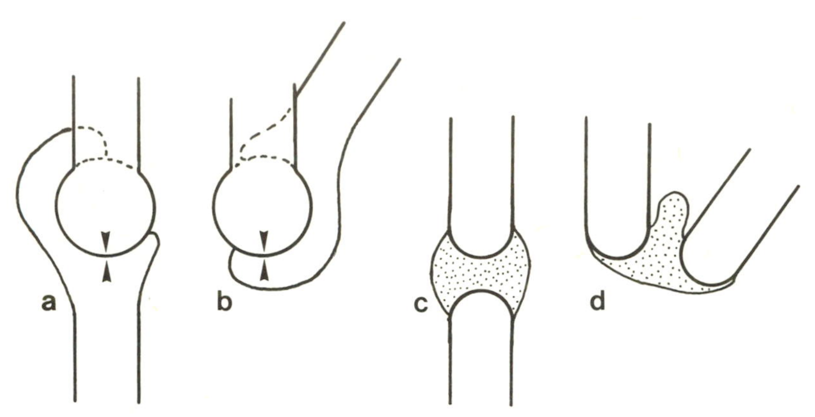

Fig. 5.17. Comparison of a synovial and a flexible joint. (*a*) Synovial joint fully extended. (*b*) Synovial joint fully flexed. In both positions there are compressive forces (arrows) between the two surfaces. These prevent disarticulation. (*c*) Flexible joint extended. (*d*) Flexible joint flexed to the same amount as (*b*). Such a large excursion requires a considerable strain in the hinge material. If the hinge were larger in volume, the strain would be less, but the compliance of the hinge material would result in a very compliant joint that would be difficult to control.

to act as scissors against the upper incisors, cropping grass and then freeing it so it can be manipulated by the tongue.

There is a great advantage in having a joint made of flexible material rather than of synovial cartilage—there is no need for the extremely delicate synovial lubrication mechanism. Against this is the difficulty that large joint excursions are not possible if large forces act across the joint. This is shown in figure 5.17. Large excursions of flexible hinges inevitably mean either very large strains in flexible links or a hinge so compliant that the movement of one bone on the other is very imprecise. In the rib cage this limitation is unimportant.

Flexible joints like intervertebral discs are good for damping out impulsive loads. If the spine were a rigid bone, then rapidly applied loads on the hip joint would be transferred with little attenuation to the base of the skull. (In a hungover state this seems indeed to be the case.) In fact, the *elasticity* of disc material greatly prolongs the time over which the force reaches a maximum; as a result, the maximum force in the bone is lower. Furthermore, the *viscosity* of the disc material results in energy being lost as heat, and so the energy transferred up the spine becomes progressively less. The other side of the coin is that the viscosity of the disc material results in all movements of the spine requiring work that is lost as heat. Synovial joints are so nearly frictionless that effectively the only work done in moving them is that required to deform the soft tissues surrounding the joint.

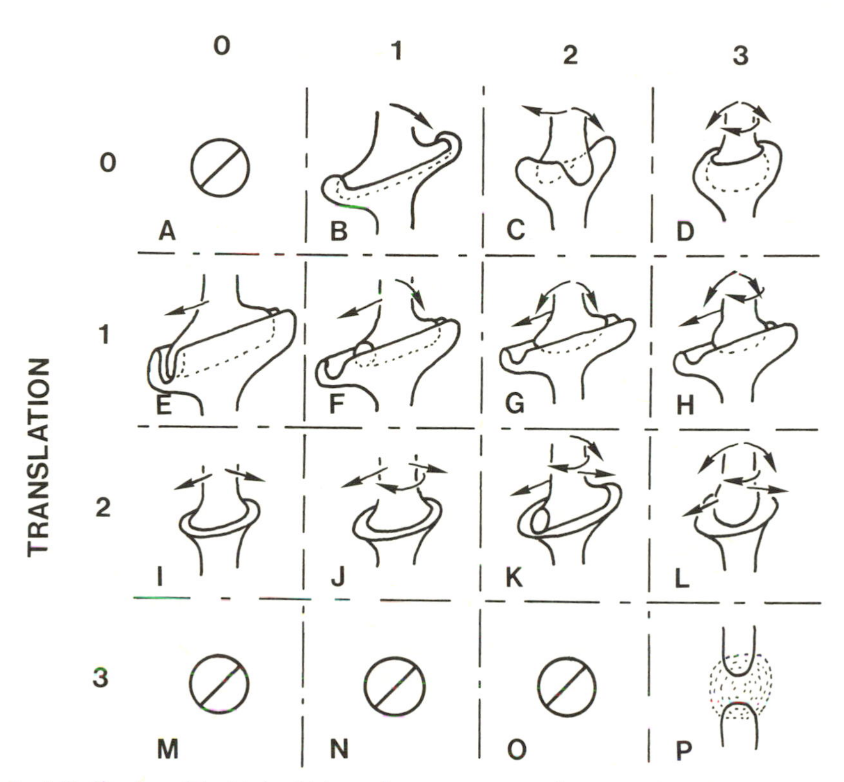

Fig. 5.18. Sketches of the kinds of joint surfaces necessary to allow up to three degrees of freedom in rotation and translation. The circle with a line through indicates that such a joint cannot exist. There are some other arrangements that would produce the same kinds of degrees of freedom.

Joints with flexible hinges inevitably have six degrees of freedom, though the amount of excursion possible may be very small in some or in all directions. Synovial joints, on the other hand, have no such necessary freedom. In fact, the great majority of synovial joints have one, two, or three degrees of freedom. Figure 5.18 shows the general types of freedom that are possible and some joints that show them. Some theoretical possibilities are not found in life, as far as I know. For instance, a joint having two translational degrees of freedom (figure 5.18.I) will almost inevitably have one degree of rotational movement. Similarly, a joint with three degrees of rotation and two of translation will be produced by a spherical bone skating around on a flat surface (figure 5.18.L). No joint I know of has this amount of freedom, along with the extremely complex cat's cradle of ligaments and muscles that would be needed to control it.

The most important effect of synovial joints on the bones is, as I have argued, the necessity for the expansion of the bone under the weak synovial cartilage. Apart from joints taking small forces, which would require little expansion, the expansion will be needed to a greater or lesser extent according to the geometry of the joint. Some joints carry low loads, and impose low stresses on the cartilage as a result. This is true, for instance, of the joint between the malleus and the incus, and between the incus and the stapes, in the middle ear. Where forces can be large, certain types of joint may necessarily impose large loads on the cartilage. In figure 5.18 the arrangements with two degrees of translational freedom, and two or three degrees of rotation (figure 5.18.K,L), have synovial cartilage surfaces meeting at a line and at a point, respectively. The stresses would be very high here, and this may be another reason, apart from the complexity of ligaments and muscles needed, why such types of joint seem not to be found. The situation can be eased by introducing accessory cartilages, like the semilunar cartilages of the knee, which take part of the load, but clearly this compounds the complexity.

On the other hand, where the joint surfaces are nearly flat, the joint pressures can be much lower, and little expansion of the ends of the bone may be needed. The small bones of ankles and wrists, which tend to have rather flat facets, show no external morphological features associated with the synovial cartilages.

5.6 Conclusion

In this chapter I have tried to show that the necessity for joints imposes very strong constraints on the design of bones. In general the higher the loads and the greater the number of degrees of freedom, the more complex and specialized must the joint become. Finally, I should like to mention the knee again. It is a structure of such mechanical implausibility that it must be designed to do something very well. Unfortunately, I do not know what that something is.

6. Bones, Tendons, and Muscles

The bony skeletons in a museum give such a good feeling for what the animals must have been like that we tend to forget what a skeleton is. We are looking only at the compression members of a structure whose tension members have rotted away. In some ways, the whole of this book can be thought of as an extended Hamlet without the Prince. Bones have no functional meaning without their muscles, tendons, and ligaments, which move them and hold them together. Contrarily, of course, muscles and tendons must always act against something comparatively unyielding, such as bone, cuticle, or water. This fact seems to be frequently ignored by people working on muscle. A colleague who knew much about the anterior byssus retractor muscle of the mussel admitted to me that he had only the haziest idea of what the byssus *was*. In fact, the viscoelastic properties of the byssus (collagenous threads that anchor the mussel to the substrate) are peculiar, and impose awkward constraints on the muscle that tightens them.

6.1 Mechanical Properties of the Tissues

Although, as we shall see, skeletons are designed to put bone in tension as little as possible, it is not immediately obvious why this should be. Clearly, if a skeleton is to be made of materials, one of which is strong in tension, the other in compression, then the skeleton should be designed accordingly. Table 6.1 shows some important physical properties of bone, tendon, and muscle. Bone has a compressive strength of about 250 MPa and a tensile strength of about 150 MPa. Tendon has a tensile strength of about 100 MPa (Elliott, 1965; Leonard, Parsons, and Adams, 1980), and it is not meaningful to talk about its compressive strength, because it is so flexible that it cannot really be loaded in compression unless it is confined laterally, a most unlifelike situation. The densities of bone and tendon are about 2,000 and 1,200 kg m^{-3} respectively, so the values of the tensile strength/density, at 7.5×10^4 and 8.3×10^4 Pa/kg m^{-3} are effectively the same. Therefore, the weight penalty for replacing tendon by bone of the same ability to resist tensile loads would be very small. Table

Table 6.1 Some Physical Properties of Bone, Tendon, and Muscle

	Bone	*Tendon*	*Muscle*
Compressive strength/MPa	250	—	—
Tensile strength/MPa	150	100	0.35*
E, Young's modulus/GPa	20	2	—
Shear modulus/GPa	5	Negligible	Negligible
Density/kg m^{-3}	2,000	1,200	1,200
$\frac{\text{Tensile strength}}{\text{Density}}$/ Pa/kg m^{-3}	7.5×10^4	8.3×10^4	2.9×10^2
$\frac{\text{Tensile strength}^2}{2E}$/ Pa	5.6×10^5	2.5×10^6	—
$\frac{\text{Tensile strength}^2}{2E \cdot \text{Density}}$/Pa/kg m^{-3}	2.8×10^2	2.1×10^3	—
Strain energy storage/J/kg	2.8×10^2	2.1×10^3	4.7*

* Taken from Alexander and Bennet-Clark, 1977

6.1 shows that muscle has an extremely low value for strength/density (and, incidentally, for energy storage/density). It would be extraordinarily expensive to resist tensile loads by using muscle. However, this is not what muscles are for, and muscles can perform work of about 250 J kg^{-1} in a single twitch, whereas bone and tendon can do no active work at all; they can only pay back some of the work done on them. It is, of course, the flexibility of tendons that makes them so suitable for transferring the force exerted by the muscles to the bones themselves. Figure 6.1 is a diagram of the human foot showing the way in which the extensor muscles of the toes originate on the shin and their tendons insert on the toes after turning through a large angle.

The tendons are forced to keep close to the bones by the two bands of the inferior extensor retinaculum. A few words about retinacula. These are bands of collagenous tissue that function to control the position of tendons as they pass from their muscle of origin to their insertion. Usually, like the inferior extensor, they prevent the tendon from straightening out. The retinacula allow the tendons complete freedom to slide, being virtually frictionless. Figure 6.2 shows the directions of the forces in the neighborhood of retinacula situated at each end of a bone. The greater the change in direction caused by the retinaculum, the greater the force on it. The force acts in the direction of the bisector of the angle made by the tendon before and after passing under the retinaculum. The geometry

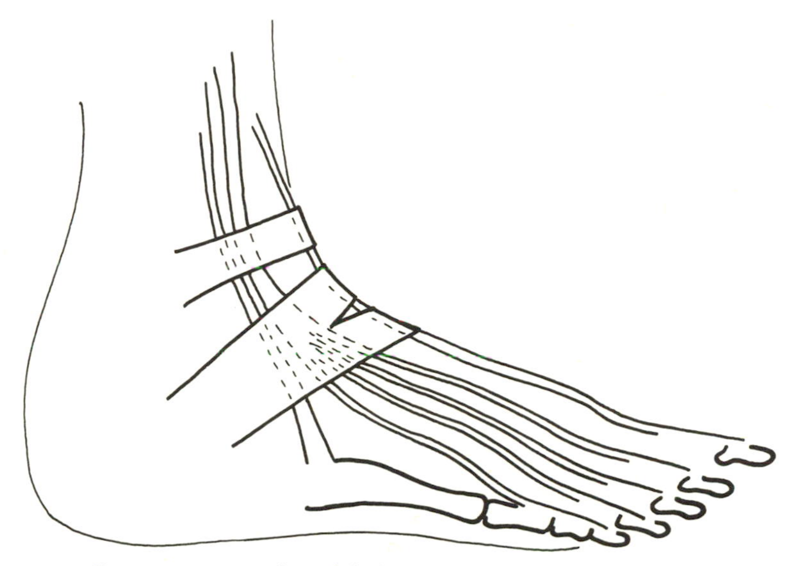

Fig. 6.1. Extensor tendons of the human toes, passing under retinacula.

of retinacula is important to the functioning of bones, and we shall return to it shortly.

The flexor tendons of complicated structures such as the foot adopt a whole variety of shapes when the foot and the toes are in various positions. It would be impossible to accommodate such excursions with a bonelike tendon. It is interesting that many arthropods do have tendons, made of cuticle, that are rigid. However, these tendons generally span only one joint, and even so they always have one or two short regions of flexible cuticle intercalated along their length to allow flexibility at critical points (figure 6.3). Some birds, such as turkeys, have ossified tendons, which work in much the same way. They are discussed below (section 6.2).

As well as being strong, tendons are flexible, yet have quite a high Young's modulus. It may not at once be clear how a material can have both qualities. In fact, tendons are flexible in bending, but quite stiff in tension. Tendons have a sigmoid stress-strain curve which is nowhere straight; that is to say, it does not have a linear region from which one can calculate the Young's modulus. However, if we take the steepest region of the curve, the modulus is about 2 GPa. We can roughly calculate the minimum radius of curvature of a tendon made of material of tensile strength 100 MPa and modulus 2 GPa, using beam theory. The maximum tensile stress in a beam of circular cross section of radius r, which was initially straight, but which has been bent into a radius of curvature

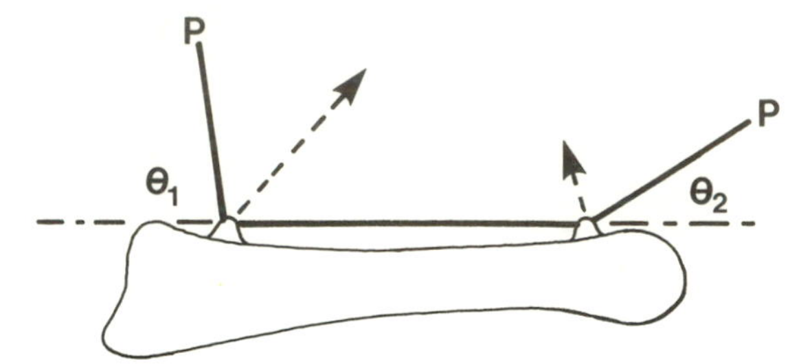

Fig. 6.2. The action of retinacula. The tendon (thick line) is under a tension *P*. The two retinacula near the ends of the bone allow a total angle change of $\theta_1 + \theta_2$. The resulting force on the retinacula is greater, the greater the change in angle they are responsible for. The vectors (with arrows) show the relative magnitudes of the forces acting on the retinacula.

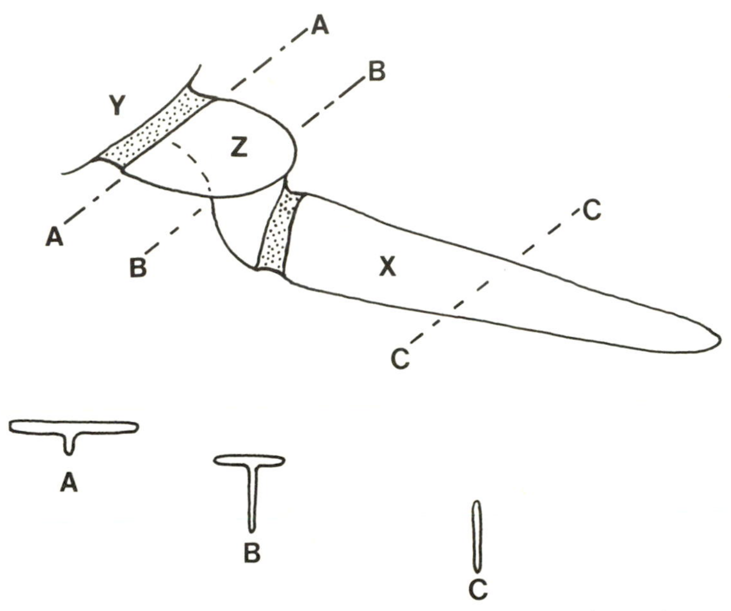

Fig. 6.3. Tendon of a joint of the American lobster, *Homarus americanus*. The muscle inserts onto the blade-shaped part on the right, *X*. This is made of rigid cuticle. The muscles have to move the rigid piece of cuticle *Y*, which is oriented at right angles to *X*. This is made possible by the intercalation of two short flexible pieces of cuticle (dotted) and the curiously shaped piece of rigid cuticle *Z*. *A*, *B*, and *C* show the section of the rigid cuticle at three levels.

R, is given by Stress $= Er/R$. Therefore, tensile strength$/E = r/R =$ 100 MPa/2 GPa $= 0.05$. This suggests that tendon cannot be bent into curves of radius of curvature less than ten times their thickness without rupturing. This is clearly not really the case: tendons can be tied into quite tight knots without damage. Something must be wrong. What is wrong is that tendon cannot be considered to be a beam. It consists of many fibrils which can shear past each other. All these fibrils cooperate to produce a fairly high modulus in tension, but in bending they do not cooperate, and the tendon can be thought of as being made of innumerable beams each with a very small cross section and therefore a small second moment of area. The same principle is made use of in wire ropes. Bone, on the other hand, is not made of independent fibrils; it has considerable shear stiffness. It is not flexible, and is stiff in both compression and bending.

Tendons attaching to bone have three somewhat different functions. One is to allow the muscle to be reasonably far away from the joint or joints it activates. Another is to make forces go around corners. We have seen how this is brought about by the use of retinacula and pulleys. The third function is to act as an energy store. The first two functions, which usually go together, require tendon to have a high stiffness and strength-to-mass ratio in tensile loading, but to be very flexible in shear. The energy storage function is best carried out by tendon with a high specific energy-absorbing capacity. This is given by a material with a high strength and *low* modulus. This last requirement runs counter to that needed for the first two functions. I have not discussed tendon elasticity for energy storage. Basically what happens is that when an animal is, say, running, it periodically has to raise and lower its center of gravity and has to accelerate and decelerate the center of gravity in the direction of motion. If, as is usually the case in running (though not in walking), the body has to be raised at the same time as it has to be accelerated, work must be done. This can be done by the muscles, but it saves energy if it can be done by the shortening of the tendon, which has been passively stretched during other stages of the locomotory cycle. This is an important energy-saving device, much used by mammals (Alexander, 1982).

6.2 Sesamoids and Ossified Tendons

Sesamoid bones are found quite frequently in mammals and reptiles (Haines, 1969). They develop, both ontogenetically and phylogenetically, in tendons in the neighborhood of joints. Usually their function, as was first clearly stated by Gillette (1872), is to ensure that the moment arm of a tendon about the center of rotation of a joint remains large at all joint angles (figure 6.4). Sesamoids occur particularly often on the plantar surfaces of feet, where it would be awkward to have large protuberances, like the olecranon of the ulna, sticking into the ground when the foot was dorsiflexed. Where there are large protuberances, there tend not to be sesamoids, and vice versa.

When a tendon passes over a bony prominence, it will be subject to both tension along its length and compression normal to its length. An

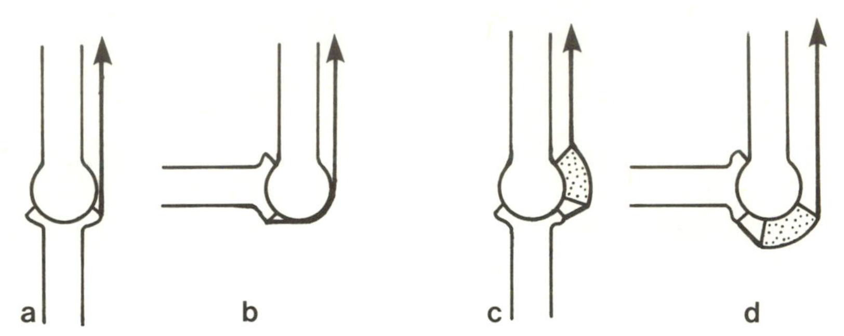

Fig. 6.4. The action of a sesamoid. (*a*) and (*b*) show a joint with no sesamoid. Although in one position (*a*) the tendon is clear of the joint surface, in another (*b*) the moment arm is smaller. (*c,d*) With a sesamoid (dotted) the tendon's moment arm is always larger. In (*c*) the sesamoid would seem likely to swing clear of the joint. Often this does happen but often, also, ligaments will keep a sesamoid closely apposed to the joint surface.

example of such a tendon is the flexor digitorum profundus where it passes over the calcaneus. This has been studied by Merrilees and Flint (1980). They found that on the side of the tendon subject to tension only, the tissue was ordinary tendon, but that on the compressive side the tissue was much more like fibrocartilage. Merrilees and Flint attribute these differences to "altered synthetic activities which are known to occur in response to changes in the physical environment . . . ." This may be so, but it is surely significant that the result of this fibrocartilage development is that when the foot is flexed, the line of action of the tendon is kept away from the center of rotation of the joint. Presumably, the changes in the tendon, described by Merrilees and Flint, when further developed, would result in the production of sesamoids.

Sesamoids do not, however, always function to keep tendons away from the center of rotation of joints. An example of this is a sesamoid embedded in the tendon at the origin of the gastrocnemius, just behind the knee, in many mammals. This sesamoid never comes in contact with the bones near it. The function of this tendon ossification is not known. Perhaps clearer is the function of the ossification of many tendons in the legs of those birds who spend much of their time walking, rather than flying. Pheasants and turkeys are examples. In these birds the tendon is replaced by bone over much of its length. Ossified tendons can presumably arise in places where the ability to turn corners, or to store energy, is not required. In turkeys, ossified leg tendons do not span two joints, and there is always a small length of ordinary tendon intercalated between ossified tendon and the bone. The movements of the joints controlled by the tendon are simple, and geometrically there are no constraints requiring the ossified tendons to bend. Therefore, the tendon does not need to turn corners. As regards energy storage, such birds as turkeys and pheasants spend most of their time walking sedately. If they need to move fast or far they take to flight, and then the legs are not used in locomotion. It may be, therefore, that ossified tendons arise when neither energy storage nor bending are required. However, there seem to be so many cases where these requirements are met and yet the tendons remain unossified that this cannot be the whole story.

6.3 Attachment of Tendons to Bone

For engineers the problem of attaching two materials of very different Young's modulus may be considerable. Usually the materials cannot be bonded to each other, because the strains each material undergoes near the interface are very different and lead to large stress discontinuities. These discontinuities in turn make adhesion difficult. Often, rather complex knots or other fastenings must be produced. In fact, there are few examples in engineering technology where a low-modulus material that is bearing a high tensile stress is attached to a high-modulus material. One example is the attachment of ropes to boats.

Bone and tendon differ in Young's modulus by about an order of magnitude, yet tendons rarely pull out of bones and also, at the naked-eye level, there is no complex knotting. A tendon appears to run straight into the bone, with only a small increase in cross-sectional area. Cooper and Misol (1970) describe the situation in the dog's patellar tendon where it inserts onto the patella from both sides, the tibial and the femoral. This is probably typical of tendinous insertions generally. They found, as had others, four regions. In order, from tendon to bone, these were:

1. Ordinary tendon.
2. A fibrocartilage region, about 300 μm wide. In this region cartilage cells appear, lying in rows in the extracellular matrix of the tendon. The cross-sectional area of the tendon increases slightly to accommodate the cells.
3. Next is a mineralized fibrocartilage region, about 200 μm wide. There is a sharp boundary between this region and the last; the mineralization does not start gradually, but appears as a clear tide-line. The cartilage cells here are usually somewhat degenerate.
4. The mineralized fibrocartilage merges imperceptibly into the rest of the bone, with no clear point where the fibers stop and the bone begins.

The orientation of the fibers from the tendon is often different from that of the generality of the bone. The collagen fibers on the surface of the bone belonging to the bone itself, the "intrinsic" fibers, are usually roughly parallel to it, whereas the penetrating tendon fibers, the

Fig. 6.5. Sharpey's fiber bone. (*a*) Sketch of surface bone, with organic material removed. Sharpey's fibers are shown by dots, with solid black representing the unmineralized core. The fibrils of mineral, which once had collagen inside them, run between the Sharpey's fibers. (*b*) Cross section of the insertion of three collagen fibrils into bone. The fibrocartilage cells are shown as ellipses between the fibrils. Incipient mineralization is shown as fine dots (this is removed along with the organic material). The fully mineralized bone and Sharpey's fibers lie beneath the transverse line.

"extrinsic" fibers, are often at a large angle to the surface. Where this happens the mineralized fibers are known as "Sharpey's" fibers. Boyde (1972, 1980) shows that if bone with Sharpey's fibers has the organic material removed chemically (is deorganified) and is then examined in the scanning electron microscope, the appearance is very characteristic. Sharpey's fibers appear as round depressions or bumps on the surface of the bone, while the intrinsic fibers, belonging to the bone proper, weave between them. If the surface is growing, the mineralization of Sharpey's fibers lags behind, and they appear as depressions (figure 6.5). If the surface is stationary, the mineralization more than catches up, and Sharpey's fibers grow out along the tendon, appearing as bumps. This was presumably the state of the bone that Cooper and Misol viewed. Sharpey's fibers are mineralized from the outside in, and so they appear to have hollow middles in deorganified specimens.

This mode of inserting tendons is certainly mechanically effective. The difference between the arrangement here and that in man-made structures is that in the latter a low-modulus material must, in some way, be attached to a high-modulus material. In the tendon insertion the low-modulus material penetrates the high-modulus one, and becomes of high modulus itself within a few μm. There are no real problems, therefore, in bonding the tendon to the bone, the only difficulty being that the grain of Sharpey's fibers and that of the rest of the bone is different. The critically important point is the continuity in the collagen fibrils right from true tendon into the heart of the bone; there is no line of weakness.

Cooper and Misol were puzzled by the need for the fibrocartilage. However, the collagen fibrils must separate from each other before they enter the bone, so that the Sharpey's fibers can be gripped by the rest of the bone. Cartilage cells are presumably a handy way of producing packing material.

Sometimes muscles seem to attach directly to bone, but in fact there is always a small intervening layer of collagen.

6.4 Devices to Reduce Tension in Bone

The ability to bear bending loads without noticeably deflecting can be thought of as bone's raison d'être. Only because of this property can the muscles act effectively (section 1.1). But, bending is dangerous, because quite small *loads* can impose large *stresses*.

Consider an idealized situation as in figure 6.6. A bone of length L, with a freely moveable joint, bears a weight P at its distal end and is prevented from rotating by the action of a tendon, exerting a force T, which inserts a distance J from the center of rotation of the joint. The value of T is PL/J. The bending moment in the bone is at its greatest at the level of the tendon insertion, and is $P(L - J)$. For a bone of symmetrical cross section the greatest stress at this position will be $P(L - J)c/I$, where c is half the depth of the bone and I is the second moment of area of the cross section. Suppose the bone has a length $(L - J)$ of 100 mm and that it is a hollow cylinder of external and internal diameter of 10 mm and 6 mm respectively. The value of c/I is $0.0117\ \text{mm}^{-3}$, the bending moment is 100 N m. Therefore, the greatest stresses are $(100P \times 0.0117)$MPa. If we take the stress that ought not to be exceeded as 150 MPa, the maximum value that P can have is 128 N, which is the force exerted by a mass of 13 kg. If, instead of being bent, the bone were supporting the same mass as a column, in compression, the stress would be $P/A = (128/50)\ \text{MPa} = 2.6\ \text{MPa}$, which is quite trivial.

The design features one might expect to find in skeletons would be that the tendons (and muscles) must be loaded in tension. The bones are

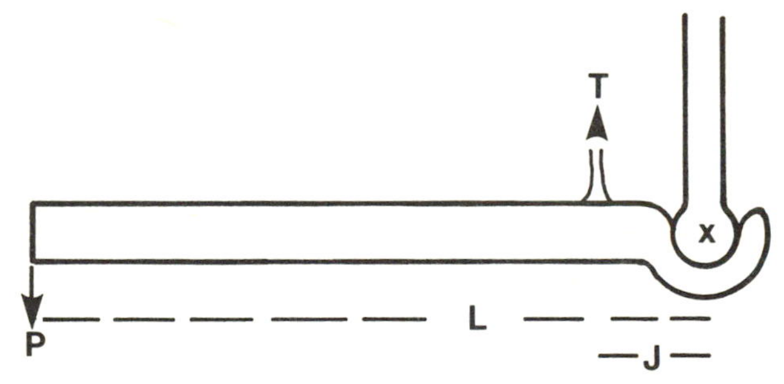

Fig. 6.6. Diagram of the effect of position of the tendon insertion relative to the joint, whose center of rotation is marked x, on the bending moment in a bone.

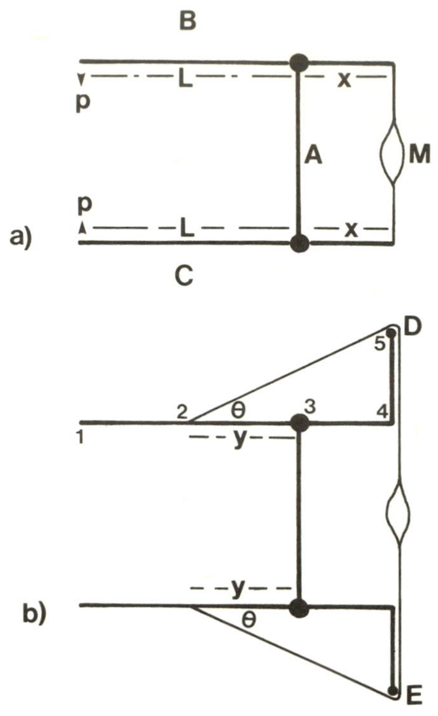

Fig. 6.7. Possible devices for reducing bending moments in bone. For explanation see text.

best loaded in compression. Things should be so arranged that the bones are loaded in bending as little as possible even though, because of the jobs skeletons have to perform, the whole structure—bones plus muscles and tendons—must be loaded in bending. It will be helpful to consider first of all a few more or less idealized situations, before considering the messy compromises that actually occur in nature.

Figure 6.7 shows a fairly common kind of situation. A bone A is joined at each end to two other bones B and C, having, to the left of the joint, a length L. The muscle M has to exert a force sufficient to balance the force p tending to collapse the structure. There will be a compressive force $p(1 + L/x)$ along A, and no bending moments. B and C will have bending moments going from zero at the point of application of the force to a maximum of pL at the joints. If these were the only forces acting on the bones, we would expect the bones B and C to be rather slender at their distal ends, becoming thicker toward the joint, and then becoming more slender again toward the muscle attachment (figure 6.8). In this way the value of Mc/I could be made constant, and the bones would have the same maximum stress all along their length. This would minimize the total weight of the bone necessary. (If some maximum stress S is produced by

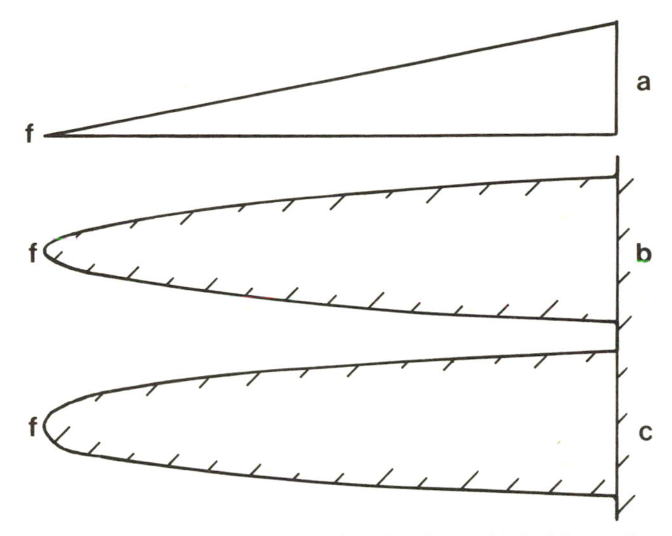

Fig. 6.8. Beams with the same maximum stress along their length. (*a*) If a light cantilever is loaded at its free end, the bending moment will increase linearly from the free end *f*. (*b*) A cantilever designed to have the same *maximum* stress all along its length. It is assumed to be solid and of *uniform width*. The depth is proportional to the square root of the distance from *f*.
(*c*) As (*b*), but now the cantilever is symmetrical about its long axis; it has a varying width equal to the depth. The depth is proportional to the cube root of the distance from *f*. Shear stresses, whose effects will be small, are ignored.

the load in some part of the bone, there is no use having more bone material in other parts of the bone reducing the local stress to less than *S*, because the bone will break at its most highly stressed point; the extra bone elsewhere will be so much useless weight. Indeed, not only will it be useless, in impact loading having bone that is lightly stressed is positively harmful, as we have seen in section 4.3.) It is true that some limb bones do show the expected change in I/c. Alexander (1975b) demonstrates this clearly with the tibia of the dog. A similar tendency can be found by recalculation of data in Minns, Bremble, and Campbell (1975) for the tibia of man. However, in both these tibiae there is a tendency for the value of I/c near the distal end to be larger than the simple calculation would predict. This is because the tibia has to expand distally to accommodate the joint surface. This leads willy-nilly to an increase in I.

On the other hand, many long bones do not show any systematic variation in I/c along their length. This is true, for instance, of the humerus and femur of nearly all mammals. Probably the reason for this is that in most long bones the loading system is both complex and variable, and the rather simple system we are considering here does not apply. In particular, if a bone of uniform cross section is loaded in torsion, it will be stressed uniformly along its length.

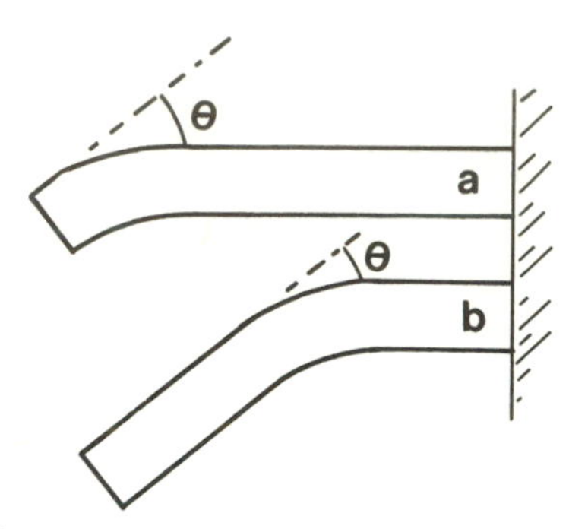

Fig. 6.9. A curvature of θ has a smaller effect on the total displacement of the end of the cantilever if it occurs near the free end (*a*) than if it occurs near the fixed end (*b*).

This suggested disposition of the bone presupposes that resistance to fracture is overridingly important. If minimizing bending deformation were the important feature, the increase in I toward the joint would be more marked. This is because the radius of curvature of a beam at any section along its length is inversely proportional to M/I. The total deflection depends more on the curvature near the joint than farther away (figure 6.9), so for a given mass of bone, more of the mass should be used to increase the value of I near the joint.

It would be possible to arrange things so as to reduce the bending moments in B and C. Figure 6.7b shows one possibility. The tendon from the muscle now loops over a slippery groove at D and E and inserts into the bones B and C at a distance y from the articulation. The distance from the insertion to the joint is constant, so any movement apart from D and E will cause the tension in the tendon and muscle to increase. In the situation shown in figure 6.7b the force in the muscle and the tendon is still PL/x. Consider only the top half of the structure. There are now forces in other directions. The tendon at its insertion at 2 exerts an upward and rightward force. Where it passes over the projection at 5, it exerts a leftward, as well as a downward, force. The resulting bending moments in the region 2 to 4 tend to bend the bone concave upward, whereas the downward forces at 1 and 5 tend to bend it concave downward. The two bending moments act counter to each other, and so the net bending moments, and stresses, in the region 2 to 4 are reduced. The previous maximum bending moment was PL; it is now $PL(1 - y \sin \theta/x)$. There is now a compressive force of $(PL/x) \sin \theta$ in the region 2 to 4.

Unfortunately for all this fine model-building, it does not seem to be that vertebrates make much use of such mechanisms for reducing bending stresses in bones. Probably the reason is that to produce a worthwhile reduction in bending stress the pulley would have to be so far from the

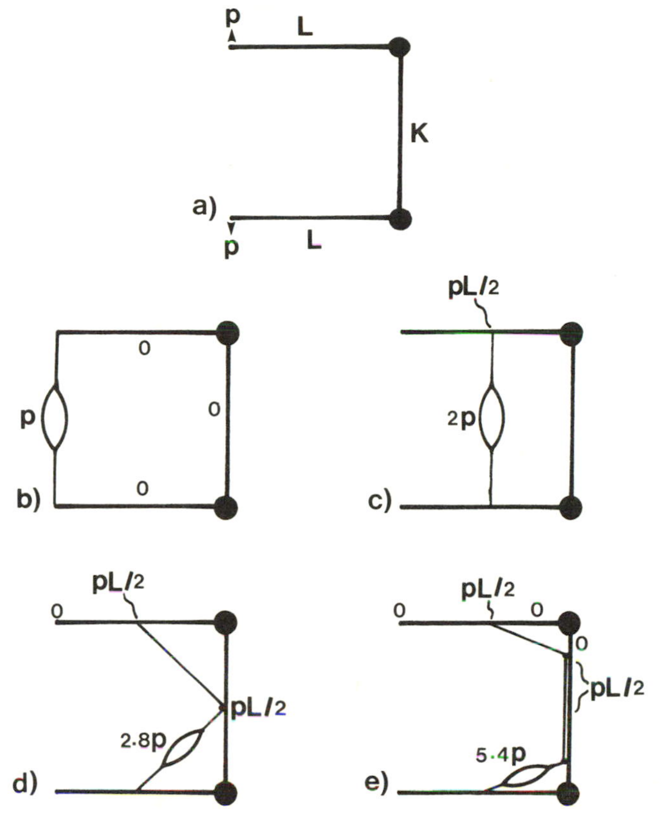

Fig. 6.10 (*a*) Two bones, of length *L*, are joined to another of length *K*. The free ends of the bone are loaded with forces *p*, tending to move them apart.
(*b*) A muscle joins the two free ends. It has to exert a force *p*. There are no stresses in the bones.
(*c*) The muscle joins the half-lengths of the outer bones. It has to exert a force 2*p*. The bending moment in the bones increases from zero at the free ends and the joints to *pL*/2 at the muscle insertion.
(*d*) The muscle joins the half-lengths of the outer bones via a retinaculum on the middle bone. The moments on the outer bones are as in (*c*). On the middle bone there is a bending moment *pL*/2 at the level of the retinaculum, but it is moment-free elsewhere. The muscle must exert a force 2.8*p*.
(*e*) The muscles join the half-lengths of the outer bones via two retinacula each 1/5 the length along the middle bone. The muscle must exert a force 5.4*p*. All the length of the middle bone between the retinacula is subjected to a bending moment *pL*/2.

axis of the bone (in figure 6.7 the distance 4 to 5 would have to be so great) that the system would be very cumbersome.

Now, consider a system that is the reverse of the first: a triplet of bones is loaded so as to straighten them out. How should muscles and tendons be best arranged to minimize the bending in the bones? Figure 6.10 shows some possible configurations. If, to take a trivial case (figure 6.10b), a

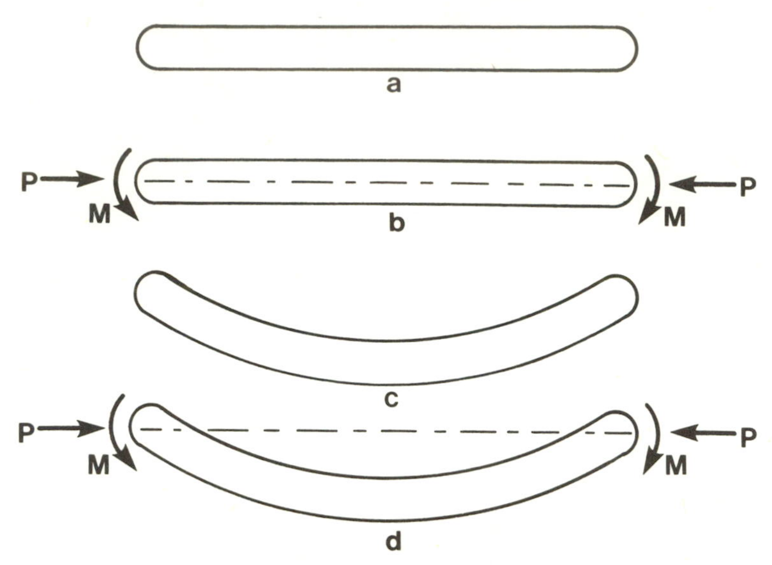

Fig. 6.11. (*a*) The unloaded bone is straight. (*b*) When an axial load P and a bending moment M are applied, their effects are additive at any point. (*c*) The unloaded bone is curved. (*d*) The force P is no longer axial: the midline of the bone deviates from the line joining the points of application of the load. This will result in the production of a bending moment, greatest in the midlength of the bone, of the opposite sense to M.

muscle spans the free ends, there is no load in the bones. However, if the idealized situation is modeling a hand grasping a bough, some room must be left for the bough and another, less trivial, configuration adopted. The other configurations are discussed in the figure legend.

Another, related device for reducing bending moments is this. A bone is subjected to a couple of value M at each end, and also a compressive load P. If we assume that the alteration in the shape of the bone caused by the bending is negligible, we can calculate the stresses by adding the stress caused by the compressive load and by the (constant) bending moment (figure 6.11). The effect of the couple can be reduced by curving the bone, because the force P will now itself have a bending moment about the bone, least at the ends and greatest in the middle. This bending moment is subtracted, if the curve is in the right direction, from the moment caused by the couple. It is possible to arrange matters so that in some parts of the bone the bending moment is completely abolished. In section 6.6 I shall discuss two curved bones, in one of which this mechanism probably works, in the other of which it probably does not.

The closer the muscles and tendons lie to the bones, the greater the bending moments in the bones, and the greater the tensile stresses in the muscles and tendons. This is in general always true, and is one of the most important features of the design of skeletons.

6.5 Why Do Tendons Run Close to Joints?

The skeletons of most vertebrates are arranged so that the muscles moving the joints usually have rather large mechanical disadvantages. For example, take the human elbow joint when it is flexed at a right angle (figure 6.12), a weight being held in the hand. The weight has a moment arm of roughly 300 mm about the center of rotation of the elbow, while the biceps brachii, brachialis, and brachioradialis have moment arms of about 55, 25, and 60 mm respectively. Similar mechanical disadvantages can be seen in most of the muscles moving the long bones of reasonably agile animals. Usually there are retinacula, which keep the tendons close to joints (figure 6.2). These act to prevent the moment arm of the muscles increasing as the joint becomes flexed.

Of course, not all muscles have large mechanical disadvantages. In man the serratus anterior draws the scapula forward. In doing so it has a

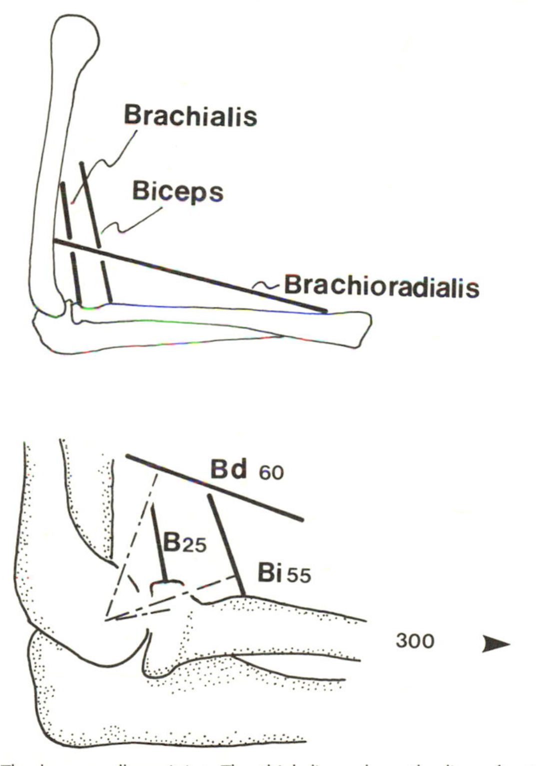

Fig. 6.12. *Above:* The human elbow joint. The thick lines show the line of action of the three main muscles resisting a weight on the hand. *Below:* Enlarged view of the elbow. *Bd:* Brachioradialis; *B:* Brachialis; *Bi:* Biceps brachii. The numbers refer to the moment arms of the muscles, and the weight, about the center of rotation of the joint, in millimeters.

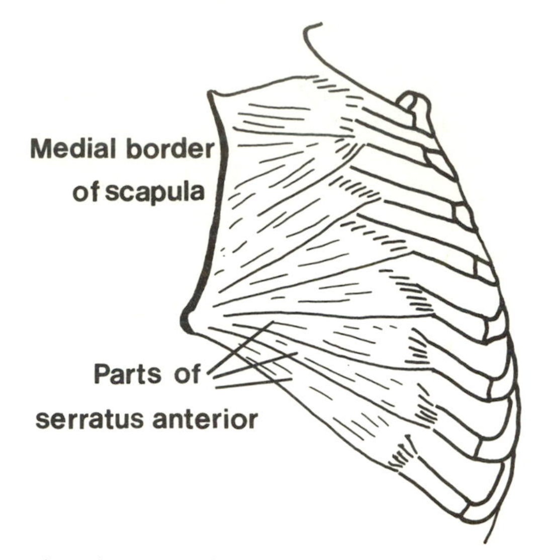

Fig. 6.13. Relationship, in man, between the serratus anterior and the scapula.

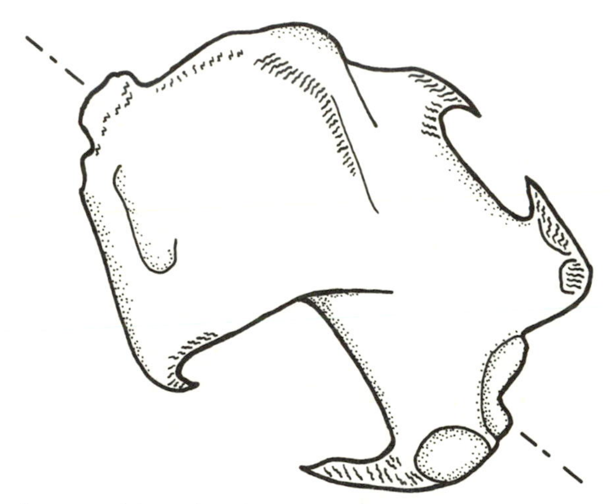

Fig. 6.14. The humerus of the mole, *Talpa*. Squiggly lines show the major muscle insertions. The long axis of the bone is shown by the interrupted line.

mechanical advantage of about one (figure 6.13). The long bones of a few vertebrates have large flanges on them; these greatly increase the muscles' mechanical advantage. The forelimb of the mole is a good example of this (figure 6.14). Although the forelimb has rather bizarre bones, the hind limb of the mole is quite standard.

6.5.1 Muscle Physiology

The selective reasons for these bony shapes have nothing to do with bone itself, but concern the nature of muscle, and a short digression into the physiology of muscle must be made. Active muscle does two separate things: it shortens and it exerts forces. The way that these two properties interact affects the shapes of bones. If a muscle is made to contract in a situation with no external load, it will shorten at its maximum possible rate. (Of course, it will really be producing a small force, because it is accelerating its own mass.) On the other hand, if the external force it has to exert is made greater and greater, there will be some load that it is unable to move, but which conversely does not actually make the muscle any longer. This load is equal to the maximum force the muscle can exert. For any muscle it is possible to construct a "force-velocity" curve showing the relationship between the force the muscle is exerting (the load it is acting against) and the velocity with which it can contract (figure 6.15a). A. V. Hill (1938), who did much of the pioneering work on muscle mechanics, showed that the force-velocity curve could be described by the equation $V(P + a) = b(P_0 - P)$, where P is the force the muscle is exerting, P_0 is the maximum force it can exert, V is the velocity of contraction, and a and b are parameters with units of force and velocity respectively. Hill's equation applies only to muscle at its resting length (the length it naturally adopts if left unconstrained). The maximum force a muscle can exert changes with its length (Gordon, Huxley, and Julian, 1966); therefore, the value of P_0 will change as the muscle shortens (figure 6.15b). The reasons for this relate to the way in which the actin and myosin in the muscle overlap. However, it turns out, fortunately for analytical purposes, that the parameters a and b hardly vary with muscle length (Abbott and Wilkie, 1953), and so the equation above can be modified to $V(P + a) = b(P'_0 - P)$, where P'_0 is the maximum force at any particular length. Hill produced theoretical explanations for his equation having the form it does. These explanations are important in studies on muscle physiology, but they need not concern us; we have to deal merely with the empirical description of the curve.

The external work done by a muscle is the product of the force it exerts times the distance it shortens. Some muscles function without doing much external work. These are the postural muscles, and muscles whose function it is to keep skeletal elements in constant spatial relationship against the action of external forces. When a chimpanzee is hanging with its hand wrapped around a branch, the digital flexor muscles are not causing any movement, and so are not doing any external work. However, they are preventing undesirable movement—the fingers do not unwrap, and so the

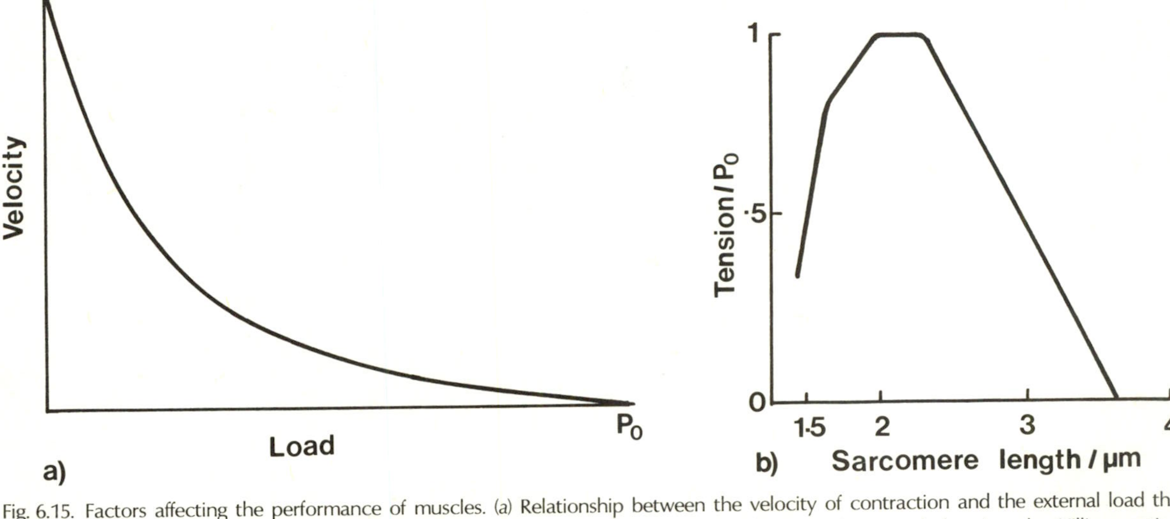

Fig. 6.15. Factors affecting the performance of muscles. (*a*) Relationship between the velocity of contraction and the external load that is being acted on. When the load is maximum (P_0), the muscle cannot shorten. The curve is a parabola, given by Hill's equation. (*b*) Relationship (idealized) between the tension a muscle can exert expressed as a fraction of P_0, and the sarcomere length. In this case the muscle is the frog semitendinosus. (Modified from Gordon, Huxley, and Julian, 1966.)

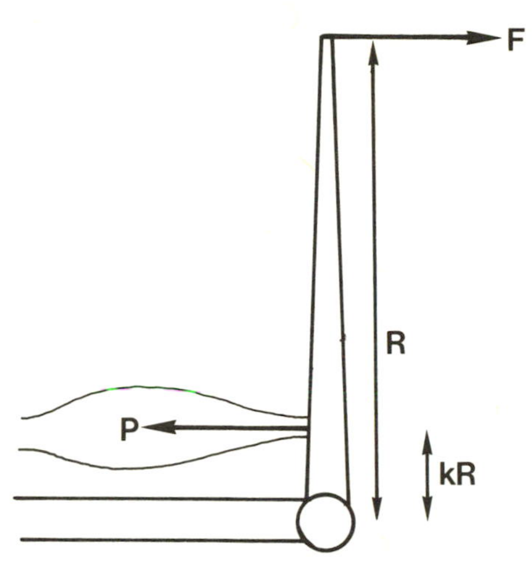

Fig. 6.16. The mechanical advantage of a muscle. See text.

chimpanzee remains suspended. Nevertheless, if an animal is to move around, its muscles must do external work.

Because power = force × velocity, the force-velocity curve can easily be transformed into a power-force curve. It turns out that for many muscles the parameter $a \sim 0.2P_0$. If so, then the power is at a maximum when the force the muscle is exerting is $0.29P_0$. Now for most muscles, particularly those involved in locomotion, the property that is likely to be selected for is their ability to move their point of application as fast as possible. The point of application will be resisting this movement with some force, and so the power output of the muscle is what, in effect, will be selected for (figure 6.16). The figure shows a muscle that is required to move the distal part of the bone at the maximum velocity. The resistance to movement is a constant force F. The force has a moment arm R, and the muscle is to be inserted where it will be most effective. Call the distance from the insertion to the joint kR. The muscle has a mechanical advantage k. Now $P = F/k$. The optimum value for P is when it equals $0.29P_0$. Therefore, $0.29P_0 = F/k$, and k should be $F/(0.29P_0)$.

This analysis shows that the greater the force the muscle is able to exert relative to the force resisting it, the nearer the point of insertion should be to the joint. Thus, if the force resisting movement is 10 N and the muscle is capable of exerting a force of 100 N, then $k = 10/(0.29 \times 100) = 0.34$; that is, the insertion should be about 1/3 of the way along the bone. If $P_0 = 1{,}000$ N, then k should be 0.034, and so on. (Note that the power exerted at the end of the bone is the same as the power at the muscle insertion. Call the velocity of the insertion V. Power of insertion = PV. Velocity of end of bone = V/k. $F = Pk$. Therefore, the power of the end of the bone = $(V/k) \times (Pk) = PV$.)

This analysis was made by Calow and Alexander (1973). They showed, in the same paper, that if instead of considering the *greatest mean velocity* that a given resisting force can be moved at over the contraction of the muscle, one considers the best arrangement for accelerating a mass to the *highest final velocity* (the requirement in jumping, for instance), the optimum position of the insertion is slightly different.

In 1974 Jack Stern carried the analysis further. Calow and Alexander's analysis was, for computational reasons, necessarily simplified. In particular, they could not allow fully for the fact that as the length of the muscle changes, the value of P_0 changes also. So, if the length of the muscle changes considerably during the course of its contraction, it will pass from some low value of P_0, through a maximum, and then decrease again (figure 6.15). Figure 6.17 shows two circumstances that will make this effect of greater or lesser importance.

The first is that the positions of the origins and insertions relative to the hinge will make the muscle change in length to a greater or lesser extent as the joint angle changes. *A*1, *A*2, and *A*3 show a fairly long muscle with its insertion close to the hinge. As the joint angle decreases, the muscle does get shorter, but not greatly so. In *B*1 the muscle is inserted farther from the hinge. As the joint angle decreases the muscle has to shorten considerably, much more, proportionally, than does the muscle in *A*1–3 to bring about the same angular rotation.

The second circumstance producing great change in muscle length is that in which more or less of the distance between origin and insertion is occupied by noncontractile tendon. In *C*1 the insertion is again close to the hinge, but at this joint angle two-thirds of the length of "muscle" is in fact tendon. As the joint angle decreases, the muscle must shorten by a large proportion of its extended length. This shortening will decrease the sarcomere length and, as shown in figure 6.15b, considerably reduce the force the muscle can exert.

This inability of muscles to exert large forces when they are greatly shortened can be easily felt in the hand. With one hand, grasp the index finger of the other and force the metacarpal-phalangeal and first interphalangeal joints into right angles in flexion. Now try to flex the finger tip. Little, and very feeble, movement is possible. This is because the flexor digitorum profundus, which originates on the ulna and has a long tendon running to the fingertip, cannot shorten any further, once two of the three joints it affects are bent into right angles.

In determining the action of a muscle around a hinge joint, there are a number of variables that must be taken into account: the distance of both the origin and insertion from the hinge; the proportion of the length of the muscle that is occupied by tendon; the amount of flexion the joint undergoes; the force resisting the movement; whether power over the

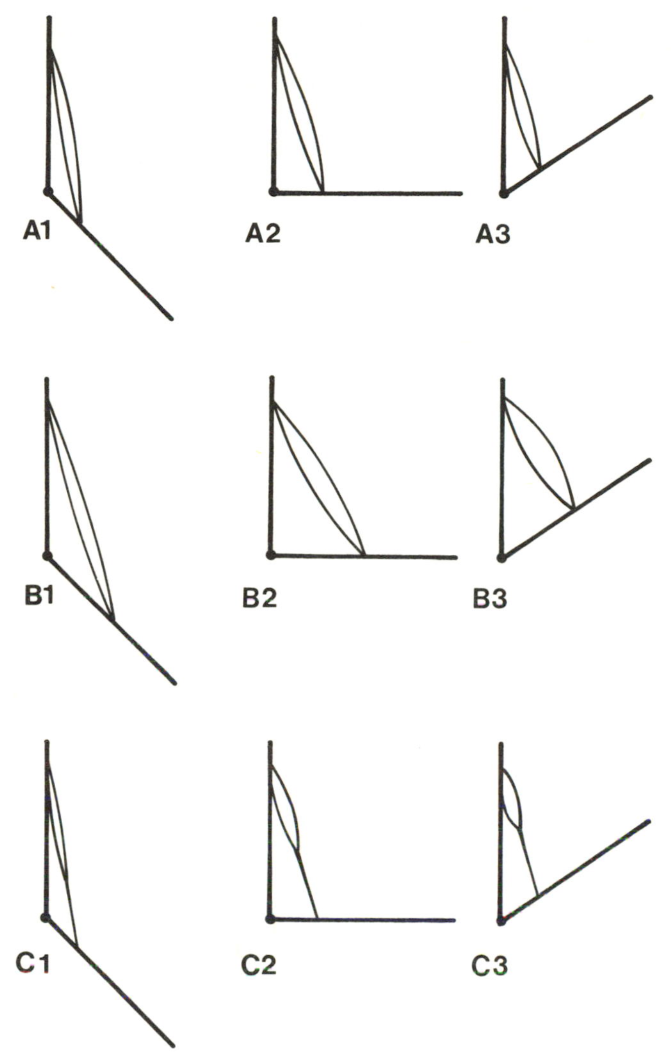

Fig. 6.17. Circumstances making a muscle, moving a joint through a given angle, contract more than a "standard" muscle. (*A*) The standard arrangement. The ratio of the distance of the joint to the origin and the joint to the insertion is called the "attachment ratio." (*B*) The muscle is inserted farther from the joint. (*C*) There is noncontractile material, tendon, in series with the muscle.

whole range of flexion is important or whether it is final velocity that is critical, as in jumpers, and so on. Obviously, the number of variables can become daunting. Stern's computer analysis (1974) attempts to take a reasonable number into account.

6.5.2 Stern's Analysis

Stern performed many analyses, and I shall merely mention a few important points here. The first, agreeing with Calow and Alexander's, is that the results are different if the time taken to complete a movement is to be minimized, or if the final velocity attained is to be maximized. In running, the total time to complete a movement is important, whereas it is the final velocity attained that is important in a jump, because this will determine the length of the jump. A sport fencer, for whom a mere touch is sufficient to gain a point, would wish to minimize the time elapsed during a lunge, to give his opponent less time in which to parry. A boxer must, notionally, settle for a compromise, a high velocity increasing the force of a blow, but a small elapsed time giving less time for blocking the punch. For any particular length of muscle, the muscle insertion should be closer to the joint (have a smaller moment arm and a greater "attachment ratio") if the aim is to achieve a high velocity at a particular point, but a greater moment arm, and therefore a lower attachment ratio, if the aim is to achieve the movement to that point in the minimum time. The reason for this is that if a muscle has a large moment about its joint, it can produce a high initial acceleration and velocity, and this will lead to high average velocities. However, the muscle will be shortening, and the force it can apply will fall off, both because the sarcomeres are shorter and because the muscle is shortening rapidly and so sliding down the curve of the Hill equation (figure 6.15a). A smaller moment arm results in a lower early velocity, and lower acceleration, but the acceleration is of longer duration.

If the contractile tissue of a muscle can be as long as desired, then it is always possible to improve the system's performance at the place where the work is done by increasing the moment arm. When the muscle is between the joint and the load, the larger the moment arm, the less the bending moments in the bone, which is advantageous. In fact, the muscles can rarely be made very long, because of anatomical constraints. The usual constraint is that if the origin of the muscle is a long way from the joint it moves, it is likely to be longer than the bone to which it should be attached. A very long biceps would have its origin on the base of the skull. There can also be difficulties if the insertion is far from the joint, particularly if the joint is concerned with grasping. Another point that comes from Stern's analysis is that smaller moment arms are increasingly favored as the excursion of the joint required becomes greater. This is because, for a given excursion, the shortening of the muscle is proportional to the moment arm. The more a muscle shortens, the farther it has to move from a sarcomere length where it can exert its greatest force.

If the total length of contractile tissue is held constant and only the length of tendon allowed to alter, it is no longer true that increasing moment arms will increase the effectiveness of muscles. As more and more of the length of the muscle plus tendon becomes tendon, the optimum attachment ratio increases, but the best position for the insertion remains about the same.

So, with muscles with a limited amount of contractile tissue there is an optimum position for the insertion. The precise position will depend on whether maximum or average velocity is being maximized, but will not be very different in these two cases. Most important, it turns out that if the muscle is reasonably long; the point of insertion should be close to the joint.

Of course, the simplifications in this discussion mean that reality is more complicated. For instance, often at the extremes of the excursion of a limb the turning moment of a muscle may be so small as to make it very difficult for it to work properly. Other muscles, with larger turning moments, must initiate the movement, the more closely inserted muscle taking over when the movement has got under way and the turning moment has increased because of the relative movements of the bones. In the hip joint of the horse (figure 6.18), the tensor fasciae latae always has a greater moment arm than the iliacus, and it is particularly well placed to initiate protraction. As the movement of protraction gets under way, the tensor fasciae latae will shorten considerably and exert less force even though its moment arm increases. However, the iliacus, which had

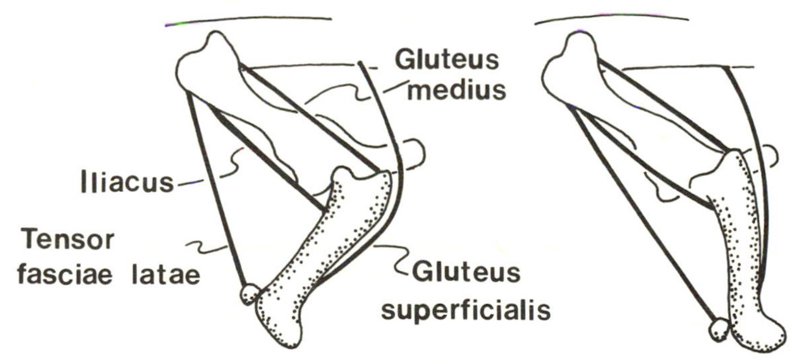

Fig. 6.18. The muscles protracting and retracting the femur of the horse. *Left,* the femur is protracted. The tensor fasciae latae is much shortened compared with its state on the right.

a very small moment at the beginning of protraction, will become effective. Similar remarks apply to the gluteus superficialis and medius.

There are some situations in which neither power nor velocity, but force, is important. In such places the insertions are far from the joints, and often the muscles are very long and straplike. A good example is the forelimb of the mole (figure 6.14). There are strong muscles, with large moment arms, acting against a large load, the earth through which the mole burrows.

The possibility of reducing bending moments in bone by altering points of insertion of the muscles that move the bone is limited in two ways. The more generally important one is because, as Stern, and Calow and Alexander, have shown, insertions need to be close to joints, and this necessarily keeps bending moments large. The other limitation is anatomical. I have mentioned already the hands of monkeys and apes, in which the tendon must lie close to the bone in order to allow room for the bough. It is a general feature of muscles that if they have long tendons these will lie close to the joints they move. This must be so, otherwise the total shortening necessary for the muscle and the tendon together would be greater than the muscle could manage. Therefore, the retinacula keeping the flexor tendons close to the phalanges of apes are what we should expect even if apes touched things with their fingertips only.

6.5.3 Ceratopsian Dinosaurs

If muscles are to have large turning moments, then the bones involved sometimes have to adopt complex shapes to allow a sufficient *length* of muscle. A good example of this is seen in the skulls of ceratopsian dinosaurs. The following few paragraphs are based on the work of Ostrom (1966). The ceratopsian dinosaurs were herbivores. Their jaws ended in a beak, presumably horny. The teeth were chisel-like, able to shear past each other with a scissors action, but incapable of grinding or chewing (figure 6.19). Their food was unknown, but it was presumably very fibrous. An advanced member of the group, *Triceratops*, had two features giving the muscle a greater turning moment than that of the more primitive *Protoceratops* (when the jaws are scaled to the same size).

1. The tooth row is farther back toward the articulation.

2. The articulation of the jaw is below the level of the tooth row, instead of being on it, thereby increasing the distance between the coronal process and the articulation. This lowering of the articulation also makes the most effective line of action of the adductor muscle flatter than it would otherwise be. This has obvious advantages if the skull is to be reasonably compact (figure 6.19).

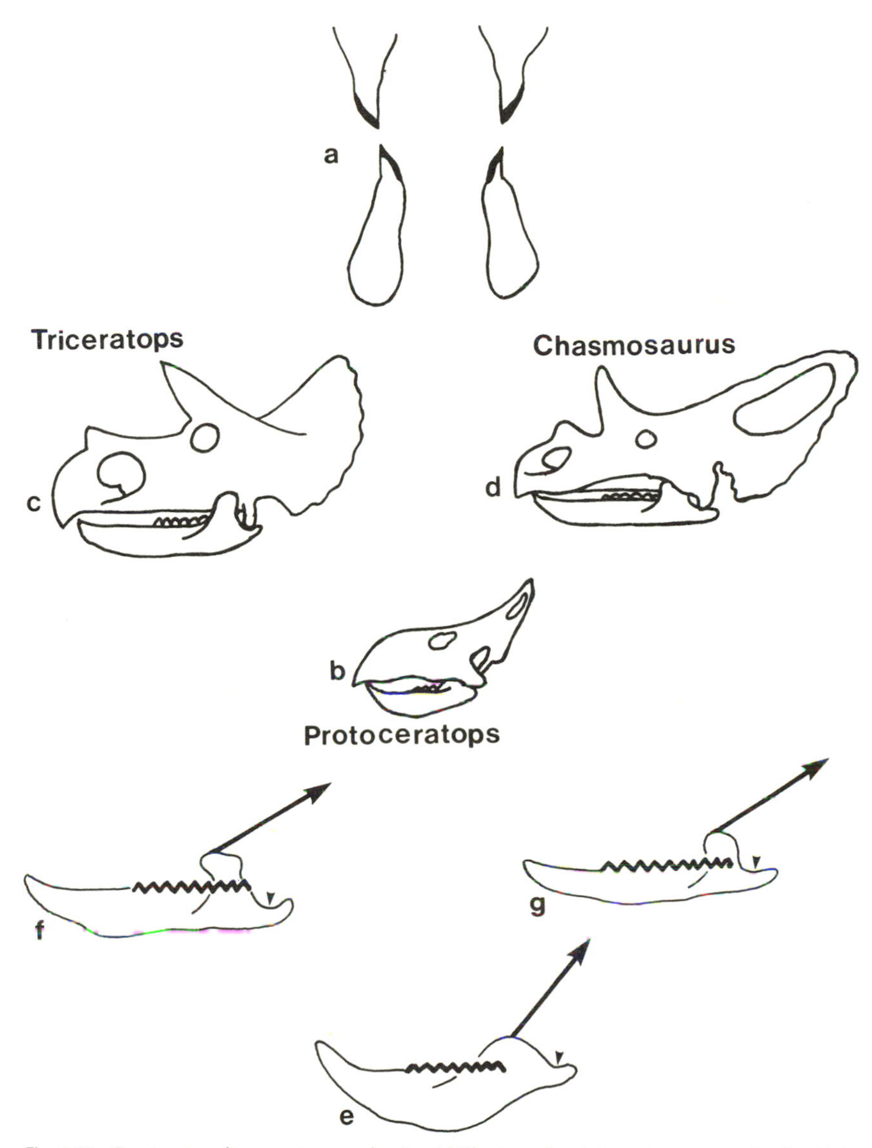

Fig. 6.19. Ceratopsian dinosaur jaw mechanics. (*a*) The jaw sliced through transversely. The thick black line represents the enamel capping of the teeth. Skulls of (*b*) *Protoceratops,* (*c*) *Triceratops,* and (*d*) *Chasmosaurus.* (*e*) Diagram of the lower jaw of *Protoceratops.* The jaw articulation is shown by the small arrowhead, the tooth row by the zigzag lines, and the direction of the principal adductor muscle by the arrow. (*f*) Diagram of the lower jaw of *Triceratops.* (*g*) Diagram of the lower jaw of *Chasmosaurus.*

These modifications of the primitive condition make the force of the bite greater by increasing the adductor moment arm. However, the great length of the adductor, making necessary the development of the frill, remains to be explained. The length of the muscle does not affect the force of the bite at any particular position of the jaw but, as Stern's analysis showed, a larger muscle allows the force to be exerted over a greater range of angular distances. If the supposedly fibrous food of the dinosaurs was bulky, the long muscle would allow a large force to be exerted as the whole of the bundle of food was sliced through.

There are two main lines of ceratopsian dinosaurs leading from the early *Protoceratops*. I have described the *Triceratops* line. The other line, leading via *Chasmosaurus* to *Torosaurus*, puzzled Ostrom. This line had very large frills, and therefore very long adductor muscles. Yet it seems that the turning moment of the muscles at the jaw was not much greater than in *Protoceratops*. Again Stern's analysis allows a resolution of this apparent paradox of great anatomical specialization with little increase in turning moment. For, with a very long muscle, a rather small turning moment (a high attachment ratio) is suited for making the final closing velocity high. I suggest that the two lines of dinosaurs chopped their food in rather different ways: the *Triceratops* line by the force of their bite, the *Torosaurus* line by opening their jaws wide and then closing them with increasing velocity, so that when they met the food they sliced through it mainly by virtue of the kinetic energy of the lower jaw. Such a mode of cutting would, incidentally, put less loading on the jaw articulation than would the more forceful bite of the *Triceratops* line.

6.6 Curvature of Long Bones and Pauwels' Analyses

I mentioned earlier that it was possible to reduce the net bending moment on a column by the device of having it initially curved and loading it axially. Long bones are quite frequently curved, so it might appear that this possibility is being realized. However, the only experimental study (Lanyon and Baggott, 1976; Lanyon and Bourne, 1979) has equivocal results. Lanyon and his co-workers looked at surface strains on the radii and tibiae of sheep. Both these bones are longitudinally curved in the anteroposterior plane. However, they show considerable differences in the greatest principal strains on the anterior and posterior surfaces, indicating considerable bending. So much was this so that one cortex was in tension, even though the bone as a whole was loaded in compression. In the tibia the tension side was the longitudinally concave side, so it is possible to suppose that the tibia is at least bent in the correct direction to minimize bending. However, in the radius the tension surface is the longitudinally convex surface. In this bone, therefore, the curvature is *enhancing* the effects of bending imposed during locomotion. It may be that at really high speeds of running, which Lanyon was unable to produce on the experimental treadmill, the pattern of stress would change. However, at the moment these studies do not confirm the adaptive function of long-bone curvature.

The hand of the gorilla acts as a hook for grasping boughs. The bones of the hand, the phalanges and the metacarpals, are curved, and this allows room for the bough. There are considerable bending moments produced by the load (the bough in this case). The action of the flexor muscles that keep the hook hook-shaped put the bones into compression (Preuschoft, 1973). However, *pace* Preuschoft, this does not in general put significant counteracting bending moments on the curved bones, because the retinacula keep the center line of the bones and the tendon parallel. But there are interosseous muscles which run from one end to the other of the metacarpals. One of their functions is to prevent the metacarpals from splaying out sideways. They also, by their contraction, put the metacarpals into compression, and this force will cause a counteracting bending moment, rather like the situation shown in figure 6.11. How important this effect is, is unknown, because the force exerted by the interosseous muscles is unknown.

This "counterbending" function of muscles has often been claimed by Pauwels and his school (Pauwels, 1980; Nachtigall, 1971) to be of general importance. Pauwels is very fond of the action of the iliotibial tract (ITT) in counteracting the tendency of the human femur to bow out laterally. Although the tract must act in this way to some extent, its magnitude is uncertain, and there is no doubt that many of the proposals that have been put forward to explain the action of muscles in preventing bending should be regarded as interesting suggestions only. We do not yet have the experimental evidence to be able to assert their correctness.

Pauwels has written many papers on the subject of the design of skeletons, nearly all of them on the human skeleton, though many papers are of general applicability. Fortunately, they have now been collected and translated into English (Pauwels, 1980). One particular concern of his is to show how the whole skeleton is designed to minimize the bending moments imposed on the bones. His procedure is to take some case of static loading of a structure, which is roughly like a limb, and to show how, as the structure is made more and more lifelike by the addition of ties, lumps, angulations, and so on, the bending moments on the bones are progressively decreased.

Figure 6.20 shows, for example, his analysis of the human lower limb when the heel is lifted off the ground. As reality is approached, the maximum stress, produced by a combination of compression and bending,

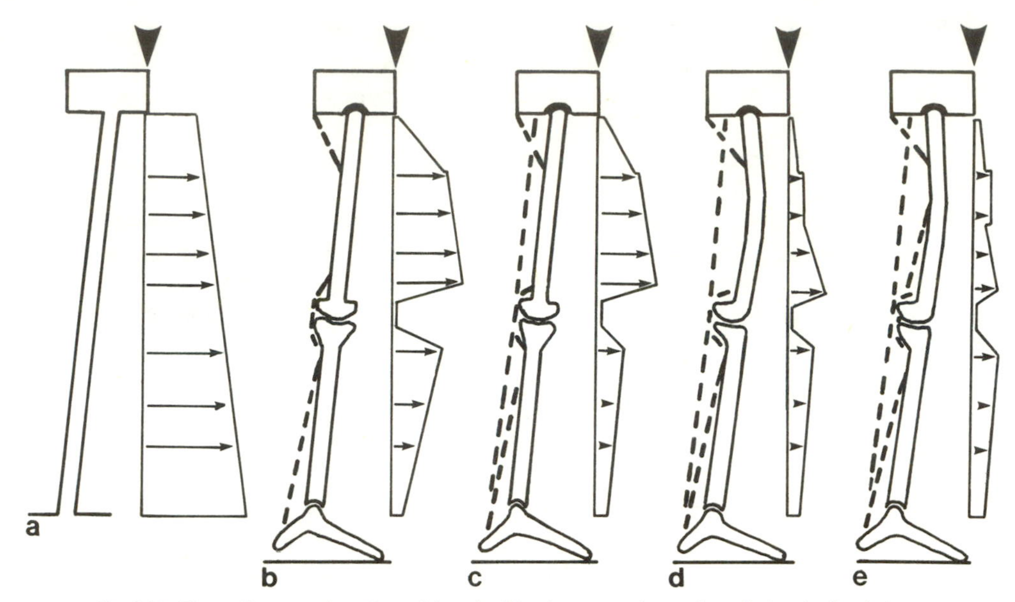

Fig. 6.20. These diagrams show Pauwels' method for demonstrating various devices in the skeleton that reduce stresses in the bone. In this case it is the human leg, in walking, just at the point where the heel leaves the ground. The stresses, shown by the polygons with the small arrowheads, are the greatest stresses in the bone at each level and are composed of bending and direct compressive stresses.

(*a*) The leg is represented by a tilted column with an eccentric load. The stress increases uniformly from top to bottom.

(*b*) Three joints appear. The system is stabilized by contraction of muscles (such as the gluteus maximus) around the hip, muscles (such as soleus) around the ankle, and by ligaments at the back of the knee, such as the posterior capsular ligament. The bending stresses in the tibia are now much reduced and, of course, there are no bending stresses across the knee joint, only compressive stresses remaining.

(*c*) Biarticular muscles are now included, such as the hamstrings, *above*, and the gastrocnemius, *below*. These could reduce still further the bending moments on the tibia and the femur. However, at this stage the structure becomes statically indeterminate, and the magnitude of the effect can only be guessed.

(*d*) The forward bowing of the femur reduces the moment arm of the load about the femur, and so reduces the bending stress in it. This effect is enhanced by a change in shape of the articulation of the femur with the tibia, so their shafts are brought forward a little.

(*e*) The action of another muscle, the short head of the biceps femori, tends to bend the bottom of the femur in the opposite sense to that produced by the main load, reducing bending stresses still further. (Modified from Pauwels, 1980).

decreases from 28 MPa to 7 MPa. Of particular interest here is the change between (*c*) and (*d*). The only anatomical differences between these two are in the shape of the femur and a concomitant change in the shape of the femorotibial joint. The femur in (*c*) is straight, but in (*d*) is bowed anteriorly. The middle of the length of the femur is now bent anteriorly by the compressive load acting between its two ends, and posteriorly by the bending moment at the hip. The combined effect is to reduce considerably the maximum tensile stress in the femur. Pauwels' results and diagrams have a seductive but specious transparency. It is necessary to understand that the actual loads borne by the ligaments and muscles have been calculated, not measured. They are the loads necessary to produce static equilibrium. However, many of the most interesting diagrams show structures that are statically indeterminate. That is to say, there are redundant members; one could cut one or more of the ties without the structure necessarily collapsing. For example, in figure 6.20b all the muscles and ligaments are needed, but in 6.20c it would be possible to cut two muscles without the structure collapsing. So, although in a general way Pauwels' calculations will show what is happening, the actual values will depend on tensions developed in muscles and in ligaments, and this will depend on the activity of the muscles and the stiffness of the ligaments.

It is also important to be clear that in Pauwels' diagrams the bones are, apart from odd protuberances, uniform in cross-sectional area along their lengths. As a result the stresses calculated for the bones at any cross section will be directly proportional to the axial loads and bending moments there. In reality, as Pauwels recognizes, the cross sections of bones, and their second moments of area, vary greatly along their length. It will almost certainly be adaptive to have bones stressed reasonably uniformly (see chapter 4), because this will minimize the mass necessary to bear the load. If these two points are borne in mind, the analyses of Pauwels are very helpful to one's understanding of how bending in bones may, to a large extent, be limited by the soft tissues.

Finally, consider the human foot. In walking, just as the big toe is leaving the ground, the front part of the foot is bearing a load of about 120% of body weight. The foot is loaded like an arch, with the body's inertial weight acting down through the tibia and fibula, and this is opposed by the Achilles tendon acting on the calcaneum, and by the ground reaction acting on the metatarsal heads and the toes. The arch is prevented from collapsing by various flexor muscles and their tendons; by ligaments running between the various individual bones, particularly the long plantar ligament; and by the plantar aponeurosis, which overlies the muscles of the foot and runs between the bottom of the calcaneum and the base of the metatarsals and the toes.

Stokes, Hutton, and Stott (1979) have analyzed the forces in the metatarsals during walking and find that, in fact, the flexor tendons and plantar aponeurosis reduce the bending moment on the metatarsals by only about 10%. The bulk of the resistance to collapse is provided by the ligaments. This calculation is based on assumptions about the forces in the tendons and aponeurosis that seem reasonable, yet it is surprising that their effect in reducing bending moments is not greater.

The values of loading that they calculate are impressive. The first metatarsal phalangeal joint has a joint force of about 80% of body weight, and the first metatarsal has a compressive force of 130% of body weight and a shearing force of about 30% of body weight. These values are for walking; they would be much higher in running.

In this chapter I have argued that bones should, if possible, be loaded in compression and not in bending. This is not usually possible to arrange because often the *function* of a bone is to produce forces at large angles to its long axis. Also tendons, which could in many circumstances reduce the bending moments on bones, are themselves constrained, by anatomical requirements and by the limited contractability of muscles, to lie very close to the bones.

6.7 The Dog's Skeleton

It would not be sensible to try to discuss a wide range of skeletal types, and instead I shall write very briefly about the dog's skeleton as an example of a mammal. I shall, however, make some comparisons between this skeleton and others. An excellent account of the mechanical design of a rather unusual type of skeleton, that of amphisbaenians (legless burrowing lizards), is given by Gans (1974).

6.7.1 The Pelvic and Pectoral Girdles

The major parts of the skeleton of the dog are the skull, the vertebral column, which supports the rib cage, and the two limb girdles, which support the fore and hind legs. The pectoral girdle is very much reduced from the condition found in reptiles. In the reptiles from which the mammals arose, and indeed in the primitive monotreme mammals, the pectoral girdle is U-shaped and rigid. This results in the head of the humerus being in a fairly constant position in space relative to the vertebral column. In more specialized mammals, such as ourselves, the pectoral girdle is reduced to two scapulae and two clavicles, the latter being attached to the sternum. The scapulae are slung in a bag of muscles and have considerable freedom, though the distance from the humeral head to the sternum is held constant by the clavicle. In mammals more specialized than ourselves, even this hindrance to movement may be lost. In the dog the clavicle is represented by a little bony plate embedded in a muscle. As a result of this reduction of the pectoral girdle, the forelimb has considerable freedom in relation to the trunk.

The pelvis, by contrast, remains rather firmly fixed to the sacrum, and little relative movement is possible. Why have these two girdles, which are rather similar in early reptiles, evolved so differently? The center of gravity of the body of the dog, and of most ground-living mammals, is nearer the forelegs than the hind legs, and so when a dog is standing still the forelegs bear a larger share of the body weight. However, this in itself would not account for the evolution of different types of suspension. When the dog is running, each leg, both fore and hind, when it strikes the ground, exerts a force tending to decelerate the body in the direction of motion and then, later in the stride, to accelerate it. However, the effects of the fore and hind legs are different. Jayes and Alexander (1978) have shown that the line of action of the legs tends to act through one point during the time the foot is on the ground. For both fore and hind limbs this point lies above the proximal joint, but is also *behind* it in the forelimb and *in front* of it in the hind limb. The result of this is that, compared with the hind limbs, the fore limbs are more concerned with decelerating the body and less with accelerating it. The forelimbs do not, of course, hit the ground while passively stretched out in front of the body; they are rotating backward at the moment of strike. Nevertheless, they do need to have a fairly efficient shock-absorbing mechanism. The hind limbs, on the other hand, have a greater responsibility for accelerating the body, and being more rigidly attached to the spine they can push the body forward effectively without the need to use muscles to prevent the hip joint from moving around relative to the center of gravity of the animal.

6.7.2 The Limbs

The proportions of the limb segments differ quite widely between mammals using different kinds of locomotion. Table 6.2 (derived from data in Gambaryan, 1974) shows this. There is a general tendency within groups for fast runners to have distal parts elongated relative to the more proximal parts. Compare the gazelle with the tapir, and the cheetah with the bear. Like the cheetah, the dog (represented in the table by the wolf) is intermediate between the bear and the gazelle, having somewhat elongated metacarpals and metatarsals. The dog is digitigrade, that is, it stands on its toes; the bear is plantigrade, standing on the flat of its feet; and the ungulates are unguligrade, standing on tiptoe.

Associated with this relative elongation of the distal elements is the removal of most muscle mass from the distal parts of the limb. For instance, the widest part of the belly of the gastrocnemius in man is 54% of the distance down the leg from hip to toe. In the dog it is 43% and

Table 6.2 Relative Lengths of the Three Segments of the Hind Limbs of Some Mammals

Species	*Femur*	*Tibia*	*Foot*
African elephant	53.5	32.4	14.1
Brown bear	44.0	36.4	19.5
Cheetah	33.6	33.6	32.8
Wolf	32.8	34.3	32.8
Tapir	35.2	28.3	36.4
Quagga	27.1	28.6	44.5
Gazelle	25.2	30.5	44.3
Red deer	25.6	30.6	43.7

Source: Derived from data in Gambaryan (1974).

in the horse 32%. The muscles of the limbs have, on the whole, the same anatomical relations in a fast-running animal and a more plodding one yet, because of the stretching out of the distal elements, the muscle mass is more concentrated near the hip. The mechanical advantage of this arrangement is clear. Suppose an animal traveling at a forward speed U has the main joint in the limb a distance h from the ground. As the limb is swung back its extremity is stationary relative to the ground. It therefore has a velocity $-U$ relative to the trunk. It must have an angular velocity $-U/h$. If the moment of inertia of the limb about the shoulder or hip joint is I, then it will have kinetic energy relative to the trunk of $U^2I/2h^2$. The energy expended in the propulsive part of the stride is proportional to I/h^2 (Alexander, 1975a). It will reduce energy expended, therefore, if legs are long, with a small moment of inertia about the proximal joint. These are, to some extent, contradictory requirements, of course. (I is not constant, as all fast running animals tuck their legs up on the recovery part of the stride; the function of this action is to reduce I.) Other things being equal I is proportional to length cubed, and therefore longer legs will not be useful from the energy conservation point of view unless I is reduced in some way. In evolution this is done by lengthening the legs while keeping the muscles confined to the proximal part of the leg.

Taylor et al. (1974) compared the cheetah, the gazelle, and the goat, which can be made to run at the same speed, and which have similar masses but dissimilar leg proportions. They found that the energy consumption of these different animals was very similar and concluded that the effect of moving the muscles proximally was not important. They say, "This suggests that most of the energy expended in running at a constant speed is not used to accelerate the limbs." This result seems to be contrary to experience—surely anyone who has introspected while sprinting will think that accelerating the limbs requires great effort. It is also not borne out by the later work of Taylor and his group referred to in chapter 4. Perhaps the cheetah, which cannot move its muscles too far proximally, has other adaptations, which overcome the effects of having a larger value of *I*. One such adaptation is the ability to flex the back much more than can the gazelle or the goat.

6.7.3 Fusion and Loss of Bones

In chapter 4 I considered how bones might become adapted so as to perform their functions with minimum mass. In these discussions I assumed that the job had to be done by a single bone. But in skeletons it is commonplace to find a set of bones all doing roughly the same thing while, in other related groups of animals, the number of bones doing the same job is smaller. This reduction is brought about, in evolutionary time, either by the fusion of bones or by their complete loss. A familiar example of this is the reduction of the number of functional metapodial bones in ungulates. In the artiodactyls there are four metacarpals in the hippopotamus, all of them of roughly the same robustness. In the pig and the cow there are also four, of which two are much reduced while in the llama there are effectively two only. In the perissodactyls, there are four metacarpals in the tapir, three in the rhinoceros, and one in the horse.

There is also a tendency in the mammals to fuse the radius and ulna, and the tibia and fibula, together, so that they become functionally one bone. The radius and ulna are fused in animals as different as the sea cow (*Trichecus*), the elephant, and the horse. The birds are well known for the tendency of the bones in the sacral region to become greatly fused, and the sacral region of mammals shows great variation in the number of vertebrae that are fused together. In the dog the radius and ulna are fused along part of their length, and the fibula, besides being very thin, is fused distally to the tibia.

The great advantage in reducing the number of bones to perform a particular function is that the amount of mass necessary for a particular strength or stiffness is also reduced. A simple example shows this. Suppose a cantilever has to support a given load over a given distance and must deflect only by some amount. Suppose also the cantilever has an overall cross section twice as deep as it is broad. What would be the effect of slitting the cantilever horizontally into two, or into four separate cantilevers stacked on top of each other?

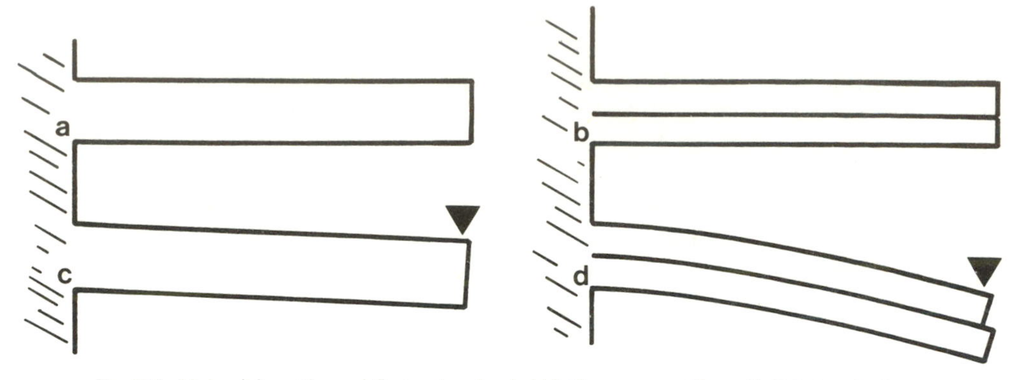

Fig. 6.21. (*a*) A solid cantilever deflects when loaded (*c*). The same cantilever, if slit in two horizontally (*b*) will deflect four times as much if loaded with the same weight (*d*).

The separate cantilevers could not support each other, because they would be free to slip past each other in shear (figure 6.21), and they would, in fact, be equivalent, mechanically, to two or four cantilevers arranged side by side, sharing the load. It is easy to show, given the conditions we have set up, that the mass necessary to produce a given stiffness is proportional to the number of sections into which the cantilever has been sliced, so cutting a cantilever into two horizontally will require it to be twice as heavy if it is to remain as stiff. (Remember, the *overall* cross section remains twice as deep as it is broad.) Slitting the cantilever will leave its resistance to load from the side unaffected, and the resistance to torsion will be reduced, though by how much cannot be simply stated.

This shows that if a set of bones is all doing the same thing, then there will be considerable weight advantages in reducing the number of separately moveable bones. But, of course, such reduction is advantageous *only* if the concomitant reduction in mobility is acceptable: as the radius becomes fused to the ulna, the mass of bone needed becomes less, but the wrist becomes unable to twist.

A survey of mammalian skeletons shows a great range in the amount of mobility allowed between different bones. (The skeletons of birds are strikingly uniform, varying mainly in the relative sizes of different bones, and little in the amount of fusion or loss.) The primitive state in mammals is to have little fusion or loss of bones. Man is interesting in that he retains the primitive condition almost entirely.

6.7.4 The Vertebral Column

Finally, in this brief survey of the dog's skeleton, we consider the functions of the vertebral column. Of all the main components of the bony skeleton the vertebral column is perhaps most able to show the division of function between hard and soft tissues, with bone taking the compressive loads and soft tissues the tensile ones. The vertebral column of the mammal has various functions: to support the head, to act as a place of attachment for the limbs, to transfer force from the limbs to the rest of the body, and to support the viscera.

The function of supporting the viscera requires that when a mammal, or any other tetrapod, is standing still, the vertebral column between the limb girdles will be loaded so as to sag, that is, it will tend to be convex downward. It might be possible to counteract this tendency, without making use of muscles attached to the column itself, by curving the vertebral column into an arch. The column would then be loaded in compression, with the viscera slung from it, and the tendency of the ends of the column to splay apart could be counteracted by the body wall musculature. Figure 6.22 is a diagram showing this. The solid circles represent the main mass to be supported. The head, which is cantilevered out on the end of the neck, is supported by tension in muscles running to the enlarged spinous processes of the anterior thoracic vertebrae. The cervical vertebrae are put into compression by this although, as shown here, any real neck is curved, and would need intrinsic musculature to stabilize it from buckling. In mammals such as the cow, which must frequently raise and lower a heavy head, there is a well-developed ligamentum nuchae running along the top of the neck (shown dotted in figure 6.22). The ligament is composed mainly of elastin, a rubbery protein. When the head is lowered, the ligament is stretched and, unlike muscle, can store the work done in stretching it as strain energy. When

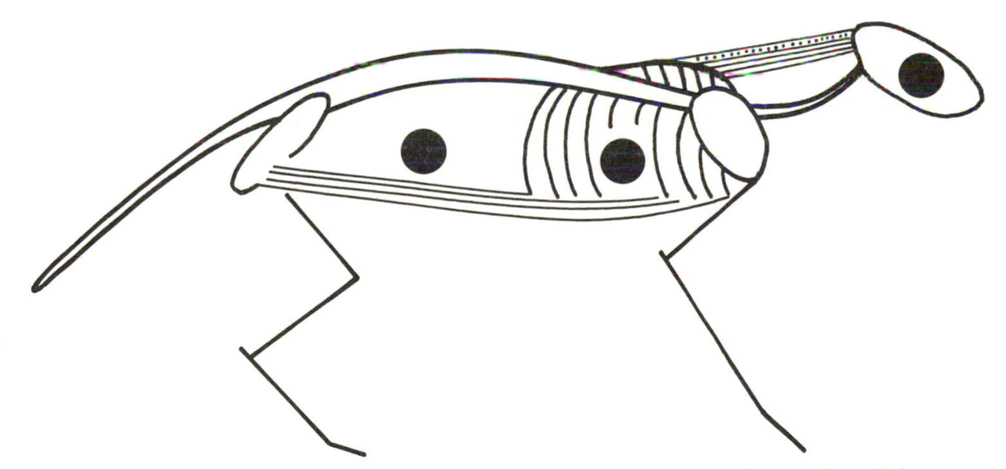

Fig. 6.22. A diagram of a tetrapod showing the complementary effects of the muscles and the spine. For explanation see text.

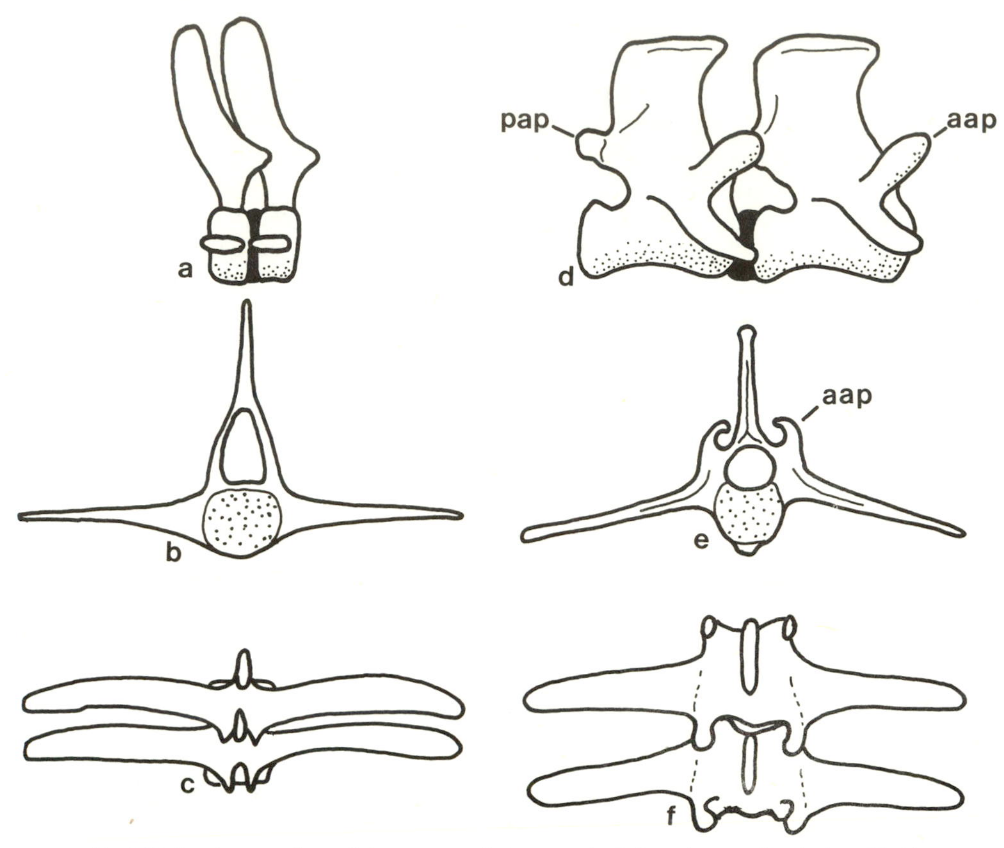

Fig. 6.23. Lumbar vertebrae of three mammals. Intervertebral discs shown solid black. In (*b*), (*e*), and (*h*) the position of the intervertebral disc is shown dotted.

the time comes to raise the head and survey the passing scene, the muscles are assisted by the force of the passive contraction of the ligamentum nuchae.

In the trunk, between the fore and hind legs, muscles running from the rib cage to the pelvis act as tension members and the curved spine as the compression member of the system bearing the weight of the viscera.

This description is greatly idealized, because real animals run around, twisting their spine to a greater or lesser amount, and most of the time do not bear much resemblance to the architectural model. In fact, there is usually a great deal of intrinsic musculature associated with the spine itself.

In locomotion different animals flex their spines to greatly differing extents, and figure 6.23 shows the effects of this on lumbar vertebrae.

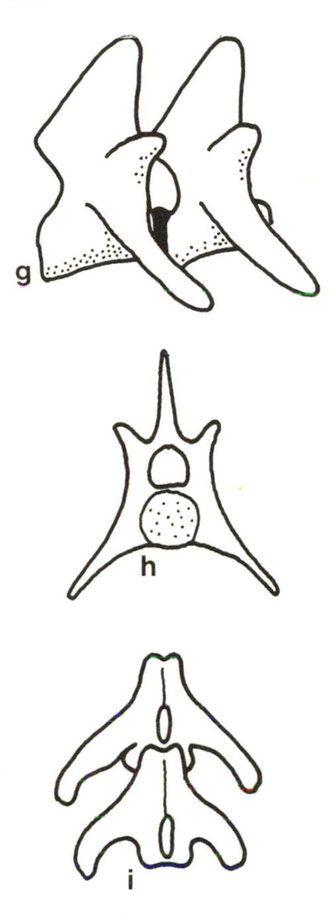

Figure 6.23 gives the right-side (*a*), anterior (*b*), and dorsal (*c*) views of lumbar vertebrae of the common porpoise, *Phocaena phocaena*; (*d*), (*e*), and (*f*) are similar views of lumbar vertebrae of the fallow deer, *Dama dama*; and (*g*), (*h*), and (*i*) are similar views of lumbar vertebrae of a dog.

The lumbar vertebrae of the porpoise are very simple. The centra are simple discs, the transverse processes stick out horizontally and directly laterally, and the spinous process, blade-shaped like the transverse processes, has a very simple articulation with the process in front. It is clear that very little relative motion of these vertebrae is possible. Virtually no flexion in the horizontal plane can occur because the transverse processes would bump into each other. If the vertebrae are to flex ventrally, the movement cannot be brought about by intrinsic muscles attached to the transverse processes, because such muscles would have no turning moment about the center of rotation of the joint. Such a movement could be brought about by blocks of muscles running beneath the vertebral column. In fact, the lumbar region of the porpoise is not capable of much flexion, but the great blocks of muscles running toward the caudal peduncle allow very forceful movement of the tail.

The lumbar region of the fallow deer is somewhat more flexible than that of a porpoise, but because of this animal's large gut the spine does not flex much during running. This is shown in the vertebrae. The transverse processes stick out almost laterally and horizontally, and there is not much room for dorsiflexion between the spinous processes. In fact, dorsiflexion is also severely limited by the articular processes. The posterior articular process (*pap*) of each vertebra fits like a peg into the socket of the overarching anterior articular process (*aap*).

The lumbar region of the dog, like that of many carnivores, is flexed and extended a great deal in locomotion. This allows the effective length of the stride to be increased beyond that allowed by rotation of the legs. The lumbar vertebrae have much more relative freedom of movement than those of the other two species I have described. There is much more space between the spinous processes. The transverse processes project ventrally, so that muscles running between them can flex the vertebrae, and the spine does not, as it were, have to rely on muscles elsewhere to do its flexion for it. This downward projection of the transverse processes is seen in other noncarnivorous mammals that flex their spines vigorously—in rabbits, for instance. The *anterior* projection of the transverse process is more difficult to explain. Dr J. Smeathers of the University of Leeds has suggested that, in vertebrae that are capable of much *lateral* flexion, the overlapping of the lateral processes keeps the moment arm of the muscles joining the processes large at all angles of flexion. They will certainly have this effect, but how important it is, is difficult to say.

6.8 Conclusion

In this chapter I have tried to describe some of the devices found in animals to take advantage of the different mechanical properties of bone, muscle, and tendon. Perhaps one should say, to mitigate the unsatisfactory features of these tissues. As always, compromises abound, and bone is particular is necessarily often stressed in tension, a mode of stressing it is not well able to withstand. As a result, bones often break. In the next chapter we look at the problem of how often they break, and whether the safety factors found in bones are adequate.

7. Safety Factors and Scaling Effects in Bones

7.1 Safety Factors

In chapter 4 we discussed the mechanical properties of whole bones. However, there was an important omission in the analysis, which we must now try to fill. We have assumed that natural selection has, for instance, specified some load that must be borne without too much deflection, or fracture. But, why is this criterion set, rather than some other? In particular, what relationship does the load that the bone is designed to bear have to the loads that are encountered in life? Essentially this is a question of safety factors.

Every improvement in some mechanical property of a bone will have a cost associated with it. We have seen that small changes in the mineralization of bone are associated with opposing changes in Young's modulus of elasticity and impact energy absorption. Similarly, an increase in the stiffness and strength of a bone, produced by an increase in mass, will be associated with the disadvantages of greater mass and the greater metabolic energy and time required to produce it.

There is little known about safety factors in bone, but we can say a few things. First, and obviously, the safety factors are insufficient to prevent bones from breaking from time to time. Many readers will have at least cracked a bone during their lifetime. Usually this will have occurred during sport and will have involved a metacarpal, metatarsal, rib, or calvicle. However, more serious fractures are not uncommon and do not all, by any means, result from "unnatural" injuries like automobile or skiing accidents, or bullets. Lovejoy and Heiple (1981), examining a large set of aboriginal Indian skeletons in Ohio, found a 45% chance of a long bone fracture in any individual. Since only 5% of these people lived beyond about 45 years, senile changes were not an important cause of fracture. Bones may break in healthy young people without any apparent accident to cause it. Throwing a baseball, hand grenade, javelin, or handball can result in fractures of a bone in the arm (Gregersen, 1971; Peltokallio and

Peltokallio, 1968; Waris, 1946). Whether these apparently spontaneous fractures are in fact the end result of slowly spreading fatigue fractures is uncertain and irrelevant. Most fatigue fractures, as discussed in chapter 2, are the result of unusual loading, or ordinary loading carried on far more often than usual, but not accidental loading. Even activity only a little more than usual can break bones. Such apparent weakness is not confined to humans. Running on the flat on grass can cause the fracture of the long bones of horses (Vaughan and Mason, 1975; Cheney, Liou, and Wheat, 1973). Greyhounds are very prone to similar fractures (Gannon, 1972; Devas, 1975). Nor is it only highly artificially selected domestic animals that suffer such injury. There is a considerable amount of anecdotal evidence from reliable observers that large wild prey animals, such as zebras and antelope, will sometimes break their legs when accelerating away from a predator.

In wild animals the breaking of a bone, particularly a long bone, is likely to threaten life. It is surprising, therefore, how many animals seem to have healed fractures. A visit to any museum is likely to show a small, but not trivial, proportion of the skeletons with healed fractures. Schultz (1939) shot 118 wild gibbons. In these there were 48 healed fractures of the major long bones and 26 healed fractures of other bones. Bramblett (1967) found that among 37 adult baboons, *Papio cynocephalus*, only 7 did not have at least 1 fracture. These studies suffer somewhat from the possibility that animals with healed fractures are easier to capture or shoot than others. This problem is absent from the study by Buikstra (1975) in which an entire social group of macaques, *Macaca mulatta*, was culled. In the 43 adults there were 17 healed fractures of long bones and clavicles. These were distributed among 11 adults.

These figures show that these primates are quite likely to break an important bone during their lifetime. We do not know the proportion of those who break a bone and survive without being hopelessly handicapped.

Of course, the high incidence of fractures does not necessarily extend to other groups. The less variable the loads imposed on a bone during a lifetime, the easier it will be to ensure that the probability of fracture will be small. A nonbony example of this is the jumping leg of the locust. Bennet-Clark (1975) has calculated the loads caused by the jump on various structures in the leg, and also the strength of these structures. He shows that the strengths are only about 20% greater than the loads. This is, in engineering terms, a negligently small safety factor. It is made possible, presumably, only because the greatest loads the leg is likely to experience are imposed on it during the jump, and these loads are more or less under the control of the locust itself. Figure 7.1 may help us to visualize the situation. The abscissa is the load a structure can withstand.

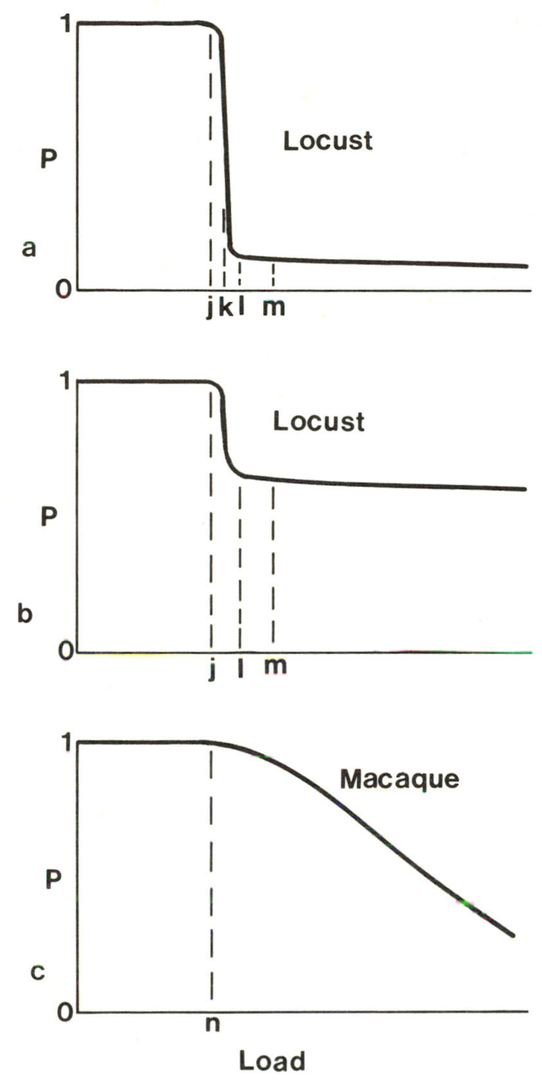

Fig. 7.1. Relationships between the load a skeletal structure can bear (abscissa) and the probability of such a loading (ordinate). (*a*) The leg of a locust, with a small probability of predation. (*b*) The leg of a locust, with a higher probability of predation. (*c*) Leg bone of a macaque.

The ordinate is the probability of the structure being loaded to that extent, *or less*, during some sensible period, such as from birth to the midpoint of the average reproductive life. In this diagram (figure 7.1a) the value *j* corresponds to the normal loads imposed during a jump. The locust is certain to load its legs to this amount. Because of slight awkwardnesses in the positioning of the leg, the nature of the ground, and so on, the leg will often undergo slightly higher loading than *j* during the jump. The probability decreases as the loads get higher. The load marked *l* is about the limit of such loads. There is, then, a rather flat tail going toward very high loading, such as would be caused by a predator munching the locust. If the leg has a strength *m*, it will be able to withstand all the loads that are likely to be imposed on it. Furthermore, increasing the strength beyond *m* is going to have little effect on the probability of failure, because any loads greater than *l* are quite likely to be *much* greater. A small safety factor *om*/*oj* is probably satisfactory from the point of view of natural selection.

On the other hand, the slightly lower strength *k* would be bad design, because the probability of being loaded beyond this point would be high (50% as drawn here). In other words, a great reduction in the likelihood of failure can be bought for a small increase in skeletal strength. Indeed, this would still be the case if the chances of being untimely eaten by a predator were quite high, as shown in figure 7.1b. If the forces are either highly predictable or irresistible, the safety factor will probably be small.

What is the likely shape of the curve for a macaque? Figure 7.1c shows a possible shape. Up to *n* are the loads imposed by running, jumping, and climbing, and the small falls that all monkeys inevitably suffer. However, above this point there is no sharp decrease in probability, because there will be a gently decreasing probability of ever more disastrous falls. Therefore, in distinction to the locust, there is no dramatic advantage in increasing the skeletal strength a small amount above that necessary to give the strength *n*.

The curves in figure 7.1 are a measure of the benefits associated with various strengths of skeletons. What are the costs? Presumably, the costs are usually some more or less complicated function of the mass of the skeleton. The cost in metabolic currency, and in currency of time taken to produce it, will be roughly linear with mass. The relative cost of actually producing the skeleton will be markedly different for different vertebrates according to their general level of metabolic activity. For a small endotherm such as a shrew, whose maintenance metabolic activity is a very high proportion of its metabolic rate, the cost of growing the skeleton will be relatively trivial. On the other hand, for an abyssal fish just ticking over on minimal energy intake, the growth of a skeleton could be a sizable proportion of its total metabolic activity. For all vertebrates the cost of *maintaining* the bony skeleton will be rather small, because bone has a low metabolic and turnover rate compared with other tissues (Thompson and Ballou, 1956). However, it will certainly be relatively lower in most teleosts than in tetrapods, because teleosts have acellular bone, whose metabolic rate must be very low indeed.

From the mechanical point of view the costs associated with the skeleton's weight or mass are likely to be complicated. Suppose we are interested only in the impact strength of the skeleton, which is often likely to be the case. If so, then the resistance to fracture can be considered to be, for any particular type of material, proportional to mass. This is because shape is rather unimportant in resistance to impact; it is the volume of bone absorbing energy that is important.

The cost of the skeleton is likely to be very different for different vertebrates according to their mode of life. In fish with swim bladders the component of the cost associated with mass as such will be proportional roughly to mass$^{2/3}$. The reason for this is that the buoyancy will decrease linearly with skeletal mass, and so swim bladder volume will have to increase linearly with skeletal mass. The drag experienced by the fish, and so the power required for swimming, is proportional to cross-sectional area, or to volume$^{2/3}$. Whales and other sea-going air breathers are in a different situation. They need lungs to breathe. If they had no bones, they would be positively buoyant. For animals needing to spend much of their time underwater, this would be disadvantageous. Therefore, an increase in skeletal mass would initially be advantageous but would, sooner or later, become disadvantageous. Some sea-going mammals, particularly the Sirenia (the sea cows), which are very slow swimmers and have lungs, which tend to give them unwanted positive buoyancy, have peculiar bones, with small marrow cavities and rather dense cancellous bone. This condition, called pachyostosis, is presumably a way of increasing the mass of the skeleton without much regard for the mechanical consequences.

For nonswimming vertebrates the mechanical component of the cost of increasing skeletal mass can be expected to be greater in the order: sedentary forms, slow runners, fast runners, fliers (figure 7.2).

Not only will the cost of increasing the mass of the skeleton, and therefore the safety factor, be greater in more actively locomoting animals, but it will also vary *within* the skeleton. This requirement can be simply exemplified in the bird. A bird, being a flying animal, will be under considerable selective pressure to have as light a skeleton as possible. Even so, this selective pressure will be greater toward the ends of the wing bones than in the axial skeleton. Consider the wing at the start of the downstroke. To accelerate the wing downward, the pectoralis major muscle will have to exert a torque. Part of this torque will be to overcome the aerodynamic load. This is inescapable, being required for flapping flight. The muscle will also have to exert a torque against the inertial forces caused by the angular acceleration of the wing. These forces will be proportional to the moment of inertia of the wing about the humeroscapular joint. If y is the distance of a narrow strip of wing from the joint, and δm is its mass, then the moment of inertia of the wing is $\sum y^2 \delta m$. Any mass of bone,

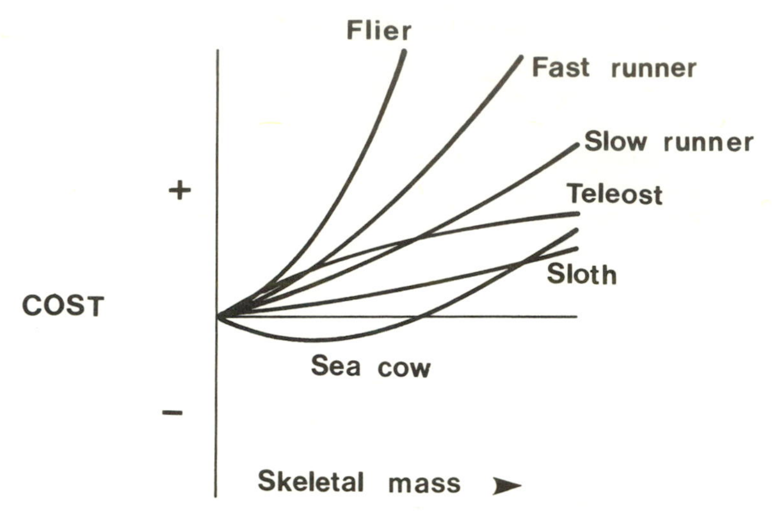

Fig. 7.2. The cost, for different types of vertebrates, of skeletons of different masses. This is very diagrammatic, serving merely to illustrate that the cost of a skeleton will differ very much according to the way of life of the animal.

or other tissue, will contribute more to the moment of inertia of the wing if it is far from the shoulder joint than if it is near to it. The greater the moment of inertia, the harder the muscles will have to work (Weis-Fogh, 1973). This is a difficulty that any fast-moving tetrapod faces, and it accounts, as we have seen in chapter 6, for the tendency, in fast movers, for the distal parts of the limbs to be lightened as far as possible. The muscles migrate proximally, and the more distal joints are controlled by long tendons.

There will, therefore, be increasing selective pressure along the limb to reduce the margin of safety of the bones. Table 7.1, derived from Vaughan and Mason (1975), is an indication of this. It records the injuries to racehorses in Britain, over a period of about two years, that resulted in the horse's death, either at once, or because it had to be destroyed. I have divided the injuries into those that involved a fall and those that did not, though these latter may have occurred when the horse landed after successfully jumping a fence. The loads on the skeleton resulting from a fall are, of course, not particularly closely related to the loads imposed during ordinary locomotion. There is a clear tendency for the accidents not involving a fall to produce fractures in the distal bones, much more than accidents involving a fall. The only surprise is the four fractures of girdle bones (one of the scapula, three of the pelvis) with no fall. A similar increase in fractures, not involving falls, from proximal to distal is seen in healthy young humans. In athletes and soldiers the tibia is much more likely to develop fatigue fractures than the femur or more proximal bones.

Table 7.1 Distribution of Fractures in Racehorses

Bones Involved	*Fall*	*Not a Fall*	*% in Fall*
1st phalanx	0	14	0
Sesamoid	0	3	0
Metapodial	7	8	47
Carpus, tarsus	3	7	30
Tibia, radius, ulna	11	5	69
Humerus, femur	10	1	91
Pelvis, scapula	6	4	60
Spine	29	0	100

Source: Derived from Vaughan and Mason.

The metatarsals in some studies are more prone to fractures than the tibia, in other studies less (Orava, Puranen, and Ala-Ketola, 1978; McBryde, 1975). Fatigue fractures are, as discussed in chapter 2, caused not by very large stresses, but by stresses merely somewhat larger or more frequent than usual. Nor do these stresses act in unusual directions.

This work on two rather different, though domesticated, mammalian species, horses and humans, shows that safety factors in limb bones are probably rather small in relation to the loads imposed during violent, but controlled, locomotion, and that these safety factors decrease distally along limbs, as we would expect.

In chapter 4 I discussed the effects of the presence of marrow on the weight advantage of having hollow bones. Even the presence of *fatty* marrow makes the advantage small. If long bones were filled with something of greater density than fat, the advantage would be even less. Fat is the least dense tissue the body produces. Hematopoetic marrow (red marrow) is denser than water. During the ontogeny of mammals about half the red marrow is eventually replaced by yellow marrow. It would obviously be mechanically advantageous to have the denser red marrow that remains concentrated in the more proximal bones. This is what happens (Ascenzi, 1976; Piney, 1922). There have been various experiments designed to explain this central-peripheral distribution of yellow marrow. These experiments tested the ideas that temperature, or blood supply, may be the controlling factor and seem to ignore the possibility that there is an important *functional* reason for the distribution of the different types of marrow. The functional, mechanical explanation seems satisfactory.

Alexander (1981) has recently tried to formalize the various factors affecting safety factors in bone and other skeletal materials. He suggests that the safety factor to be found in a bone can be determined by knowing two variables. One is the ratio of the cost of growing and using a bone of a particular "designed" strength, stiffness, or whatever, to the cost to the animal of its failure. The other variable is a combination of the variability of the loading that will be imposed on the bone and of the quality control of the construction of the bone. Before showing how these variables may interact, we must say a little more about them.

If an engineer were to design a structure in, say, an airplane to have a particular strength, it would be fairly easy to quantify its costs. The cost would include the manufacturing cost, and the effect that the mass of the structure would have on the performance of the airplane. Notice that the costs are not in any simple way measured in the same currency; one is in dollars, the other in performance.

In general, of course, the greater the strength of the structure, the greater will be its mass, and so the greater the cost, whether measured in dollars or performance. The cost of failure of the structure would not be the same for all structures. In an airplane if the catch retaining a blanket locker door breaks, the cost would be trivial, whereas if the main wing spar were to break, the airplane would probably be destroyed. For a manufacturing company it might be possible to forecast the cost of an airplane crashing, in dollars, but such a quantification would clearly be highly imperfect, depending as it would on such factors as how full the airplane was, the cupidity of lawyers, and so on. When trying to apply concepts such as safety factors to animals, biologists are, in one way, in an easier position than engineers, because the currency of cost is rigidly defined; it is Darwinian fitness. As this controls the way that bone, and indeed everything, has evolved, it is worthwhile saying more about the concept of fitness.

Suppose a population of animals consists of two types, A and B, which differ genetically. The average number of offspring produced by an A individual (counted from the moment of its own fertilization) is, say, Ra, and the average number produced by B individuals is Rb. Suppose that Ra is greater than Rb. Then the *fitness* of the two types is defined as A: fitness = 1; B: fitness = Rb/Ra. Notice that fitness says nothing about the *number* of offspring successfully produced. The chance of survival to sexual maturity of a fertilized codfish egg might be only 0.0001, while a fertilized mammal egg might have a 0.5 chance of surviving. Even so, in a fairly stable, sexually reproducing population, the average value for R will be about 2. Fitness, however, will vary between 1 and 0, depending on how well the opposition is doing.

Notice also that the concept of fitness has no implications about why one type is fitter than another. Indeed, if the two types do not differ much in their characteristics, the difference in fitness in any generation may to a considerable extent be determined by chance. The reason for the need for this rather neutral definition of fitness is that fitness, defined in this way, allows one to calculate the way that genes will spread, or fail to spread, through populations. Most differences between species are heritable and are the result of different genes spreading through populations of species.

The currency, therefore, determining whether changes in the safety factor of bones will, or will not, spread through a population is simply whether the change enhances, or does not enhance, the competitive ability, relative to other members of the species, of the bearers of the change to produce offspring in the next generation.

All the costs and benefits of a particular safety factor are expressed, in Alexander's 1981 formulation, in the term $(G + U)/F$, the costs of growth and use divided by the cost of failure. In the airplane example above, this could not be readily done, because the different costs are measured in different currencies. In animals there is in theory no problem, because all costs are measured in terms of Darwinian fitness. Of course, in reality actually determining the effects of various characters on Darwinian fitness can be very difficult indeed.

As well as the concepts of cost and benefit, the concept of the safety factor is inevitably also different in biology and in engineering. It is, really, too grand a concept for biology. Human designers may be thought to go about designing in one of two ways. One is to have a rather precise idea of the loads that are to be imposed on the designed structure, and then to design the structure so that, theoretically, it will bear, say, twice that load without failing. This kind of procedure is used for car bodies, airplane frames, dams, bridges, skyscrapers, and so on. Any refining of the process depends on a better understanding of the imposed loads and better stress analysis. The other method is to design something that "looks about right," and then to see how it does in practice. If too many customers come back complaining that the soles of their shoes have cracked, or that the handle of the deep freeze has broken off, the design can be improved. If, on the other hand, no one ever complains, the designer may think that he has produced an object that is too good. Material, and costs, could be saved by making the object a little weaker. Natural selection, I am sorry to say, acts in the latter way. Structures will be increased or decreased in strength until an acceptable probability of failure is attained—that which will maximize Darwinian fitness.

Alexander suggests that the maximum loads imposed on a population of bones, up to some point of interest like production of offspring, can be considered to have a log normal distribution. That is, if the values of load are expressed using a log scale, the distribution of load against probability of that load will have a normal distribution. I use "normal" here in the formal, statistical sense. Alexander recognizes that the main advantage of this distribution is its mathematical convenience, but it does in fact describe many loading distributions quite well (Bompas-Smith, 1973), though probably not, in fact, some that I discussed earlier in the chapter.

Alexander defines a safety factor (arbitrarily, for we are dealing with natural selection, not a written specification) as having a value of 1 if the probability of failure is 0.5. If the safety factor defined in this way is called n, the probability of failure with any particular safety factor is called $P(n)$, then

$$dP(n)/dn = (-1/nV2\pi) \exp[-(\ln n)^2/2V^2]$$

where V is a measure of the variability of the loading and of the strength of bony structures of a particular mass. Once this is accepted, it is possible to make calculations on the values of the safety factors one should find in bone if natural selection is working in this way. Figure 7.3 shows the results of calculations of safety factors for bones loaded in bending. (The way in which the mass and, therefore, the cost increase with safety factors will vary according to the loading mode that is important, see chapter 4. However, the general shape of the distribution will not be greatly affected by the loading mode.)

Figure 7.3a has several features of interest. As might be expected, the safety factor is high if the cost of failure is high relative to the cost of $(G + U)$ and, at the same time, the loads imposed are variable. This latter condition is important because, although by definition a safety factor of unity gives a 0.5 chance of survival, a large value for V will mean that to have a high, rather than a moderate, chance of survival, the safety factor must be much more than unity.

If the cost of failure is small, and the variability of loading is large, then the safety factor should become small. Indeed, the diagram shows a region, dotted, where it is not worthwhile keeping a bone at all, and it should disappear. Figure 7.3b shows the probability of failure with a particular safety factor. Note that having the same safety factor does *not* mean that the probability of failure is the same; that depends on the variability of loading as well. Shown on these diagrams are points that Alexander suggests are roughly where different bones should lie. T refers to the limb bones of terrestrial mammals; G to monkeys and arboreal apes such as gibbons. S refers to the antlers of stags. The stag's antler is not critical for its survival, though it does help it to reproduce in any one season. It is a heavy bone, and therefore its cost of use and production are fairly high compared with the cost of its failure. Since it is used in fights, the variability of loading on it will be very large. These together imply a position on the top right of the diagram. The frequency of fracture in one population is known (Clutton-Brock et al., 1979), and so it is reasonable to place the stag on the 0.3 chance of failure line.

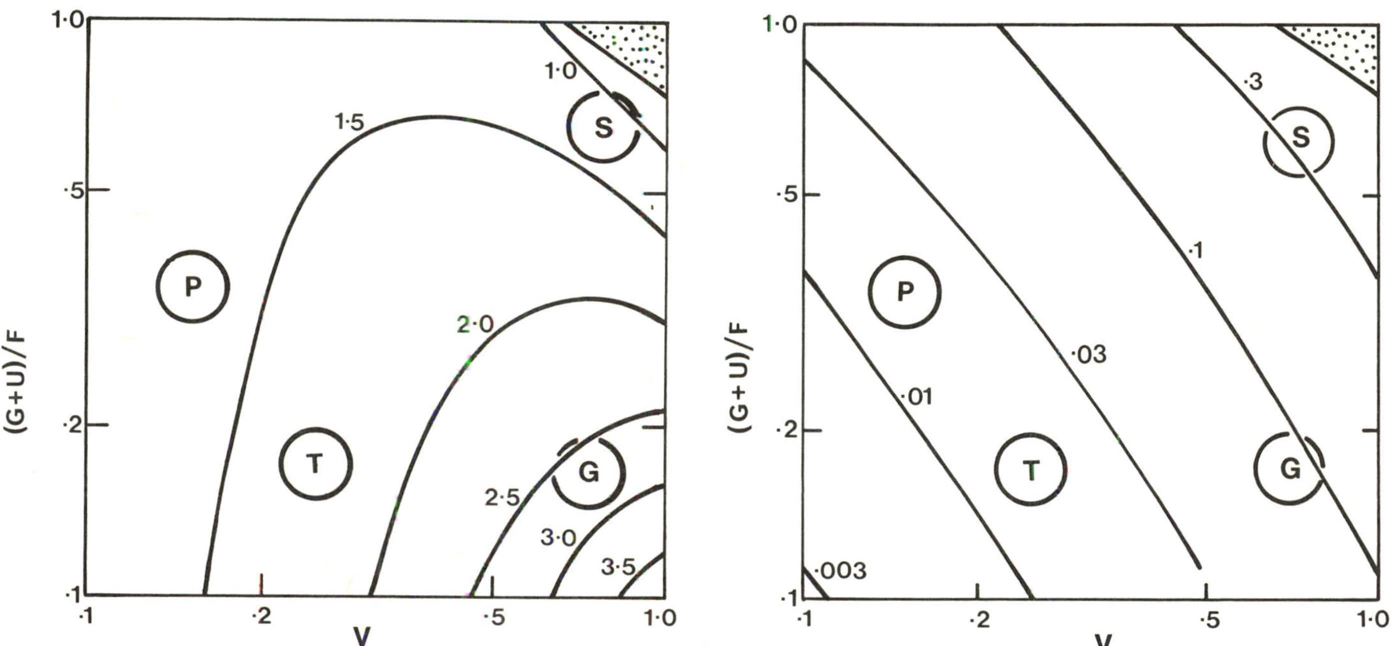

Fig. 7.3. Diagrams, slightly modified from Alexander (1981), showing the safety factors, and probability of fracture, that may be expected in different bones as a result of the action of natural selection. Ordinate: (Cost of growth + use)/cost of failure. Abscissa: variability of load and of the bone's response to load. *S*: Stag's antler; *G*: Gibbon's long bone; *T*: Terrestrial mammal's long bone; *P*: Pterodactyl's wing bone. (*Left*) The contours represent safety factors. (*Right*) The contours represent probability of fracture.

In terrestrial mammals the cost of growing and using a bone will be about the same as in gibbons, as will the cost of failure, which will be high in each case. Perhaps, for an arboreal animal, the cost of increased mass will be slightly higher than for a land-based one, if only because for animals in trees the ability of the terminal branches to bear the animals' weight without excessive bending may be important. In general the earth is more unyielding (though I have noticed my small son is better able to traverse bogs than I am). On the other hand, for arboreal animals, which cannot limp from tree to tree while their bones heal, the cost of failure of a long bone is probably higher. Terrestrial mammals and gibbons should, therefore, have about the same position on the ordinate. However, because of falls, the *variability* of loading in an arboreal animal is likely to be much higher than in one living on the ground all the time. Therefore, the gibbon should be farther to the right on the diagram, and should have a higher value for its safety factor. An important point to notice is that, although the gibbon will have a higher safety factor than a land-based mammal, it nevertheless will have a higher frequency of failure. This is partly because of the arbitrary definition of safety factor, but it also does show that some ways of life are more hazardous than others, and greater risks have to be accepted.

I have added another point, P, to Alexander's diagram. This represents the limb bones of pterosaurs. In chapter 4 I showed that they were extremely thin-walled, fragile, and pneumatized. Their thin-walledness made them liable to local buckling, and they were obviously very much minimum weight structures. Pterosaurs had large wings and were very light (Bramwell and Whitfield, 1974). This very low wing loading was essential to their mode of flight. Therefore, although the cost of breaking a limb for a pterosaur would be as high as, or even higher than, for a gibbon, the cost of making the respective bones a few percent heavier would be much greater for the pterosaur. So, the ratio $(G + U)/F$ would also be greater. On the other hand, the flight behavior of pterosaurs was so sedate that the variability of loading would have been less than for a ground-living mammal, and much less than for a gibbon. Therefore, it is reasonable to place the point for pterosaurs above, and to the left of, the terrestrial mammals. We shall never know the frequency of fractures in pterosaur bones, but if I have put the point anywhere near the right place, then although the safety factor for the limb bones would have been very low, the probability of fracture would not have been much higher than for a terrestrial mammal.

There is not a great deal of information about the magnitude of safety factors found in nature. The difficulty is not that of finding out the strength of bone, but rather of determining the stresses imposed on bones during locomotion. There are two ways of approaching this problem. One is to calculate what stresses should be present in the bone, by making use of force plate or other data external to the bone. Another, described later, is to measure directly the strains in the bones during locomotion.

Alexander (1977a) investigates the stresses that may be found in a 70 kg antelope when it is running flat out. He is able to talk of "an antelope" because he makes clear in this paper that various antelopes that he investigated showed such precise allometric relationships in all the relevant variables that it is not necessary to tie the discussion down to a particular individual or even species.

The argument below is a simplified version of Alexander's.

1. The sum of the vertical forces on the four feet must, over time, average the weight of the animal, Mg.

2. Because the forefeet are nearer the center of gravity than are the hind feet, they must bear more of the weight. Other evidence shows the ratio to be about 3 to 2. The force on each forefoot is 0.3Mg, that on each hind foot 0.2Mg.

3. The feet can support weight only while they are on the ground. Observations using film show that the duty factor (the proportion of time during a stride for which a foot is on the ground) for the forefeet is 0.2, and 0.18 for the hind feet.

4. The average vertical force on a hind foot is therefore time-averaged force/duty factor = 0.2Mg/0.18. The shorter the time the foot is on the ground, the greater the force it will have to exert in order to produce the time-averaged force necessary to support the animal.

5. This force will not be constant while the foot is on the ground. Geometrical considerations show that the maximum force, with which we are here concerned, is about $\pi/2$ times the average force. Therefore, the maximum force is $0.2\text{Mg} \cdot \pi/(0.18 \times 2)$. This force is exerted when the direction of the force relative to the ground is nearly vertical, and has a negligible horizontal component. The *bending* forces on the bone inclined at an angle θ to the vertical will be $F \cdot \sin \theta$ where F is the force. Film analysis shows the tibia to be inclined at an angle of 50°, so the bending force is $(0.2 \times 70 \times 9.81 \times 0.766)/(0.18 \times 2) = 9.18\text{N}$.

6. The length of the tibia and the second moment of area of the various cross sections along its length are known. From this can be calculated the greatest bending stresses. Halfway along the bone these are 140 MPa.

7. The force on the ground will also have a longitudinal component $F \cdot \cos \theta$. Also, because the leg is not collapsing around its joints, muscles must be holding the joint angles steady. These forces can fairly easily be calculated from a knowledge of the geometry of the limb. The cross-sectional areas of the bones are known at different levels. It turns out that

Table 7.2 Maximum Calculated Stresses in Mammal Bones During Strenuous Activity

Animal	*Activity*	*Mass/kg*	*Tensile Stress/MPa*	*Compressive Stress/MPa*
Dog	Long jump	36	60–80	100
Dog	High jump	36	60	80
Kangaroo	Hopping	7	60	90
Wallaby	Hopping	11	65	90
Antelope	Galloping	70	(80–150*)	
Buffalo	Galloping	500	35–95	60–115
Elephant	Running	2,500	45–70	55–85

* This is the mean of the maximum tensile and compressive stresses, and is not quite comparable with the other measures.

the longitudinal compressive stress in the tibia is 14 MPa. This must be added algebraically to the bending stresses. It will, of course, tend to reduce the tensile stresses.

8. The maximum compressive stress is calculated to be about 150 MPa, the maximum tensile stress 130 MPa.

9. The yield stresses of bovine bone are roughly 170 MPa and 280 MPa in tension and compression respectively. There is no reason to think that antelope bone will be very different.

10. According to these calculations, therefore, the actual strength of the bone material exceeds the loads imposed during fast galloping by 30% and 90% in tension and compression.

Alexander and Vernon (1975) have obtained values for the stress in the tibia of a wallaby running on a treadmill and also on a force platform. The calculations are more direct than the previous ones, because the use of the force platform allows the ground reaction force to be determined directly. The values of the greatest stresses in the bones vary from about 60–110 MPa in tension and 90–150 MPa in compression. The wallaby was running on a rather short treadmill; in the wild it could presumably hop faster. In doing so it would have imposed greater stresses on its bones. Alexander et al. (1979a) have extended this work to mammals of great size, using cine film but not, of course, force platform data. They find that the stresses in the bones during reasonably fast locomotion are about the same in large animals as in small ones. Table 7.2 shows this.

It is interesting that ordinary, though fast, locomotion induces stresses that are roughly the same in animals that differ in mass by a factor of 350. Although we have rather little comparative information, it seems that the mechanical properties of the compact bone of different animals do

Table 7.3 Safety Factors in Various Bones and Tendons

		Bone		Tendon
Animal	*Locomotion*	*Tension*	*Compression*	*Tension*
Dog	Jumping	2–3	2.8	1
Kangaroo	Hopping	3	3.2	1–2
Kangaroo	Pulling on lead	1.6	1.9	
Buffalo	Galloping	1.8–5	2.5–5	1.8–6
Elephant	Running	2.5–4	3.3–5	
Horse	Galloping	4.8	4.9	
Horse	Pacing	4.4	4.2	
Man	Weightlifting		1–1.7	
Ostrich	Running	2.5	2.6	2.6
Goose	Flying		6	
Pigeon	Flying	4–5.6		1.6
Sunfish	Feeding		8	

Source: Taken from Alexander (1981).

Note: Assumed yield strengths: Bone, tension 172 MPa, compression 284 MPa; tendon, tension 84 MPa. For details of calculations and authors, see Alexander's paper.

not differ greatly (chapter 2). Alexander (1981), using the kind of analysis discussed above, has tabulated the safety factors for bones involved in a range of strenuous activities (table 7.3). He also does the same for tendon. Of course, the reliability of the results varies greatly. For instance, analyses for the horse, by two different workers, suggest that the safety factor in pacing (in which the legs on one side are in phase, but are 90° out of phase with those on the other side) is about 4.3, while in galloping, a much more strenuous gait, there is another estimate of a safety factor of 4.8. However, the experiments of Rubin and Lanyon (1982) show that peak stresses in the radius of a horse become less when the gait changes from the trot to the canter, so the findings about the pace and the gallop may be correct. The general impression is that safety factors vary from the barely adequate to around 5 or 6. The higher values are probably the result of the animals' inability to run hard enough in the experimental conditions.

Another method of estimating safety factors is to measure the strains in the bones during locomotion. Strain gages are attached to the bones, the animals are allowed to recover, and they then run while the strains in the bone are measured. Lanyon and his co-workers have been foremost in this attempt. Table 7.4 shows the results of such tests. This table is pleasing because it shows results that are quite consistent with each other, and also with the values tabulated by Alexander.

Table 7.4 Safety Factors in Various Bones, Measured from Strain Gages

Animal	*Bone*	*Activity*	*Safety Factor*
Horse	Radius	Trotting	3.0
Horse	Tibia	Galloping	2.7
Horse	Metacarpal	Accelerating	2.9
Dog	Radius	Trotting	3.5
Dog	Tibia	Galloping	4.0
Goose	Humerus	Flying	3.0
Cockerel	Ulna	Flapping	4.0
Sheep	Femur	Trotting	3.9
Sheep	Humerus	Trotting	3.9
Sheep	Radius	Galloping	3.8
Sheep	Tibia	Trotting	4.0
Pig	Radius	Trotting	3.5
Fish	Hypural	Swimming	2.9
Macaque	Mandible	Biting	3.9

Source: Taken from Rubin and Lanyon (1982).

Note: I have increased the safety factors by a factor of 1.26 from those given in Rubin and Lanyon's paper to make the results comparable with table 7.3. For details of calculations and authors, see Rubin and Lanyon.

This fairly close relationship between load and strength must mean that the build of the bones is actually determined by the forces imposed during locomotion. This statement does not distinguish between determination over evolutionary time and determination during a lifetime. It is probable that the build of bones is roughly established by the genes of an animal, but the precise relationship between build and imposed load is established during life. There is much evidence that the build of a bone can be altered by loads imposed on it, so it is *not* the case that animals have a particular build of their bones and simply limit their own locomotory activity to what their bones are capable of bearing, though obviously elephants do not behave like month-old lambs.

It is as if the osteogenic mechanism has a means of measuring strains in the bone, and can increase or decrease the amount of bone to keep reasonable safety factors. However, consideration of some bones, such as the human cranium, that are capable of bearing loads *much* greater than those to which they are normally subjected shows that this cannot be the whole story. These possible control mechanisms are discussed in the next chapter.

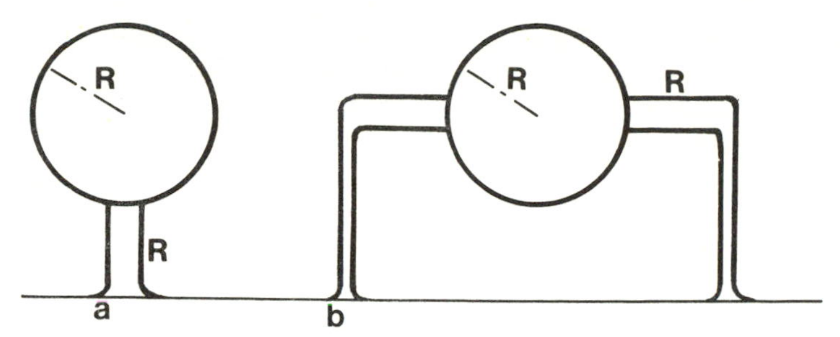

Fig. 7.4. Two hypothetical animals. (*a*) A monopod, body of radius *R*; length of leg *R*. (*b*) Animal with four legs (two not shown). Body radius *R*; length of proximal segment of legs *R*.

7.2 Size and Shape

7.2.1 Scaling

Animals that are closely related to each other, but are of different size, tend also to be of different shape. This is a well-known fact whose significance has been debated for many years. It is of importance in the understanding of bones, and we shall spend some time examining it.

It is best to start by considering a hypothetical animal. It will have a spherical body of radius R, of density D, supported on a single, solid cylindrical bony leg whose length is also R (figure 7.4a). The bone has strength $X Pa$, and the stress in the bone must not exceed this value when the animal is standing on one leg. Ignore the mass of the leg, and also suppose that it is sufficiently stiff for us to be able to ignore the possibility of buckling.

What should be the leg's cross-sectional area if it is not to be crushed? Force/Area $= X$. Force $= (4/3)\pi R^3 Dg$. Therefore, area $= 4\pi R^3 Dg/3X$, and the radius of the leg is $r = \sqrt{4R^3 Dg/3X}$. Since $(4/3)Dg$ is a constant, we can say $r \propto R^{1.5}$. Therefore, in order for the stress in the leg to remain constant, as the body increases in radius, the cross-sectional area of the leg must increase in proportion to the volume of the body, and so the radius of the leg must increase in proportion to the 1.5 power of body radius. As the body increases in size, the leg will get relatively thicker.

This well-known result is very artificial, particularly as no account is taken of bending, which is the loading mode in which bone is most in danger. So, consider the stresses in another fairly unlikely animal: a spherical animal supported by four solid cylindrical limbs whose first segment is held out horizontally. The radius of the body is R and the length of the first limb segment is also R (figure 7.4b). What should happen to the radius of the limbs if the stress at the root is not to exceed XPa?

Each limb supports 1/4 of the total mass, so the bending moment at the root, where the limb leaves the body, will be $\pi R^4 Dg/3$. The maximum stress at the root will be: Bending moment × radius/second moment of area. If the radius is r, and the stress is to be X, then $r^3 = (4/3)(R^4 Dg/X)$. Again, since $(4/3)Dg$ is a constant, $r^3 \propto R^4$; $r \propto R^{4/3}$.

One can go on for a long time thinking up more or less reasonable animals and calculating appropriate allometric relations for their skeletons. One of the main problems is that it is possible to be extremely ad hoc about the way the length of the limb element varies in relation to body length. In the examples above I have assumed that the limb length varies proportionally to body length. In fact, there is no particular reason for thinking this should be the case.

7.2.2 Elastic Similarity

McMahon (1973) proposed a relationship which he held to be generally true and which had the great virtue of simplicity. In what follows I shall spend some time producing evidence that McMahon's ideas are not generally true, but even if the reader accepts these arguments, this should not detract from the great importance of McMahon's contribution, for it is against his ideas that all later ones must be measured.

McMahon's basic idea is that organisms are designed so that the *deflections* they undergo are limited, not the stresses they bear. As we saw in chapter 4, a slender bone may collapse by Euler buckling even though its compressive strength is not even approached as it starts to collapse. Take the case of a cylindrical column. How tall can it be made before it collapses by buckling under its own weight? The collapse length is $0.79(E/D)^{1/3}d^{2/3}$, where d is the diameter of the column, E is Young's modulus, and D is the density. One can achieve greater lengths than this by, for instance, making the column taper. However, it turns out that the length is always proportional to the two-thirds power of some characteristic diameter, such as the diameter halfway up. To take a related example, suppose we have a branch of a tree that is growing in length without getting any wider. As it grows, it will droop more and more until its end is actually nearer the trunk than when it was shorter. If the proportional drooping, or deflection d/L, is to remain constant as the branch grows, it can be shown that the length should increase as diameter to the two-thirds power again or, in other words, the diameter should be proportional to the 1.5 power of the length.

McMahon calls structures in which the relationship between the deflections and the length are the same "elastically similar." It turns out to be very generally true that elastically similar structures usually scale so

Table 7.5 Allometric Exponents for Bones

Group	*Humerus*	*Ulna*	*Femur*	*Tibia*	*Metatarsal*	*Forelimb*	*Hind Limb*
Bovidae	.63	.62	.76	.65	.61	.64	.62
Cervidae	.79	.97	.79	.83	.80	.85	.76
Suidae	.54	.54	.58	.41	—	.52	—
Artiodactyla	.66	.63	.75	.65	.68	.67	.65
Perissodactyla	.60	.20	.90	.28	.53	.54	.87
All ungulates	.65	.62	.72	.60	.68	.67	.60

Note: Values of the exponent x in the equation Length $\propto$ Breadth^x.
Expected value: elastic similarity, 0.67; geometric similarity, 1.00.

that $d \propto L^{1.5}$. This is so even when we do not simply take solid cylinders, but more bonelike structures with marrow cavities, and so on. McMahon argues that it is a general property of organisms that all body segments scale in this way, so that the trunk as well as the limbs will obey $d \propto L^{1.5}$. He gives two examples where this has been shown to be the case, in studies on macaques and Holstein cattle. This idea about the proportions of animals is, if true, obviously very important; it seems that the masses of the individual parts of the body should maintain the same relative values as the animals increase or decrease in size.

There have been several attempts to test McMahon's ideas. The first was by himself; he investigated the limb bones of adult ungulates (McMahon, 1975). The results are shown in table 7.5.

The ungulates as a whole agree rather well with McMahon's predictions, although the perissodactyls (horses, tapirs, etc.) do not. McMahon attributes their lack of agreement, reasonably, to the very different locomotory patterns of the rhinoceroses and the horses. The artiodactyls are more similar in their locomotion. Furthermore, the range of size in the perissodactyls is small.

7.2.3 Geometric Similarity

If organisms of different size are simply scale models of each other, they are said to show geometric similarity. Alexander (1977a) examined the relations between body mass and various bone sizes in bovids. (McMahon had not known the masses of the animals whose skeletons he measured.) The bovids ranged in size from the dikdik (4.4 kg) to the oryx (176 kg). The fit in these cases is good. However, there are many studies in which the fit is much less good. Howell and Pylka (1977) found in the femora

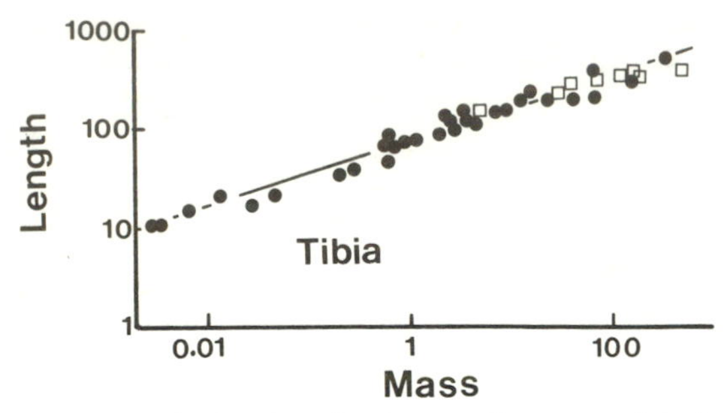

Fig. 7.5. Relationship between the length of the tibia, in millimeters, and the body mass, in kilograms, of various mammals. The fitted line for all points has a slope of 0.32. The bovids (open squares) are fitted by a slope of 0.22. (From Alexander et al., 1979b.)

of bats that elastic similarity would fit in vampires, which scuttle around on the ground a good deal, although the size range was too small for Howell and Pylka to be very confident about this. However, for the great majority of bats, which hang upside down, the femora obey geometric similarity more closely. Howell and Pylka suggest that this is because the femur, being hardly loaded in compression in life, is not loaded in the kind of way considered by McMahon in developing his hypothesis. However, Norberg (1981) in an extensive survey also considered the wings of bats, and found that more than half the species showed a reasonable fit to geometrical similarity, and that none showed a good fit to elastic similarity.

Prange, Anderson, and Rahn (1979) have examined the scaling in birds' femora and humeri. They measured birds ranging in mass from hummingbirds (0.0036 kg) to the swan (10.8 kg). Their results do not confirm McMahon's hypothesis. In fact, it is very difficult to find any fairly simple hypothesis that will account for the humerus results, which show the bone getting relatively *thinner* as it gets longer. Prange and his colleagues are at a loss to explain these findings.

Later, Alexander and his co-workers made a large survey, the results of which are impossible to reconcile with elastic similarity (Alexander et al., 1979b). McMahon had by chance chosen the Bovidae (because they show a good range of size and have a reasonably uniform mode of locomotion) as a subject for his detailed measurements. Alexander et al. measured homologous bones from insectivores, primates, lagomorphs, rodents, fissipede carnivores, an African elephant, and various artiodactyls. The overall equation for all species, and indeed the equations within most of the orders, have exponents close to those expected if the animals were obeying geometric similarity (figure 7.5, table 7.6).

Of course, the mammals being considered here have an enormous range of body mass, about six orders of magnitude, from the pygmy shrew, *Sorex*

Table 7.6 Allometric Exponents for Bone Dimensions for Running Mammals and Some Other Animals

Group	Range of Masses (kg)	Bone Length	Bone Diameter	x in: $L \propto D^x$	Skeleton Mass
Geometric similarity		.33	.33	1	1.00
Elastic similarity		.25	.38	.67	1.00
Insectivores	.003–.6	.38	.39	.97	
Primates	.3–20	.31	.38	.82	1.05
Primates	.6–60	.34	.39	.87	
Rodents	.01–2	.33	.40	.82	
Carnivores	.6–150	.36	.40	.90	
Cats	2–130				.99
Bovidae	4–500	.26	.34	.76	
All mammals	.02–7,000				1.13
All mammals	.006–7,000				1.09
All mammals	.003–2,500	.35	.36	.97	
Running birds (leg bones)	.1–40	.36	.40	.90	1.01
All birds (leg bones)	.003–80	.36			1.07
Spiders	.00005–.001				1.13

minutus, 2.9 g to the African elephant, *Loxodonta africana*, 2.5 tonnes. This implies that the modes of locomotion are very different, and so it could be argued that the allometric relations proposed by McMahon could not be expected to hold. In fact, the exponents for bone diameter agree better with elastic similarity than with geometric similarity. However, the main thrust of the idea of elastic similarity is that bones (and segments of the body as a whole) should get relatively much stockier as body mass increases. In particular, the length of bones should scale to the 2/3 power of bone diameter. Geometric similarity, of course, should make the exponent 1. Table 7.6 shows that although bone diameter scales to body mass with an exponent of about 0.38, length scales to diameter to the 0.82 to 0.97 power (the bovids excepted, as usual). These findings mean that the predictive and explanatory power of McMahon's hypothesis is markedly reduced. Another problem remains, however: it does seem strange that geometric similarity is adhered to, because this should imply that larger bones will have larger stresses imposed on them. However, the analysis of Alexander et al. (1979b) shows this is not so. There is some general thickening of the skeleton with body size, and this is shown in the exponent for skeletal mass, which is slightly more than unity; for the mammals it is 1.13. Although this is not a large exponent, the range of body masses is great, and the skeleton is 4% of the mass of the shrew and 14% of the mass of the elephant. Rubin and Lanyon have emphasized the role that behavioral adaptations may play; larger animals may, by simply behaving more sedately, not expose their bones to such large forces, relatively, as do small animals (Lanyon and Rubin, 1983; Rubin and Lanyon, 1982).

7.3 Conclusion

This chapter has spent almost as much time considering *how* we might approach the adaptiveness of the sizes and shapes of bones as actually doing so. This is not surprising, because the ways of looking at bone I have talked about are fairly new, and the experimental evidence is skimpy. Alexander's theoretical, observational, and experimental work has provided us with most of what we know. McMahon's concept of elastic similarity seemed, for a few years, to be the basis for a great illumination, but unfortunately too many counterexamples have appeared for it to be now any more than a starting point for speculation.

It seems clear that vertebrate skeletons are designed in some way to keep the safety factors within bounds. In this chapter I have discussed the relative importance of strength, and of weight, which is an inevitable concomitant of strength, in animal skeletons. This is the kind of trade off between costs and benefits that selection should optimize over evolutionary time. In the next chapter we pass on to the question of how adaptations may come about during a single lifetime.

8. Construction and Reconstruction

8.1 The Need for Feedback Control

In the previous chapter I discussed the way in which the build of bones and the mechanical properties of bone material are adapted to the requirements of the animal. I did not discuss how this adaptation was brought about. Lanyon and Rubin (1983) have an interesting discussion of this topic. Bones have quite intricate shapes, yet they also have a general size and build that seems to be adapted to the loads they bear. It is well known that bed rest (Minaire et al., 1976) and weightlessness, as in space flight (Vogel et al., 1977; Morey and Baylink, 1978), lead to a reduction in bone mass and reduced growth rate of bone. The morphological changes resulting from weightlessness reduce the strength, stiffness, and energy-absorbing ability of the bones (Spengler et al., 1979). Similarly, intensive loading of a bone can result in an increase in mass. The arm bones of professional or obsessive tennis players are larger on the racket-holding side than on the other (Jones et al., 1977), and middle-aged runners have more mineral in their bones than do sedentary people (Dalén and Olsson, 1974).

It is inconceivable, on the other hand, that the shapes of bones could result *merely* from mechanical adaptation to loads placed on them. This was shown to be the case by many experiments starting in the 1920's. For instance, Murray and Huxley showed in 1925 that a small fragment of the limb bud of a chick embryo, grafted into the chorioallantois of another, older embryo, developed into a recognizable femur with a head even though there was no pelvis with which it could articulate. Niven (1933) showed in the chick that mesenchyme that would normally develop into a patella would develop into a recognizable patella even in vitro. These and many other very interesting experiments are described by Murray (1936) in a book that is a classic of clear scientific writing.

Bones, therefore, develop at least partly without reference to the loading system. However, their final build is dependent in some way on the mechanical environment in which they find themselves, either during development or in maturity. I shall barely discuss the intrinsic, genetically determined, development of bone; the subject is too vast. Nevertheless, be clear that the form of bones, lying latent in the genes, is the result of

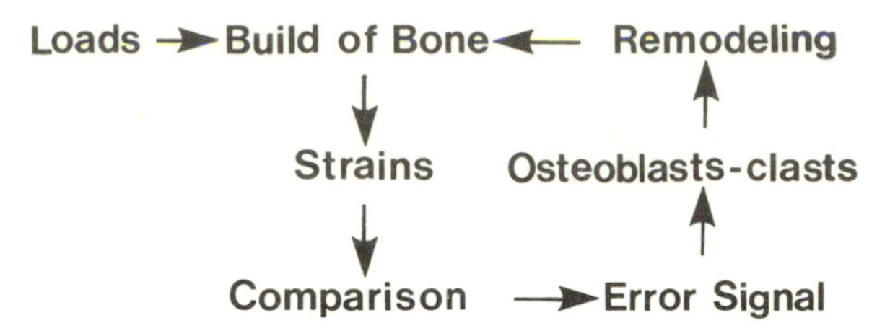

Fig. 8.1. The kind of remodeling system that must operate in bone. The strains come from the interaction of the loads applied to the bone and its build. The error signal stimulates bone-remodeling cells, and their action results in the remodeling of the bone, which changes its build and, in turn, the strains produced by loads.

natural selection acting in the past on mechanically functioning skeletons. The question I shall be concerned with is: how is the shape and size of bones affected by the forces acting on them during life? Unfortunately, it will become clear that there are many many gaps in our understanding.

One may diagram the *kind* of thing that must happen (figure 8.1). Loads are applied to a bone. This results in strains in the bone, which are measured in some way and compared with some allowable or desirable strain. (I use "strain" rather than "stress" because, in general, stresses are not directly measurable.) If the actual strains deviate from the desirable strain, either at all or by some threshold amount, bone is added to surfaces or removed from them. This reconstruction alters the build of the bone so that the same forces will now result in different and, if the reconstruction has been correct, more acceptable strains in the bone. Although there is argument about this (Frost, 1973), it is probable that bone will, sooner or later, actually be removed unless information is given to the remodeling system making it adaptive for it to remain in place or to be increased in amount.

Only two experiments, as far as I know, have been performed in which the strain in the living bone has been monitored before and after remodeling. Both experiments were by Lanyon and his colleagues. Goodship, Lanyon, and McFie (1979) removed part of the ulna of pigs (figure 8.2). This removal resulted in roughly a doubling of the peak strain on the radius during walking, as measured by strain gages attached to the surface of the bone, compared with the strains on the control side. An explosive growth of the bone followed, the bone became rounded rather than elliptical in cross section, and the difference in peak strain became less. By the end of three months the strain on each side was about the same. This seems a good example of the control system involved when remodeling works properly. The other experiments (Lanyon et al., 1982) on the forelegs of sheep showed that the strain at the end of a year was somewhat lower in the radius whose ulna had been removed than in the radius on the other side.

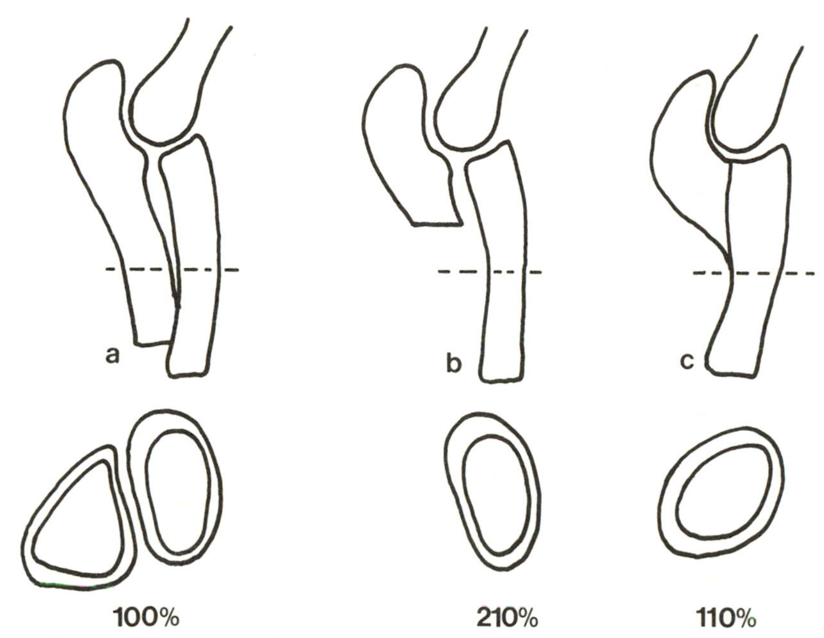

Fig. 8.2. An experiment of Goodship, Lanyon, and McFie (1979). (*a*) Normal limb of a pig. The ulna is present. The strain on the radius in walking, shown at the bottom, is taken as 100%. Interrupted lines show the plane of the section. (*b*) Ulna partially removed. Strain more than doubled. (*c*) After three months the radius has changed shape, and the strain is about the same as on the control side.

If bones are subjected to an altered loading regime, there are two ways they might respond: by altering the build, or achitecture, of the bone, or by altering the quality of the bone tissue in some way. Woo and others (1981) have looked for both these responses in bone. In one experiment they studied the effect of saving a dog's femur from stress by screwing a plate to it, so that its stiffness was artificially increased. After a year, strips of bone were taken from control animals, as well as from animals that had had attached to them either stiff plates made of cobalt chrome alloy or much less stiff plates made of graphite fibers in a polymethylmethacrylate matrix. The maximum bending moment and the area under the load-deformation curve (a measure of energy absorbed before failure) were measured for strips of standard length and width (their depth, however, was determined by the thickness of the cortex). Woo found that the control, nonplated strips were stronger and absorbed more energy than strips from the stiff-plated bones, but that the strips from the flexible-plated bones were barely affected, behaving just like the controls. However, calculation of the Young's modulus and tensile strength of the bone *material* shows these to have been unaffected. The cortical thickness had been reduced in the stiff-plated femora, and this accounted for all the change in the specimen properties observed.

The contrary experiment, of increasing the level of activity well above normal, was performed on pigs. The control animals were allowed to do what pigs like doing, which is nothing much and a lot of eating. The experimental animals were trained to trot for about 40 kilometers a week, and this was kept up for 8 months. Again, strips were tested, and marked differences found in the load the strips could bear and the energy they were able to absorb; these properties were enhanced in the exercised animals. However, there was again no difference in the Young's modulus, in the tensile strength, or in the mineral or organic constitution of these strips.

In these long-term experiments, therefore, it seems that the bone laid down, if bone mass is increased, or the bone remaining, if bone is removed, is very similar to the original bone. This is what one would expect, because there is no reason for supposing that a different type of bone material would be more suitable for dealing with the altered loading regime. Of course, if bone is laid down during remodeling, its constitution will be different initially from that of the original bone because new bone has a lower mineral content than older bone.

8.2 The Nature of the Signal

There are two main questions to be answered about adaptive remodeling. (*a*) How are the strains in the bone measured? (*b*) How, the strains having been measured, can appropriate action be taken?

The strain-related messages must be read and acted upon locally. Although bone resorption or deposition, mediated through parathyroid hormone or calcitonin, may affect some parts of the skeleton more than others, there seems to be no good evidence that these effects are localized differently according to the differing mechanical needs of various parts of the skeleton. It is very difficult, indeed, to see how blood-borne hormones could be centrally programmed to act at some times more on one bone, at another, on another This is not to say, of course, that PTH or calcitonin may not be necessary for remodeling. However, their efficacy must be modulated locally.

8.2.1 Electrical Effects

For some years now electrical effects of some kind have been most favored by investigators as being the likely mode of information transfer between bone strains and the cells in bones. There is, indeed, a great body of evidence that electrical phenomena can alter remodeling and fracture healing (Brighton, 1977; Becker, 1979). These studies show that the application of potential differences across bone, which causes a current flow, results in new bone formation taking place to a greater extent than in the absence of such potentials. It is also known that electrical potentials can be developed in bone by deforming it. The actual mechanism producing these potentials is still in some doubt, and the matter is one of intense research activity. Starkebaum, Pollack, and Korostoff (1979) give a useful set of references. There is a tendency to avoid the issue at the moment, if at all possible, by calling the potentials "stress-generated potentials" (SGP). For our purposes the mechanism producing the potentials is not of central importance, but I shall briefly describe two candidates.

Classical piezoelectricity. A crystal made of anions and cations will have no net charge, nor will the ends be polarized with respect to each other. However, many crystals have a lattice structure such that, when they are strained, there is a net separation of charges. This results in a potential difference being set up between opposite ends of the crystal. This happens because the crystals have no center of symmetry. The effect is called the direct piezoelectric effect. The reverse effect, in which a potential set up across the crystal causes strain in the crystal, also exists, but does not concern us here.

Apatite has a center of symmetry and, so, does not exhibit piezoelectricity. Dry collagen does show it, however, though wet collagen barely does, probably because when wet it has little shearing stiffness. However, it is possible that wet collagen, stiffened by mineral, could be responsible for the SGP of bone.

Streaming potentials. If a solid surface in contact with a polar liquid such as plasma has a surface charge, then the fluid will have a concentration of oppositely charged ions near the surface. Many of these ions will be firmly bound to the surface, others less so. If the fluid is made to flow with respect to the surface, the less rigidly bound ions will be made to move, and such a net movement of charge is a current; therefore, a potential difference must appear between the upstream and the downstream parts of the fluid. Eriksson (1976), in particular, argues that this is an important part of the SGP of wet bone. This view is contested by others (Bur, 1976; Korostoff, 1979).

Johnson et al. (1980), though not championing streaming potentials, have produced evidence that classical piezoelectricity cannot be responsible for all SGP in bone. They point out that bone at 98% relative humidity, the relative humidity at which many previous experiments have been carried out, is very different from fully saturated bone in its electrical properties, having a much higher resistivity. They claim the size of the

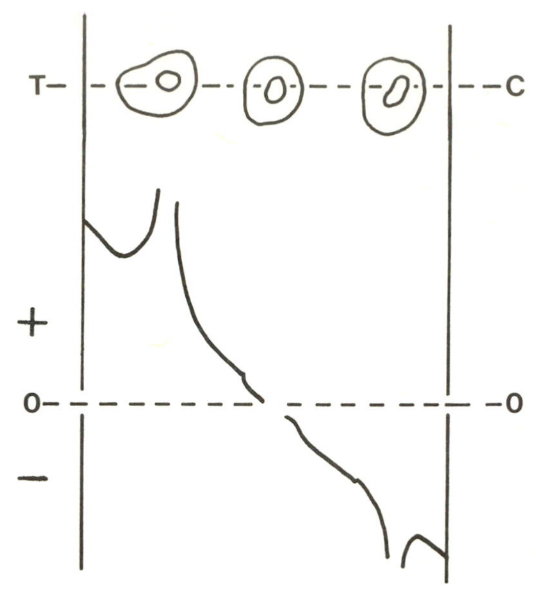

Fig. 8.3. Diagram of the electrical potential gradient in a bending specimen. *Above:* Three Haversian systems are shown in section. There is a bending moment so that one side is in tension *T*, the other in compression *C*. *Below:* Diagram showing how irregularities in the potential are associated with Haversian systems. The potential goes from positive on the tensile side to negative on the compressive side.

SGPs in wet bone is far too high, and the rate at which the SGPs decay is far too low, for them to be attributable to classical piezoelectricity. At the moment the question of the source of SGPs is unanswered.

In a uniform beam subjected to pure bending the strain increases linearly with distance from the neutral axis. The SGP should be proportional to the difference in strain between any two points. Therefore, for a uniform beam the SGP can be calculated between any two points if the potential difference between the two outermost surfaces can be measured. However, as was foreseen (McElhaney, 1967), the situation in bone is complicated because it cannot be considered to be uniform. Starkebaum, Pollack, and Korostoff (1979), using a microelectrode on human bone at 98% relative humidity, found that the potential distributions, though in general behaving in the expected way, showed great changes over very short distances. Most interestingly, these changes were associated with Haversian systems (figure 8.3). The effects of tension or compression on SGPs seem to be accentuated in Haversian systems compared with the interstitial lamellae. The effects seem to be actually confined to the Haversian system itself, as if by the cement sheath surrounding the system.

The significance of these elegant experiments is not clear. The variations in SGP were recorded from the surface of the test specimens, and it is possible that conditions deep in the bone substance may be different. Also, as mentioned above, the work of Johnson and his co-workers suggests that bone at 98% relative humidity may be mechanically similar, but electrically very dissimilar, from fully saturated bone. It may be that these local irregularities in the field, even if they occur in life, are not significant in signaling, because they are confined to the Haversian systems. The morphology of bone is such that the resting cells, lying on the surface, which can become osteoclasts or osteoblasts and which are, therefore, responsible for initiating remodeling, are always outside cement sheaths. If the anomalous effects in Haversian systems are confined to them, the signaling system should be unaffected. Nevertheless, these experiments are important in showing that SGPs are far from being the smoothly varying potentials that it is all too easy to imagine.

If a specimen of wet bone is loaded in bending, the potential difference that appears is transient. The tension side will become positively charged with respect to the compressive side. The peak difference depends upon both the strain and the strain rate (Lanyon and Hartman, 1977) and is of the order of 5 to 10 mV. When the bone is held at a constant deformation, the potential decays quite quickly, going to near zero in about two seconds (Cochran, Pawluk, and Bassett, 1968). If the bone is unloaded, so that it returns to its original shape, a potential of equal magnitude and opposite sign appears, and decays as quickly.

8.2.2 Direct Measurement of Strain

There have been suggestions that strain could be measured more directly than through the mediation of electrical effects. For instance, Jendrucko et al. (1976), making some reasonable assumptions about Poisson's ratio for bone and the bulk modulus of water, showed that the hydrostatic stresses in the osteocytes could be quite high. They suggest that these transient high pressures may in some way be a stimulus to remodeling. Bassett (1976) pointed out that remodeling was mediated through cells on free surfaces, not by the osteocytes. This is not to say that hydrostatic stresses are not important in the life of osteocytes (indeed, large tensile strains in the bone might cause the cell contents to boil), but that it is unlikely that they are the signal for remodeling. If they are, the enclosed cells must signal to the cells on the surface, which do the remodeling.

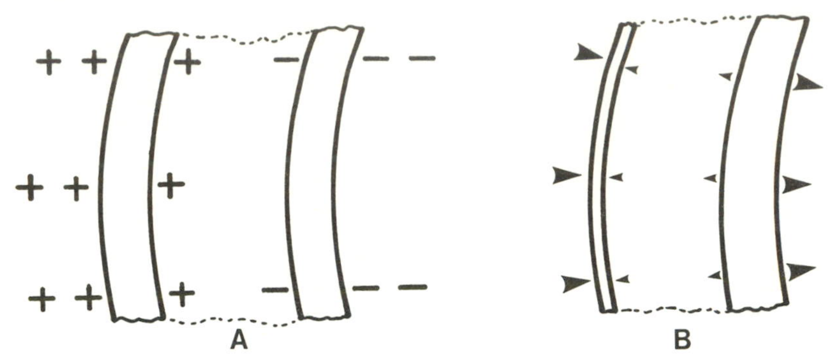

Fig. 8.4. The consequences of Bassett's hypothesis. (*A*) The strains in an anatomically badly bent, hollow bone, loaded in compression. The strains are progressively less positive from the convex side to the concave one. (*B*) The arrows show the direction of bone remodeling, which will be more intense on the periosteal side than on the endosteal surface. The resulting asymmetrical thickness of the cortical walls is shown.

8.3 How Does Bone Respond to the Signal?

Suppose, then, that bone cells can in some way determine the local state of strain. How are they to react adaptively to this information? Studies on bone remodeling have been bedeviled by Wolff's law. Wolff (1892) stated (to paraphrase him) that bony structures remodeled so as to fit them to their function. He wrote a great deal more, of course. The unfortunate thing is that, for many workers, it seems only necessary to show that bone is adapting, invoke Wolff's law, and depart, conscious of a day's work well done. No thought is given as to how the bone remodels in an adaptive fashion.

However, in more recent times people have tried to produce testable models of the mechanisms of adaptive remodeling. Some years ago Bassett (1965, 1971) suggested that bone cells lay down bone in regions of compressive strain, and erode it from regions undergoing tensile strain. As it stands this cannot be correct, because it would lead to a hollow long bone that was anatomically bent and loaded in compression longitudinally becoming thinner-walled on one side of the marrow cavity and thicker-walled on the other (figure 8.4). Nevertheless, this model has been taken as a basis for experimentation and thought.

However, in 1964 Frost had produced a more satisfactory model. He suggested that bone cells responded to *changes* in *curvature* of surfaces. If, on the application of load, a surface became more concave, bone would be laid down on it. If it became more convex, it would be eroded. Frost suggested a number of additional features for his model, but we need not

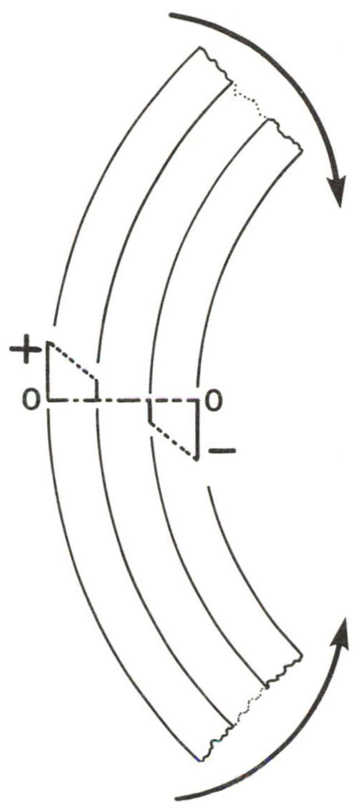

Fig. 8.5. Section of a hollow bone loaded in bending (deformations are grossly exaggerated). The strain in the section *O–O* is shown by the dotted line. The bone cells could measure the *change* of strain with depth.

consider them at the moment. This model does get over the major geometrical problem of Bassett's model, but it has another difficulty. In 1968 I pointed out that it would be extraordinarily difficult for cells to measure changes of curvature of surfaces. However, with the aid of canaliculi penetrating into the bone substance, it would probably be rather easy for them to measure changes of strain with depth (Currey, 1968). If this were so, a bone cell should lay down bone on its local surface if, on application of load, the strains deep to it became more tensile with depth. It should remove bone if the strains became less tensile with depth (figure 8.5). Unfortunately, the model of Frost would work only if the bone were never loaded in net tension—that is to say, if it is loaded in tension as well as being bent or twisted. There seems to be no signaling mechanism, relying solely on local information, that can give the right type of reconstruction if it is possible for bone to be loaded both in net tension and in net compression. In one mode or the other the reconstruction would go in the wrong way, making the situation worse, not better.

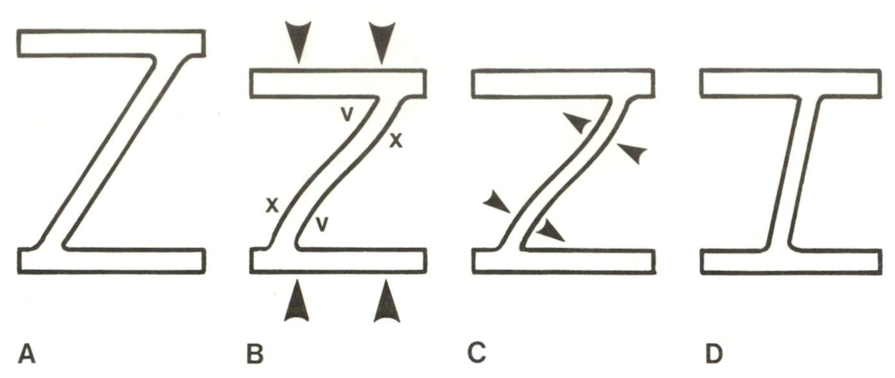

Fig. 8.6. Remodeling of cancellous bone. (*A*) The original shape: two sheets joined by a diagonal strut. (*B*) A load (large arrows) is applied. In the strut some surfaces become more convex (*x*), others more concave (*v*). (*C*) The direction of remodeling according to Frost's hypothesis. (*D*) The shape after some remodeling.

I proposed that there was some integrating mechanism, possibly nervous, that indicated to the local cells the sign of the overall strain. This seemed rather clumsy at the time, and shortly afterward Oxnard (1971) showed, by anatomical analysis, that bony structures can very rarely be loaded in tension overall. I must admit, however, to being pleased to see that the analysis of Hayes and Snyder (1981), discussed at length in chapter 4, does show that the human patella may well be loaded in net tension *pace* Oxnard's analysis.

Frost has gone on to elaborate his ideas in a series of papers and books. His book of 1973 is a fairly comprehensive account of his theory. Many of his ideas seem somewhat farfetched. This is because Frost attempts to subsume all remodeling phenomena under his theory, a laudable aim, but almost certainly premature. Even so, Frost's ideas come from an inventive mind and clinical experience and do not deserve to be ignored in the way they have been. Recently, Frost has produced a thought-provoking and interesting model to account for adaptive growth in hyaline cartilage (Frost, 1979). I shall not consider it further here.

Frost's ideas work well in explaining remodeling in cancellous bone. In chapter 4 I discussed the fact that the trabeculae in cancellous bone are arranged more or less along the lines of principal strain. In this way bending stresses are effectively eliminated. The way in which this could come about has been analyzed by Pauwels (1973). Figure 8.6 shows the deformation occurring in a strut, set diagonally between two sheets, which are being forced toward each other. There will be changes of curvature, which could be removed by remodeling of the Frost type. A portion of cancellous bone that had spaces set at random in it would come to be regularly and effectively arranged in relation to the loads put on it (figure

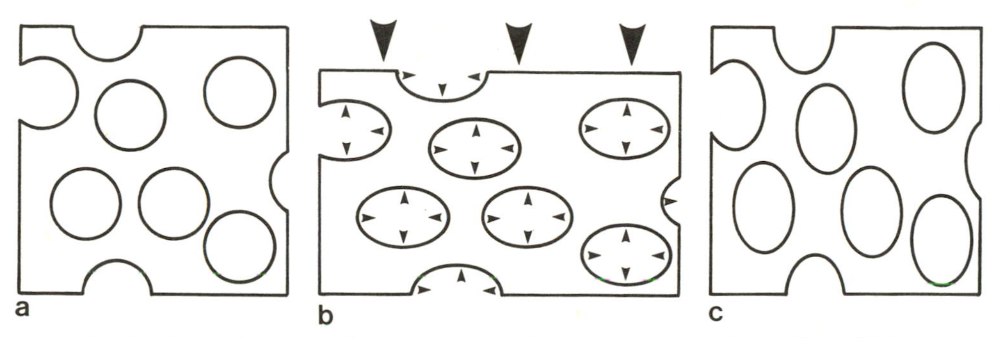

Fig. 8.7. (*a*) A portion of cancellous bone with circular spaces set more or less randomly. (*b*) On loading, the bone deforms (shown greatly exaggerated here). Surfaces that become more convex are eroded, those that become more concave are added to. The direction of reconstruction shown by small arrowheads. (*c*) After reconstruction, the bone has a better shape to resist the loads.

8.7). Indeed, as discussed in chapter 4, it looks as if such a fairly simple method of control produces structures that are good approximations to structures of minimum weight.

One situation in which Frost's model would produce useless remodeling is when a bone is loaded in bending (figure 8.8). An originally straight bone would remodel into an arch, but this would leave the stresses almost unchanged. However, this is a small objection because it must be very unusual for a bone to be loaded only in bending. Almost always there will be an axial compressive component as well. The axial component will deflect the bone, and adaptive remodeling will act to cancel out the effect.

Frost is extremely ingenious in his arguments and tries to explain the shapes of actual bones in terms of his model. Take, for example, his treatment of vertebral centra. He produces the following morphological facts. The centra of young people are only slightly inwaisted (figure 8.9) and have a fairly dense mass of trabeculae, whereas the centra of aged people are more obviously inwaisted and have a much sparser mass of trabeculae. He explains it thus. When a centrum is compressed, it will bulge sideways because the trabeculae inside bulge on being compressed. The sideways bulging produces an increase in convexity on the outer surface of the compact bone, causing that surface to be eroded, and an increase of concavity on the inner surface, causing that surface to be added to. The resultant inwaisting will continue until the tendency of the trabeculae to bulge the cortex outward will just balance the tendency of the longitudinal compression to make the cortex bend inward (figure 8.9). When this happens, there will be no change in curvature of the cortex, the cortex will not be exposed to bending strain, and remodeling will stop. In older

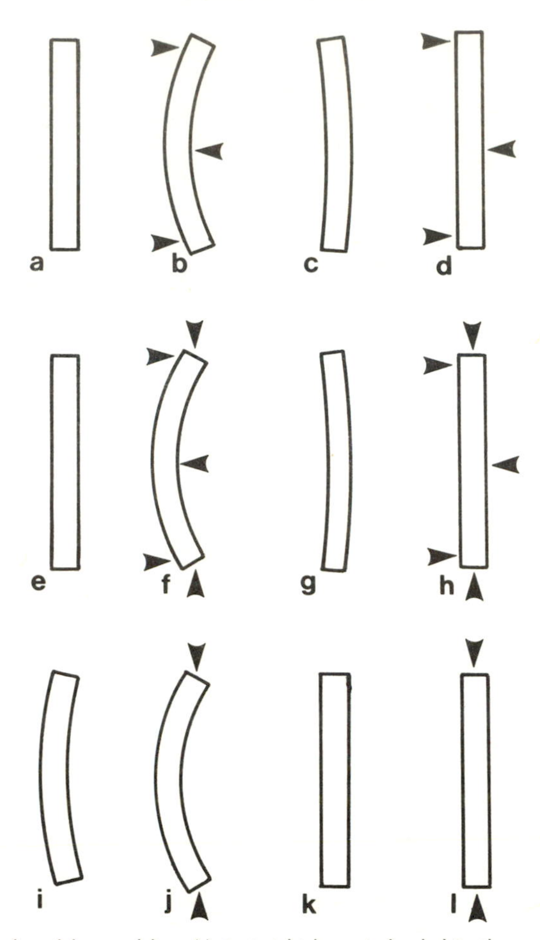

Fig. 8.8. Futile and useful remodeling. (*a*) A straight bone is loaded in three-point bending (*b*). Remodeling alters its shape (*c*), so that when it is loaded again (*d*) it becomes straight. However, this will not materially affect the strains in the bone. (*e–h*) If an axial load is imposed as well as bending, the resulting remodeling *is* adaptive because, when the bone straightens under the influence of the bending loads, the axial load will have no bending moment, as it does in (*f*). It will be advantageous to (*i*), an anatomically bent bone subject to an axial load (*j*), to remodel because it will not, in its remodeled state (*k*), be subject to bending when (*l*) loaded axially.

people there are fewer trabeculae. "As a result, more vertical shortening occurs under the same compressive loads. This increases the lateral bulging tendency, which requires more inwaisting to neutralize it." (Frost, 1973). This analysis would seem at first sight to explain inwaisting. However, as an explanation it goes so far beyond the known facts as to be misleading.

Suppose the vertebral centrum is initially straight sided. Will it bulge when compressed? It will do so if either of two things is true: the ends,

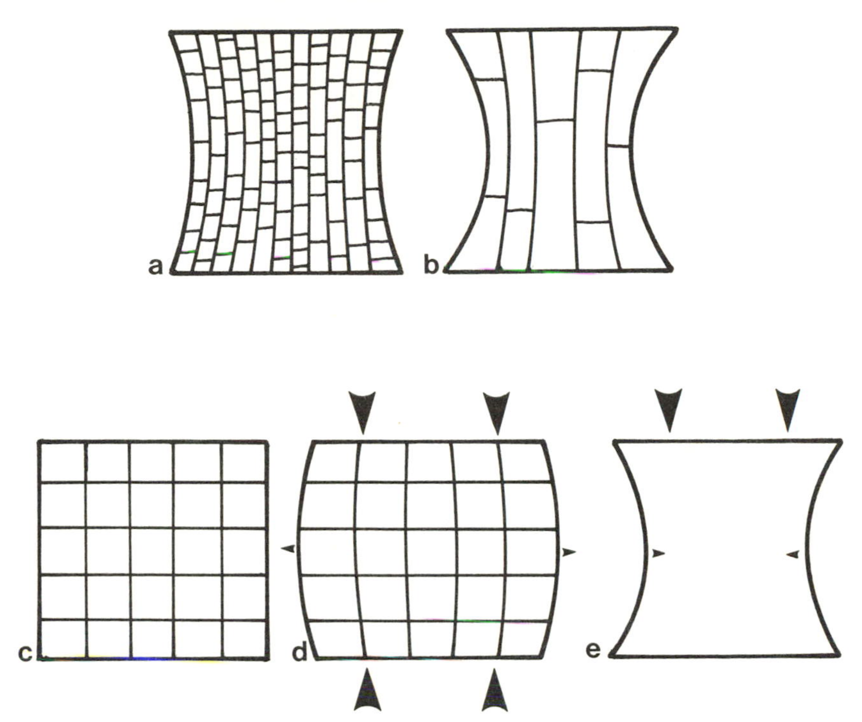

Fig. 8.9. (*a,b*) Caricatures of young and old centra. (*c–e*) Frost's idea of the function of inwaisting. A straight-sided centrum will bulge outward on being axially loaded. Inwaisting will produce a contrary tendency to bulge inward.

but not the middle, are constrained from expanding laterally (this is the usual reason for the barreling of compression specimens) or if there is a change in Poisson's ratio through the bone so that it is greater in the middle of the length of the bone than at the ends. It is not at all likely that either of these conditions is true in centra. The centrum is like a test piece compressed between the two intervertebral discs. The discs will be much more compliant in the radial direction than the subchondral bone at the ends of the centrum. Therefore, there is no reason to suppose that the ends are noticeably restricted in the radial direction. There is also no reason to think that the effective Poisson's ratio of the trabeculae is greater nearer the middle of the centrum than at the ends. Indeed, in old people's centra the lateral struts in the cancellous bone tend to disappear before the longitudinal ones. This will reduce the bulging effect of the cancellous bone, so it would seem that the appropriate remodeling of the centrum would result in it being more straight sided, rather than more inwaisted.

Such theoretical discussions are all very well, but of course when we look at what happens in real life, the situation is not very clear cut. Heřt,

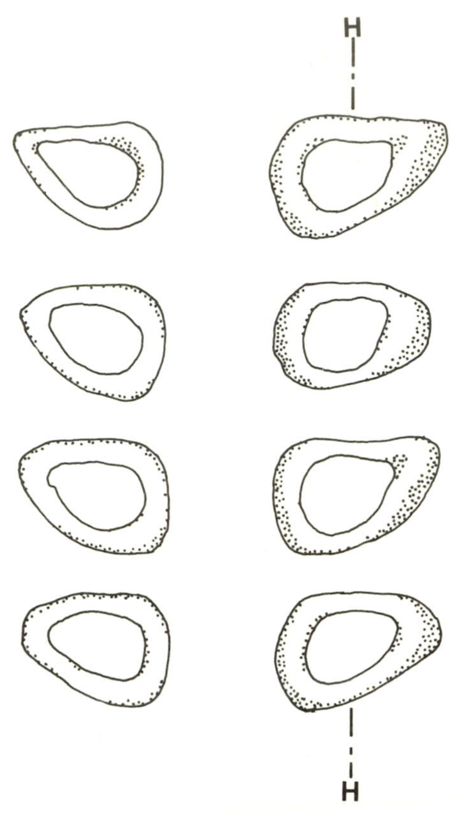

Fig. 8.10. An experiment of Lišková and Heřt. The left-hand sections are controls; the right-hand sections are the corresponding bones, from the same rabbit, that have been loaded intermittently in bending about the *H–H* axis. New bone shown by stippling.

Lišková, and Landgrot (1969) and Lišková and Heřt (1971) performed some experiments in which the tibiae of rabbits were subjected to continuous or to intermittent bending loading for many months. These were beautifully designed experiments; the force on springs producing the bending was occasionally altered so that the maximum bending stress in the bone remained effectively constant, at about 60 MPa, as the bone grew. The experiments showed that the constant stress had very little effect on the bone over many months. In the experiments involving intermittent loading, bending stresses of about the same magnitude as in the static experiments were applied at about 0.5–1 Hz, for from one to three hours a day. The results were quite clear; there was considerable extra bone deposited, both periosteally and endosteally in the experimental bone compared with the cortical bone (figure 8.10). Bone was laid down almost equally on the tension and on the compression side. This, of course, is not in accord with the predictions of the models discussed above, which are that bone should be lost from the subperiosteal surface that was in tension and from the endosteal surface that was in compression.

One of the problems of inferring the correct remodeling that should take place is that the state of loading is often not known. Wright and Yettram (1979) used finite element analysis on two structures in which the kind of remodeling occurring is well known: a malaligned femur, loaded longitudinally, and a tooth subjected to constant, though tiny, loading from an orthodontic device. The analysis of the femur produced few surprises. However, the model of the teeth was strange, because the changes in curvature of the alveolar bone induced by the cantilever loading of the tooth depended on the grain of the bone, that is to say on its elastic anisotropy. When the grain of the bone was modeled as being placed parallel to the long axis of the tooth, the changes in curvature produced were of the opposite sign from those produced in the femur, if adaptive remodeling were to result. However, if the grain of the bone were normal to the long axis of the tooth, the curvature changed sign, and became consistent with the curvature requirements of the Frost model.

If the computer model reflects reality, the situation is rather disturbing, because it suggests that relatively small changes in the mechanical characteristics of bone could profoundly affect the kind of remodeling produced. Also, note the contrast between the results of the experiments of Heřt and his colleagues, showing that constant force did not produce remodeling, and the experience of countless orthodontists who make alveolar bone melt away by the application of small, constant loads. Indeed, the contrast is so stark that one is tempted to think that the control systems in the bone around teeth are quite different from those in "ordinary" bone.

I described above the experiment of Goodship and colleagues, which showed remodeling reducing strain to acceptable levels. In osteoporosis this does not happen. Osteoporotic vertebrae, in particular, can collapse "spontaneously" when the loading is little, if at all, more than that imposed in day-to-day activities. It would be possible to attribute this lack of adaptive balance between loads and bone erosion to a general decay associated with aging. But bone erosion is an active process, requiring the activity of osteoclasts. It is possible that the feedback system is deranged in some way?

Because the SGP declines quite quickly after the bone is loaded, the rate of strain is important in determining the magnitude of the potential evoked. Lanyon and Hartman (1977) investigated the potential produced in the radii of sheep during normal locomotion. Measuring both the strains and the potentials at the same time, they found, as might be expected, that the maximum strain in the bones increased with speed of locomotion. However, they also found that the size of the stress-generated potentials increased much more rapidly than the strains themselves with speed. At a brisk trot the strains were obvious, as were the SGPs. However, when the sheep walked slowly, although the strains could be clearly recorded, the SGPs were "small and irregular, with no recognisable strain-linked patterns." In other words, strain *rates* must be high for SGPs to reach reasonably high values.

Lanyon and Hartman suggest that, as activity in old people decreases, the strain rates imposed on the bones fall to a level at which the resulting SGPs are small and incapable of initiating bone deposition. If, as seems to be the case, bone is removed unless it is actively protected, this level of SGP will result in bone being removed even though the stresses and the strains in the bone become higher and higher as a result. O'Connor et al. (1981) have shown that bone remodeling is indeed much more sensitive to strain *rate* than to maximum strain.

The general thrust of the work of people like Lanyon is that bones are adapted to the loads exerted on them, so that the strains they experience are brought to some reasonable level—in fact, that they have some safety factor. If this is accepted, the subtlety of the process becomes apparent, because different bones in the same skeleton will need to have different strains as the "desired" level. This is because the variability of loading is different for different bones. Consider the human skeleton. If the control system in the tibia produces a strain of 0.002 during day-to-day locomotion, a sensible safety factor may well result. But if the bones in the vault of the skull were remodeled so that the day-to-day strains were 0.002, what would the effect be? The top of the human skull, which is only trivially loaded by muscle action, would have to be very thin indeed if normal activities were to produce strains as great as 0.002. But a very thin skull would be useless against the occasional sharp blow to which the skull is subjected. In fact, therefore, the vault bones must usually experience extremely small strains. One can suppose either that the control system in some bones of the skull has a different set point from that in other bones, or that the bones here do not have a mechanically controlled structure, but instead merely grow to a particular thickness.

The auditory ossicles are subjected to extremely small forces, and are correspondingly lightly stressed. Yet the control system must ensure that they do not melt away through disuse osteoporosis.

The case of deer antlers is even more striking, because while the antlers are growing they are covered by a periosteal "velvet." This is richly supplied with blood, and the deer seem to do their best to prevent the antler from banging into things. When growth has finished the velvet is shed, and no more growth is possible. Only then are the antlers loaded in earnest, in fights. The whole growth process seems especially designed to prevent any adaptive remodeling taking place, yet deer can fight fairly successfully, and their antlers do not break all the time, nor yet are they overdesigned; they seem in fact to have a sensible safety factor (chapter 7).

We are left with the conclusion that there must be some way of modifying, or even overriding, the control system in particular bones and even in different parts of the same bones. The system is subtle, indeed.

Another aspect of this subtlety concerns bones that are liable to fail by buckling. If a bone is long and slender, it may collapse suddenly by Euler buckling (chapter 4). The problem for a signaling system designed to prevent such buckling is that the strains at any point in the bone give little information about whether the bone is near collapse. In a column loaded very nearly along its long axis (there is bound to be *some* deviation of the line of action of the force from the long axis), the local stress may become large only just before failure occurs. If, however, more bending is imposed on such a slender bone, two things happen: (*a*) the load that will cause collapse decreases, but (*b*) the greatest stresses in the bone give a better warning of the likelihood of collapse. Lanyon and Rubin (1983) suggest that some bones may be loaded in bending when they could be loaded in direct compression. This would seem to be bad skeletal design. They suggest that such bending may increase "the perfusion of the tissues" or maintain a "healthy flux of strain generated potentials." However, it may be that imposing apparently gratuitous bending on slender bones may not directly help the bones resist the loading, but may help the remodeling system to respond more effectively by giving a greater warning of collapse.

Another problem for local signaling systems is to produce useful remodeling when *deformation* of the bone must be restrained in some way. If, for instance, a bone is loaded as a cantilever, then the deformation of the free end depends much more on the strains near the fixed end than on those near the free end (figure 6.9). It would, therefore, be adaptive to have quite low strains near the fixed end, while the strains near the load could be higher without affecting the total deformation greatly. For the remodeling system to work properly in such a bone would require the local signal producers to have information about the loading system as a whole. It is very difficult to see how this could be effected.

8.4 The Precision of Response

It is possible to quantify the precision with which adaptive growth and remodeling is carried out by examining the actual asymmetry of bones that should be similar. The argument goes like this: We know that paired bones can quite readily become asymmetrical if their habitual use differs.

Table 8.1 The Values of *D*, and Their Standard Deviations, for Various Bones of Gulls and Hens

	Bone	Gulls N	Gulls Mean of D	Gulls S.D.	Hens N	Hens Mean of D	Hens S.D.
Load	Humerus	12	−1.4	2.1	10	−1.3	10.8
	Radius	12	0.9	3.9	12	2.3	7.6
	Ulna	12	1.5	3.2	12	−1.0	5.0
	Tibia	11	−0.1	3.9	11	−1.1	5.3
	All	47	0.24	3.9	45	−0.23	7.3
Work	Humerus	12	1.5	5.8	10	−6.1	33.0
	Radius	12	−1.6	8.0	12	3.9	18.4
	Ulna	12	−0.6	7.1	11	−0.2	7.7
	Tibia	11	−4.4	8.1	11	−1.8	18.7
	All	47	−1.20	7.4	44	−0.84	20.6

Note: Maximum load borne, and work to fracture.

This was shown by the example of the passionate tennis players. Therefore, paired bones are not constrained to be symmetrical by any inherent genetical effect. Paired bones will tend to be symmetrical because the construction and reconstruction system in each will be responding to the same kind of loading. Examining the asymmetry of paired bones will give, therefore, a maximum value for the *lack* of precision of control. It is a maximum value because some of the asymmetry will be the result of different loading histories on the two sides.

Recently I, with Al Jayes and Neill Alexander of Leeds, have examined the asymmetry between paired limb bones of the lesser black-backed gull, *Larus fuscus*. These wild birds (shot on a city garbage dump) depend a great deal on their flying ability, and there is no reason to think there should be any side-to-side asymmetry in their locomotion.

For each pair of bones we calculated the value $D = 200(L - R)/(L + R)$, where R and L are the values for the property being measured for the right and left bones respectively. This value essentially gives the difference between the two sides expressed as a percentage of the mean. We tested the humerus, radius, ulna, and tibia, and the properties we examined were the bending load borne in three-point bending and the work shown under the load-deformation trace in this three-point bending. The results are shown in table 8.1.

The *mean* value for D is always very low, the greatest value being 1.2%. This shows that the gulls are not right or left handed or footed. Of more interest are the figures in brackets. These are the standard deviations of

Table 8.2 Standard Deviations of *D* for Various Bones

Animal	*Bone*	*Test*	*S.D. of D for Load or Torque*	*Work*	*Reference*
Man	Femur	Bending	6.2	40.5	1
Dog	Fibula	Torsion	16.2	36.6	2
Dog	Tibia	Torsion	4.2	—	3
Rabbit	Humerus	Torsion	34.2	63.0	4
Rabbit	Tibiofibula	Torsion	9.4	17.8	5
Rabbit	Radius	Bending	5.6	—	6
Gull	Various	Bending	7.3	20.6	7
Hen	Various	Bending	3.9	7.4	7

Sources: Mather, 1967; 2. Puhl et al., 1972; 3. Strömberg and Dalén, 1976; 4. White et al, 1974; 5. Paavolainen, 1978; 6. Henry et al., 1968; 7. Currey, Jayes, and Alexander.

D, and show the average asymmetry of the bones, without respect to left and right. The statistical distribution of *D* is very close to normal, so the standard deviation is an appropriate parameter to discuss. The overall standard deviation for load is only 3.90%. This implies that in only one pair in twenty will the two sides differ by more than 8%. Similarly for work, only one pair in twenty will differ by more than 14%. These are remarkably low values, and show that for an animal such as a gull the control system determining the build of the bones, and through that their mechanical properties, does indeed exert a very precise control. These figures are, of course, upper values for actual asymmetry, because inevitably the testing procedure introduces some random effects, which must tend, on average, to increase the apparent differences between the two sides.

We also examined battery hens, and one can see from table. 8.1 that in these birds, which are reared in cramped conditions, and on which selection for skeletal symmetry must be low, the difference between members of paired bones is greater. Table 8.2 shows results calculated from the work of other authors. Some of these values are much higher than ours, and probably to some extent represent greater experimental error. Anyhow, our results for the gull (the only undomesticated animal tested, on which natural selection is likely to be acting most rigorously) show what the control system is capable of in an animal with strong selection for symmetry.

At the other extreme of selection one can instance the Sirenia, the sea cows. These are mammals that live in the sea all their lives, and swim around very slowly. As I discussed in chapter 7, in such animals there must be little selection for saving weight in the skeleton which, indeed, through pachyostosis, acts as ballast to counteract the buoyant effect of the lungs. As a result, there would also be little selection for symmetry. Petit (1955) mentions that the asymmetry of the sirenian skeleton is the most striking thing about it.

8.5 The Function of Internal Remodeling

So far in this chapter I have discussed the erosion of bone from, and its deposition onto, the endosteal and periosteal surfaces. I have not discussed internal remodeling, that is, the formation of new Haversian systems. In chapter 2 I mentioned the possible adaptive reasons for the production, or presence, of Haversian systems. Haversian systems seem to have no straightforward mechanical advantage over primary bone. It is possible, however, that they could reorient the grain of the bone if altered loading systems make the present grain inappropriate; they could repair micro-cracks; they could remove dead bone; and they could have a function in calcium or phosphorus metabolism.

8.5.1 Changing the Grain

It is true that, where the imposed forces change in direction in relation to the grain of the bone, Haversian remodeling can alter the grain of the bone adaptively. This is seen under large muscle insertions and during fracture repair (Vasciaveo and Bartoli, 1961). Enlow (1975) argues persuasively that the Haversian remodeling seen under muscle insertions results in muscles having firm attachments to the bone even when the muscle insertion is migrating during growth, and also during erosion of the bone surface when the shape of the bone is being altered. Compact coarse-cancellous bone is often badly oriented, and replacing it with well-positioned Haversian systems might be mechanically advantageous. However, the usual orientation of Haversian systems makes it certain that improving the grain of bone is not their only function. Quite simply, many of the Haversian systems developing in long bones do not materially alter the grain of the bone, because they have a mainly longitudinal, though very gently spiraling, course, and the blood channels they replace have the same orientation (Cohen and Harris, 1958). Furthermore, in animals such as man that undergo many generations of remodeling, there does not seem to be any general change of direction between earlier and later generations of systems. Certainly, the longitudinal orientation of the systems in long bones is appropriate for the force systems usually acting, but no more

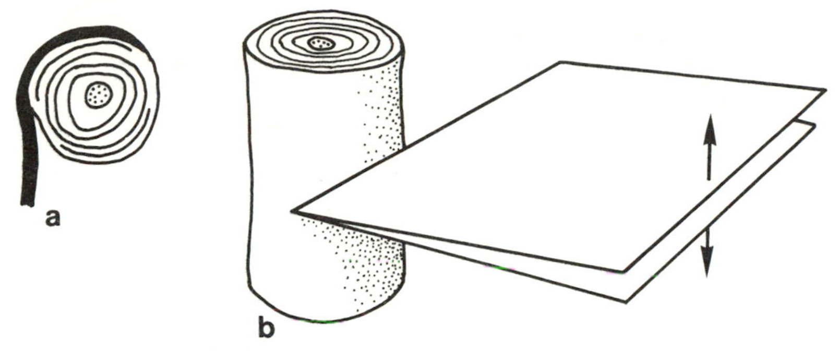

Fig. 8.11. (*a*) A microcrack, shown black, will tend to get caught by a cement sheath around a Haversian system, seen in cross section, central cavity stippled. (*b*) However, most dangerous cracks have the orientation shown here, with tensile forces (arrows) producing cracks traveling in the plane normal to the axis of the Haversian system.

appropriate than that of the primary laminae they replace. The only effect that might possibly be important occasionally is the reduction in the anisotropy of the bone. Fibrolamellar bone seems to be rather weak when loaded in tension radially to the long axis of the bone; Haversian bone is somewhat less weak. However, long bones can be loaded in this way only with great difficulty and artificiality, and it can hardly be important in life.

8.5.2 Taking out Microcracks

What of the suggestion that Haversian systems form at the end of microcracks and thereby blunt them? Martin and Burr (1982) have suggested that it could happen like this. A crack enters the vicinity of a Haversian system. The crack will tend to travel around the cement sheath, because this is a zone of weakness. The bone inside the system is now relieved of stress (figure 8.11). As a result the local bone cells will become osteoclasts and destroy the bone around them—the usual response to a marked reduction of stress in bone. When the cavity becomes filled in, a new Haversian system will have appeared. This suggestion is ingenious, but seems to run into geometrical difficulties. A dangerous crack front in a long bone is likely to travel in a radial direction and to be oriented normal to the long axis of the bone. Haversian systems will be oriented in general with their long axis normal to such a crack front, and so there will be no great tendency for the crack front to travel around the cement sheath (figure 8.11). However, this idea is important in emphasizing the role that the stresses, or lack of them, near a crack front may have in inducing reconstruction locally.

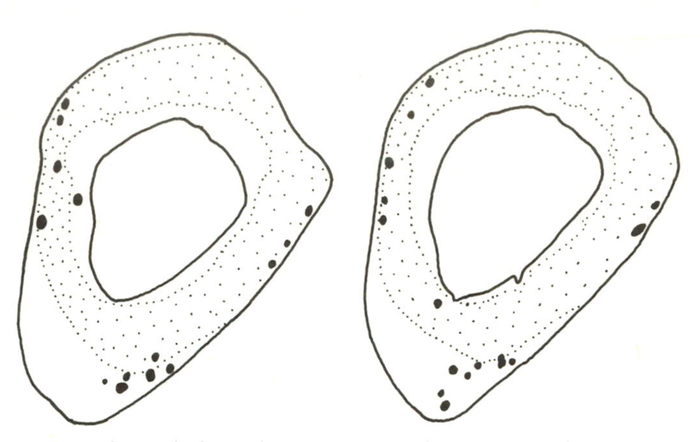

Fig. 8.12. Sections of paired ulnae of a cat. Fine stippling: Haversian bone; Black splotches: Haversian systems in process of forming.

A considerable difficulty for the idea that Haversian systems function to take out cracks is the distribution of these systems within the bone. They are often remarkably symmetrical between right and left bones. Figure 8.12 shows diagrams taken from microprojection of the midshaft of the right and left ulnae of a mature cat. (I chose one section of the pair before the other ulna had been sectioned so as not to compare sections that looked similar. These two sections are, however, typical of several samples I took from cats.) The distribution of Haversian bone in the two sections is very similar, and even some of the erosion cavities seem to be in equivalent places. It is surely stretching things too far to suppose that microdamage was occurring with such symmetry in the two bones as to necessitate such symmetrical remodeling. Marotti (1961) also showed in the bones of adult dogs that there is a close similarity in the amount of Haversian remodeling that occurred in equivalent regions of paired bones. There is no doubt (Enlow, 1963, 1975; Currey, 1968) that Haversian systems quite often form to replace dead bone. However, the topographical difficulties associated with supposing that microcracks are important also apply to any suggestion that the replacement of dead bone is an overridingly important function of Haversian systems.

It is certainly true that when bone is loaded heavily, it undergoes more remodeling than when it is not (Churches and Howlett, 1981; Heřt, Přibylová, and Lišková, 1972; Bouvier and Hylander, 1981). This, of course, would accord with the idea that the systems may be repairing microcracks. But it is the *distribution* of remodeling that argues against this hypothesis. Heřt, Přibylová, and Lišková loaded rabbit tibiae in bending about the anteroposterior plane and found Haversian remodeling in the regions that had been loaded by bending stresses, the sides, but not in the regions near the neutral axis, the front and back. So far so good. However, Heřt's picture shows that the remodeling was going on not where the stresses induced by bending would be greatest, near the subperiosteal surface, but beneath this, about 1/2 to 3/4 of the way from the endosteal to the subperiosteal surface. It is a general finding that Haversian remodeling does not go on in the regions most highly stressed; instead it seems to occur as if the reconstruction were going according to some program. The distribution of remodeling in different species of vertebrates is instructive. In general, in the lower vertebrates, that is, all vertebrates except the mammals and birds, there is rather little Haversian remodeling. The dinosaurs and some of the larger mammal-like reptiles do, however, show many Haversian systems. If one were to hold to the idea of Haversian systems functioning by eliminating microcracks, one could explain this distribution of Haversian remodeling either by supposing that reptiles and amphibians do not suffer fatigue cracks or that they are unable, for some physiological reason, to remodel. It is indeed possible that reptiles suffer less microcracking in their bones because, being generally sluggish, the penalties to them for having overdesigned bones, with great safety factors, would be less than for active mammals or birds. In the mammals and birds extensive internal remodeling seems to go with size of bone rather than anything else. Small birds have very narrow cortices, and remodeling is almost absent. Ostriches, on the other hand, have extensive Haversian bone. In mammals, the primates and carnivores and artiodactyls show much remodeling, the smaller mammals rather little. The distribution of remodeling in the artiodactyls is interesting because one finds in a cross section of, say, the femur that most of the bone is still primary fibrolamellar bone, with occasional isolated Haversian systems, but that in one part of the section the bone has been completely remodeled. Usually this remodeling is under a place where large muscles insert. This general distribution of remodeling in birds and mammals does not correspond well with what one feels is likely to be the distribution of more or less highly stressed bone.

Finally, in nearly all tetrapods that show remodeling, the remodeling that does take place seems usually to be more intense toward the marrow cavity than toward the subperiosteal part of the bone (Atkinson and Woodhead, 1973; Bouvier and Hylander, 1981). This is like the remodeling in Heřt's rabbits, with little remodeling in the most highly stressed regions. Bouvier and Hylander gave monkeys hard or soft diets and observed more Haversian systems in the mandibles of the hard-diet group. The remodeling was sparse in the subperiosteal region, however. In commenting on this they suggest that, because of the way the mandible grows, the subperiosteal bone was younger, and therefore had had less time to develop fatigue cracks and to be remodeled. Against this evidence that internal remodeling is not associated with high stresses is the rather indirect evidence I discuss in chapter 7: that the distribution of remodeling does seem to be related to the safety factors that one would expect in bones in a particular skeleton. Remodeling is less in the distal limb bones, which should have lower safety factors.

8.5.3 Removing Dead Bone

If bone cells die, the bone tissue around them can be considered as dead. It is by no means clear in what ways dead bone is less effective than living bone. However, there is some evidence that Haversian systems may occasionally form where cells have died (Currey, 1960). However, this idea of a revitalizing function runs into the same kinds of problems as does that of removing microcracks—that is, the distribution of remodeled bone within bones, between bones in the same animal, and between species does not seem like the distribution that would result from the death of cells.

The final explanation for the phenomenon is that remodeling has a physiological function, either in improving the blood supply of bone or by being part of the calcium or phosphate metabolism in the body.

8.5.4 Improving the Blood Supply

The first of this pair of explanations can be readily dismissed. The replacement of fibrolamellar bone by Haversian bone reduces the density of the blood supply (Currey, 1960; Vasciaveo and Bartoli, 1961). Furthermore, it reduces it in a particularly deleterious way: it cuts off the bone in the interstitial lamellae from the nearest blood channels by interposing the cement sheath, across which no or few canaliculi pass. There is anatomical evidence that the presence of a few Haversian systems in a small volume of bone induces further remodeling because of the death of cells in the interstitial lamellae (Currey, 1960).

8.5.5 Mineral Homeostasis

The second of the pair of explanations obviously does have some foundation. Bone is a convenient source of mineral. Laying birds often develop a mass of bony trabeculae in the marrow cavity of some of their long bones over a short length of time before egg laying starts. This bone is removed, and the calcium is made available for the eggshell. (Because the shell is made of calcium carbonate the phosphorus in the bone is not needed, and must be disposed of.) It would not be sensible to attribute other than an entirely incidental mechanical function to these trabeculae. Ruth (1953) showed that a kind of crude Haversian remodeling could be induced in rats, which do not normally have Haversian systems, by giving lactating mothers a calcium-free diet. The mothers suffer calcium stress, and cavities appear in the bones. When the mothers are later fed a calcium-rich diet, the cavities are filled in with lamellar bone.

A difficulty in accepting calcium regulation as the main function of internal remodeling is the distribution of remodeling between species. Why should young American adults, who can surely very rarely have been in negative calcium balance, have massive Haversian remodeling in their bones, whereas many small mammals and birds, which must be nutritionally near imbalance quite often, do not? Since internal remodeling has quite bad mechanical consequences, and since most animals must be in calcium balance in the long run, it seems strange, if the function of remodeling is to tide animals over short-term imbalances, that this function is not carried out by mechanically otiose bone, like the medullary spongiosa of laying birds.

In summary, there are several explanations for the phenomenon of Haversian remodeling, and there is some evidence that they all may be valid for some remodeling. However, I think that remodeling does seem so mechanically disadvantageous, according to our present knowledge, that either we are missing some important mechanical advantage or the calcium metabolism explanation has another dimension of which we are unaware.

8.6 Conclusion

The study of adaptive remodeling, both internal and external, is one of great activity at the moment. Most of the interesting experiments are difficult because they have considerable logistical problems. Imposing excessive loads is technically (and ethically) difficult, and small animals, such as mice, are not usually suitable. But the time scale of growth in larger animals is quite large, and experiments often have to be carried on for months. There is, furthermore, the difficulty that the operation to impose the loading pin, plate, or whatever is to be used to overload or underload the bone will itself tend to produce a reactive bone growth, and controls have to be extremely carefully thought out. As a result of these difficulties, it is perhaps not surprising that apparently similar experiments have quite dissimilar results. I have not emphasized these differences in this chapter because they are at the moment merely confusing, and we can perhaps hope for a consensus to appear soon. However, a reading of Churches and Howlett, Carter et al., Lanyon, and Woo, in Cowin (1981) will show that some considerable reconciliation must be brought about before such consensus emerges!

A final point must be emphasized: the interaction between the genetic endowment of the cells concerned with remodeling and the strain imposed on the bone must be complex. It must be complex because, in the mature skeleton, the kinds of stresses imposed on bones will differ from place to place. Some bones, such as vertebrae, function to bear compressive loads. Some, such as long limb bones, function through resistance to bending. Other bones, like much of the vault of the human skull, normally bear very little stress but must be strong in an accident. Some bones, like long bones, must be straight as they bear bending loads, while still others, like the lower jaw, must be complex and curved to bear bending loads.

To subsume such different relations between stress and function under any very simple law seems a futile activity.

References

Abbott, B. C., and Wilkie, D. R. 1953. The relation between velocity of shortening and the tension-length curve of skeletal muscle. *Journal of physiology* (London) 120:214–23.

Alexander, R. McN. 1975a. *Biomechanics.* London: Chapman and Hall.

———. 1975b. Evolution of integrated design. *American zoologist* 15:419–25.

———. 1977a. Allometry of the limbs of antelopes (Bovidae). *Journal of zoology* (London) 183:125–46.

———. 1977b. Terrestrial locomotion. In: *Mechanics and energetics of animal locomotion*, ed. R. McN. Alexander and G. Goldspink, pp. 168–203. London: Chapman and Hall.

———. 1981. Factors of safety in the structure of mammals. *Science progress* 67: 109–30.

———. 1982. *Locomotion of animals.* Glasgow: Blackie.

Alexander, R. McN., and Bennet-Clark, H. C. 1977. Storage of elastic strain energy in muscle and other tissues. *Nature* 265:114–17.

Alexander, R. McN., and Jayes, A. S. 1978. Optimum walking techniques for idealized animals. *Journal of zoology* (London) 186:61–81.

Alexander, R. McN., and Vernon, A. 1975. The mechanics of hopping by kangaroos (Macropodidae). *Journal of zoology* (London) 177:265–303.

Alexander, R. McN., Maloiy, G.M.O., Hunter, B., Jayes, A. S., and Nturibi, J. 1979a. Mechanical stresses in fast locomotion of buffalo (*Syncerus caffer*) and elephant (*Loxodonta africana*). *Journal of zoology* (London) 189:135–44.

Alexander, R. McN., Jayes, A. S., Maloiy, G.M.O., and Wathuta, E. M. 1979b. Allometry of the limb bones of mammals from shrews (*Sorex*) to elephant (*Loxodonta*). *Journal of zoology* (London) 189:305–14.

Andrews, E. H. 1980. Fracture. In: Vincent and Currey, 13–35.

Ascenzi, A. 1976. Physiological relationship and pathological interferences between bone tissue and marrow. In: *The Biochemistry and physiology of bone*, ed. G. H. Bourne, vol. 4, pp. 403–44. New York: Academic Press.

Ascenzi, A., Benvenuti, A., and Bonucci, E. 1982. The tensile properties of single osteonic lamellae: technical problems and preliminary results. *Journal of biomechanics* 15:29–37.

Ascenzi, A., and Bonucci, E. 1967. The tensile properties of single osteons. *Anatomical record* 158:375–86.

———. 1968. The compressive properties of single osteons. *Anatomical record* 161: 377–88.

———. 1976. Mechanical similarities between alternate osteons and cross-ply laminates. *Journal of biomechanics* 9:65–71.

Ascenzi, A., Bonucci, E., and Bocciarelli, D. S. 1967. An electron microscope study on primary periosteal bone. *Journal of ultrastructure research* 18:605–18.

Ascenzi, A., Bonucci, E., Ripamonti, A., and Roveri, N. 1978. X-ray diffraction and electron microscope study of osteons during calcification. *Calcified tissue research* 25:133–43.

Atkinson, P. J., and Woodhead, C. 1973. The development of osteoporosis. A hypothesis based on a study of human bone structure. *Clinical orthopaedics and related research* 90:217–28.

Bassett, C.A.L. 1965. Electrical effects in bone. *Scientific American* 213 (October): 18–25.

———. 1971. Biophysical principles affecting bone structure. In: *The Biochemistry and physiology of bone*, ed. G. H. Bourne, vol. 3, pp. 1–76. New York: Academic Press.

———. 1976. Comment on theoretical evidence for the generation of high pressure in bone cells by R. J. Jendrucko *et al. Journal of biomechanics* 9:485.

Becker, R. O. 1979. Significance of electrically stimulated osteogenesis—more questions than answers. *Clinical orthopaedics and related research* 141:266–74.

Behiri, J. C., and Bonfield, W. 1980. Crack velocity dependence of longitudinal fracture in bone: *Journal of materials science* 15:1841–49.

Bennet-Clark, H. C. 1975. The energetics of the jump of the locust *Schistocerca gregaria. Journal of experimental biology* 63:53–83.

Berthet-Colominas, C., Miller, A., and White, S. W. 1979. Structural study of the calcifying collagen in turkey leg tendons. *Journal of molecular biology* 134: 431–45.

Biewener, A. A. 1982. Bone strength in small mammals and bipedal birds: do safety factors change with body size? *Journal of experimental biology* 98:289–301.

Bompas-Smith, J. H. 1973. *Mechanical survival: the use of reliability data. London:* McGraw-Hill.

Bonfield, W. 1981. Mechanisms of fracture in bone. In: Cowin, 163–70.

Bonfield, W., and Behiri, J. C. 1983. Fracture mechanics of bone—evaluation by the compact tension method. In: *Biomedical engineering*, ed. S. Saha, pp. 343–47. Elmsford, N.Y.: Pergamon.

Bonfield, W., and Grynpas, M. D. 1977. Anisotropy of Young's modulus of bone. *Nature* 270:453–54.

Borden, S. 1974. Traumatic bowing of the forearm in children. *Journal of bone and joint surgery* 56-A:611–16.

Bouvier, M., and Hylander, W. L. 1981. Effect of bone strain on cortical bone structure in macaques (*Macaca mulatta*). *Journal of morphology* 167:1–12.

Boyde, A. 1972. Scanning electron microscope studies of bone. In: *The biochemistry and physiology of bone*, ed. G. H. Bourne, vol. 1, pp. 259–310. New York: Academic Press.

———. 1980. Electron microscopy of the mineralizing front. *Metabolic bone disease and related research* 2 (Supplement):69–78.

Boyde, A., and Hobdell, M. H. 1969. Scanning electron microscopy of lamellar bone. *Zeitschrift für Zellforschung* 93:213–31.

Bramblett, C. A. 1967. Pathology of the Darajani baboon. *American journal of physical anthropology* 26:331–40.

Bramwell, C. D., and Whitfield, G. R. 1974. Biomechanics of *Pteranodon. Philosophical transactions of the Royal society of London* 267-B:503–81.

Bright, R. W., Burstein, A. H., and Elmore, S. M. 1974. Epiphyseal-plate cartilage. A biomechanical and histological analysis of failure modes. *Journal of bone and joint surgery* 56-A:688–703.

Brighton, C. T., ed. 1977. Symposium on bioelectrical effects on bone and cartilage. *Clinical orthopaedics and related research* 124:2–143.

Brookes, M. 1971. *The blood supply of bone.* London: Butterworths.

Bryant, J. G. 1983. The effect of impact on the marrow pressure of long bones *in vitro. Journal of biomechanics* 16:659–65.

Buikstra, J. A. 1975. Healed fractures in *Macaca mulatta:* age, sex, and symmetry. *Folia primatologica* 23:140–48.

Bur, A. J. 1976. Measurements of the dynamic piezoelectric properties of bone as a function of temperature and humidity. *Journal of biomechanics* 9:495–507.

Burstein, A. H., Currey, J. D., Frankel, V. H., and Reilly, D. T. 1972. The ultimate properties of bone tissue: the effects of yielding. *Journal of biomechanics* 5: 34–44.

Burstein, A. H., Zika, J. M., Heiple, K. G., and Klein, L. 1975. Contribution of collagen and mineral to the elastic-plastic properties of bone. *Journal of bone and joint surgery* 57-A:956–61.

Calow, L. J., and Alexander, R. McN. 1973. A mechanical analysis of a hind leg of a frog (*Rana temporaria*). *Journal of zoology* (London) 171:293–321.

Carter, D. R., Caler, W. E., Spengler, D. M., and Frankel, V. H. 1981a. Fatigue behavior of adult cortical bone—the influence of mean strain and strain range. *Acta orthopaedica Scandinavica* 52:481–90.

———. 1981b. Uniaxial fatigue of human cortical bone. The influence of tissue physical characteristics. *Journal of biomechanics* 14:461–70.

Carter, D. R., Harris, W. H., Vasu, R., and Caler, W. E. 1981. In: Cowin, 81–92.

Carter, D. R., and Hayes, W. C. 1976a. Fatigue life of compact bone. I Effects of stress amplitude, temperature and density. *Journal of biomechanics* 9:27–34.

———. 1976b. Bone compressive strength: the influence of density and strain rate. *Science* 194:1174–76.

———. 1977a. The compressive behavior of bone as a two-phase porous structure. *Journal of bone and joint surgery* 59-A:954–62.

———. 1977b. Compact bone fatigue damage. I Residual strength and stiffness. *Journal of biomechanics* 10:325–37.

———. 1977c. Compact bone fatigue damage—microscopic examination. *Clinical orthopaedics and related research* 127:265–74.

Carter, D. R., Hayes, W. C., and Schurman, D. J. 1976. Fatigue life of compact bone. II Effects of microstructure and density. *Journal of biomechanics* 9:211–18.

Carter, D. R., Schwab, G. H., and Spengler, D. M. 1980. Tensile fracture of cancellous bone. *Acta orthopaedica Scandinavica* 51:733–41.

Carter, D. R., and Spengler, D. M. 1978. Mechanical properties and composition of cortical bone. *Clinical orthopaedics and related research* 135:192–217.

Cavagna, G. A., Saibene, F. P., and Margaria, R. 1964. Mechanical work in running. *Journal of applied physiology* 19:249–56.

Chamay, A. 1970. Mechanical and morphological aspects of experimental overload and fatigue in bone. *Journal of biomechanics* 3:263–70.

Chapman, D. I. 1981. Antler structure and function—a hypothesis. *Journal of biomechanics* 14:195–97.

Cheney, J. A., Liou, S. Y., and Wheat, J. D. 1973. Cannon-bone fracture in the thoroughbred horse. *Medical and biological engineering* 11:613–20.

Churches, A. E., and Howlett, C. R. 1981. The response of mature cortical bone to controlled time-varying loading. In: Cowin, 69–80.

Clutton-Brock, T. H., Albon, S. D., Gibson, R. M., and Guinness, F. E. 1979. The logical stag: adaptive aspects of fighting in red deer (*Cervus elaphus* L.). *Animal behaviour* 27:211–25.

Cochran, G.V.B., Pawluk, R. J., and Bassett, C.A.L. 1968. Electromechanical characteristics of bone under physiologic moisture conditions. *Clinical orthopaedics and related research* 58:249–70.

Cohen, J., and Harris, W. H. 1958. The three-dimensional anatomy of Haversian systems. *Journal of bone and joint surgery* 40-A:419–34.

Collins, R., and Kingsbury, H. B. 1982. On the mechanism of lubrication of human articular joints. In: *Osteoarthromechanics*, ed. D. N. Ghista, pp. 93–160. New York: McGraw-Hill.

Cook, J., and Gordon, J. E. 1964. A mechanism for the control of crack propagation in all-brittle systems. *Proceedings of the Royal society of London* 282A: 508–20.

Cooper, R. R., and Misol, S. 1970. Tendon and ligament insertion. A light and electron microscopic study. *Journal of bone and joint surgery* 52-A:1–20; 170.

Cowin, S. C., ed. 1981. *Mechanical properties of bone*. AMD-vol. 45. New York: American Society of Mechanical Engineers.

Currey, J. D. 1959. Differences in the tensile strength of bone of different histological types. *Journal of anatomy* 93:87–95.

———. 1960. Differences in the blood-supply of bone of different histological types. *Quarterly journal of microscopical science* 101:351–70.

———.1962. Stress concentrations in bone. *Quarterly journal of microscopical science* 103:111–33.

———. 1964a. Three analogies to explain the mechanical properties of bone. *Biorheology* 2:1–10.

———. 1964b. Metabolic starvation as a factor in bone reconstruction. *Acta anatomica* 59:77–83.

———. 1965. Anelasticity in bone and echinoderm skeletons. *Journal of experimental biology* 43:279–92.

———. 1967. The failure of exoskeletons and endoskeletons. *Journal of morphology* 123:1–16.

———. 1968. Adaptation of bones to stress. *Journal of theoretical biology* 20:91–106.

———. 1969. The mechanical consequences of variation in the mineral content of bone. *Journal of biomechanics* 2:1–11.

———. 1975. The effect of strain rate, reconstruction and mineral content on some mechanical properties of bovine bone. *Journal of biomechanics* 8:81–86.

———. 1979a. Mechanical properties of bone with greatly differing functions. *Journal of biomechanics* 12:313–19.

———. 1979b. Changes in the impact energy absorption of bone with age. *Journal of biomechanics* 12:459–69.

Currey, J. D., and Brear, K. 1974. Tensile yield in bone. *Calcified tissue research* 15:173–79.

Currey, J. D., and Butler, G. 1975. The mechanical properties of bone tissue in children. *Journal of bone and joint surgery* 57-A:810–14.

Dalén, N., and Olsson, K. E. 1974. Bone mineral content and physical activity. *Acta orthopaedica Scandinavica* 45:170–74.

Davidge, R. W. 1979. *Mechanical behavior of ceramics*. Cambridge: Cambridge University Press.

Day, W. H., Swanson, S.A.V., and Freeman, M.A.R. 1975. Contact pressures in the loaded human cadaver hip. *Journal of bone and joint surgery* 57-A:302–13.

De Ricqlès, A. 1977. Recherches paléohistologiques sur les os longs des tétrapodes VII (deuxième partie, fin). *Annales de paléontologie* 63:133–60.

———. 1979. Quelques remarques sur l'histoire évolutive des tissus squelettiques chez les vertébrés et plus particulièrement chez les tétrapodes. *Annales biologiques* 18:1–35.

Devas, M. 1975. *Stress fractures*. Edinburgh: Churchill Livingstone.

Dickerson, R. E., and Geis, I. 1969. *The structure and action of proteins*. Menlo Park: Benjamin.

Driessens, F.C.M., van Dijk, J.W.E., and Borggreven, J.M.P.M. 1978. Biological calcium phosphates and their role in the physiology of bone and dental tissues. 1. Composition and solubility of calcium phosphates. *Calcified tissue research* 26:127–37.

Dunegan, H. L., and Green, A. T. 1972. Factors affecting acoustic emission response from materials. In: *Acoustic emission*, STP 505, pp. 100–112. Philadelphia: American Society for Testing and Materials.

Elliott, D. H. 1965. Structure and function of mammalian tendon. *Biological reviews* 40:392–421.

Enlow, D. H. 1963. *Principles of bone remodeling*. Springfield: Charles C. Thomas.

———. 1969. The bone of reptiles. In: *Biology of the reptilia*, ed. C. Gans, vol. 1, pp. 45–80. New York: Academic Press.

———. 1975. *A handbook of facial growth*. Philadelphia: W. B. Saunders.

Enlow, D. H., and Brown, S. O. A comparative histological study of fossil and recent bone tissues. *Texas journal of science* part 1 (1956) 8:405–43; part 2 (1957) 9:186–214; part 3 (1958) 10:187–230.

Eriksson, C. 1976. Electrical properties of bone. In: *Biochemistry and physiology of bone*, ed. G. H. Bourne, vol. 4, pp. 329–84. New York: Academic Press.

Evans, F. G. 1973. *Mechanical properties of bone.* Springfield: Charles C. Thomas.

Evans, F. G., and Bang, S. 1967. Differences and relationships between the physical properties and the microscopic structure of human femoral, tibial and fibular cortical bone. *American journal of anatomy* 120:78–88.

Evans, F. G., and Lebow, M. 1957. Strength of human compact bone under repetitive loading. *Journal of applied physiology* 10:127–30.

Fedak, M. A., Heglund, N. C., and Taylor, C. R. 1982. Energetics and mechanics of terrestrial locomotion. 2. Kinetic energy changes of the limbs and body as a function of speed and body size in birds and mammals. *Journal of experimental biology* 79:23–40.

Frasca, P. 1981. Scanning-electron microscopy studies of "ground substance" in the cement lines, resting lines, hypercalcified rings and reversal lines of human cortical bone. *Acta anatomica* 109:114–21.

Frasca, P., Harper, R., and Katz, J. L. 1977. Collagen fiber orientations of human secondary osteons. *Acta anatomica* 98:1–13.

———. 1981. Strain and frequency dependence of shear storage modulus for human single osteons and cortical bone microsamples—size and hydration effects. *Journal of biomechanics* 14:679–90.

Freeman, M. A. 1979. *Adult articular cartilage.* London: Pitman Medical.

Frost, H. M. 1960. Presence of microscopic cracks in vivo in bone. *Henry Ford hospital bulletin* 8:25–35.

———. 1964. *The laws of bone structure.* Springfield: Charles C. Thomas.

———. 1973. *Bone modeling and skeletal modeling errors.* Springfield: Charles C. Thomas.

———. 1979. A chondral modeling theory. *Calcified tissue international* 28:181–200.

Frost, N. E., Marsh, K. J., and Pook, L. P. 1974. *Metal fatigue.* Oxford: Oxford University Press.

Galante, J., Rostoker, W., and Ray, R. D. 1970. Physical properties of trabecular bone. *Calcified tissue research* 5:236–46.

Gambaryan, P. P. 1974. *How mammals run: anatomical adaptations.* New York: Halsted Press.

Gannon, J. R. 1972. Stress fractures in the greyhound. *Australian veterinary journal* 48:244–50.

Gans, C. 1974. *Biomechanics: an approach to vertebrate biology.* Philadelphia: Lippincott.

Ghista, D. N., ed. 1982. *Osteoarthromechanics.* New York: McGraw-Hill.

Gibson, L. J., and Ashby, M. F. 1982. The mechanics of three-dimensional cellular materials. *Proceedings of the Royal society of London* 382-A:43–59.

Gillette, E. P. 1872. Des os sesamoides chez l'homme. *Journal d'anatomie et physiologie* 8:506–38.

Glimcher, M. J. 1976. Composition, structure, and organization of bone and other mineralized tissues and the mechanisms of calcification. In: *Handbook of Physiology*, ed. G. D. Aurbach. sect. 7, vol. 7, pp. 25–116. Washington: American Physiological Society.

Goodship, A. E., Lanyon, L. E., and McFie, H. 1979. Functional adaptation of bone to increased stress. *Journal of bone and joint surgery* 61-A:539–46.

Gordon, A. M., Huxley, A. F., and Julian, F. J. 1966. The variation in isometric tension with sarcomere length in vertebrate muscle fibres. *Journal of physiology* (London) 184:170–92.

Gordon, J. E. 1976. *The new science of strong materials.* London: Penguin.

Gottesman, T., and Hashin, Z. 1980. Analysis of viscoelastic behaviour of bones on the basis of microstructure. *Journal of biomechanics* 13:89–96.

Gould, S. J., and Lewontin, R. C. 1979. The spandrels of San Marco and the Panglossian paradigm: a critique of the adaptationist programme. *Proceedings of the Royal society of London* 205B:581–98.

Gregersen, H. N. 1971. Fractures of the humerus from muscular violence. *Acta orthopaedica Scandinavica* 42:506–12.

Griffith, A. A. 1920. The phenomena of rupture and flow in solids. *Philosophical transactions of the Royal society of London* 221A:163–97.

Haines, R. W. 1969. Epiphyses and sesamoids. In: *Biology of the reptilia*, ed. C. Gans, vol. 1, pp. 81–115. New York: Academic Press.

Haldane, J.B.S. 1953. Animal populations and their regulation. *New biology* 15:9–24.

Halstead, L. B. 1974. *Vertebrate hard tissues.* London: Wykeham.

Hancock, J. R., ed. 1975. *Fatigue of composite materials*, STP 569. Philadelphia: American Society for Testing and Materials.

Hancox, N. M. 1972. *Biology of bone.* Cambridge: Cambridge University Press.

Harris, B. 1980. The mechanical behaviour of composite materials. In: Vincent and Currey, 37–74.

Hasselbacher, P., ed. 1981. The biology of the joint. *Clinics in rheumatic diseases* 7: (no. 1) 1–287.

Hayes, W. C., and Snyder, B. 1981. Toward a quantitative formulation of Wolff's law in trabecular bone. In: Cowin, 43–68.

Heglund, N. C., Cavagna, G. A., and Taylor, C. R. 1982a. Energetics and mechanics of terrestrial locomotion. 3. Energy changes of the centre of mass as a function of speed and body size in birds and mammals. *Journal of experimental biology* 79:41–56.

Heglund, N. C., Fedak, M. A., Taylor, C. R., and Cavagna, G. A. 1982b. Energetics and mechanics of terrestrial locomotion. 4. Total mechanical energy changes as a function of speed and body size in birds and mammals. *Journal of experimental biology* 97:57–66.

Henry, A. N., Freeman, M.A.R., and Swanson, S.A.V. 1968. Studies of the mechanical properties of healing experimental fractures. *Proceedings of the Royal society of medicine* 61:902–6.

Heřt, J., Kučera, P., Vávra, M., and Voleník, K. 1965. Comparison of the mechanical properties of both the primary and haversian bone tissue. *Acta anatomica* 61:412–23.

Heřt, J., Lišková, M., and Landgrot, B. 1969. Influence of the long term, continuous bending on the bone. *Folia morphologica* 17:389–99.

Heřt, J., Přibylová, E., and Lišková, M. 1972. Microstructure of compact bone of rabbit tibia after intermittent loading. *Acta anatomica* 82:218–30.

Hill, A. V. 1938. The heat of shortening and the dynamic constants of muscle. *Proceedings of the Royal society of London* 126B:136–95.

Horsman, A., and Currey, J. D. 1983. Estimation of mechanical properties of the distal radius from bone mineral content and cortical width. *Clinical orthopaedics* 176:298–304.

Howell, D. J., and Pylka, J. 1977. Why bats hang upside down: a biomechanical hypothesis. *Journal of theoretical biology* 69:625–31.

Hughes, J., Paul, J. P., and Kenedi, R. M. 1970. Control and movement of the lower limbs. In: *Modern trends in biomechanics*, ed. D. C. Simpson, pp. 147–79. London: Butterworths.

Huiskes, R., and Chao, E.Y.S. 1983. A survey of finite element analysis in orthopedic biomechanics: the first decade. *Journal of biomechanics* 16:385–409.

Huiskes, R., Janssen, J. D., and Sloof, T. J. 1981. A detailed comparison of experimental and theoretical stress-analysis of a human femur. In: Cowin, 211–34.

Hukins, D.W.L. 1978. Bone stiffness explained by the liquid crystal model of the collagen fibril. *Journal of theoretical biology* 71:661–67.

Hull, D. 1981. *An introduction to composite materials.* Cambridge: Cambridge University Press.

Inglis, C. E. 1913. Stresses in a plate due to the presence of cracks and sharp corners. *Transactions of the institute of naval architects* 55:219–30.

Jackson, S. A., Cartwright, A. G., and Lewis, D. 1978. The morphology of bone mineral crystals. *Calcified tissue research* 25:217–22.

Jayes, A. S., and Alexander, R. McN. 1978. Mechanics of locomotion of dogs (*Canis familiaris*) and sheep (*Ovis aries*). *Journal of zoology* (London) 185: 289–308.

Jendrucko, R. J., Hyman, W. A., Newell, P. H., and Chakraborty, B. K. 1976. Theoretical evidence for the generation of high pressure in bone cells. *Journal of biomechanics* 9:87–91.

Jeronimides, G. 1980. Wood, one of nature's challenging composites. In: Vincent and Currey, 169–82.

Johnson, M. W., Chakkalakal, D. A., Harper, R. A., and Katz, J. L. 1980. Comparison of the electromechanical effects in wet and dry bone. *Journal of biomechanics* 13:437–42.

Jones, H. H., Priest, J. D., Hayes, W. C., Tichenor, C. C., and Nagel, D. A. 1977. Humeral hypertrophy in response to exercise. *Journal of bone and joint surgery* 59A:204–8.

Katz, J. L. 1971. Hard tissue as a composite material. 1 Bounds on the elastic behavior. *Journal of biomechanics* 4:455–73.

———. 1981. Composite material models for cortical bone. In: Cowin, 171–84.

Kempson, G. E. 1972. The tensile properties of articular cartilage and their relevance to the development of osteoarthrosis. In: *Orthopaedic surgery and traumatology.* Proceedings 12th International meeting of the society for orthopaedic surgery and traumatology, Tel Aviv pp. 44–58. Amsterdam: Excerpta Medica.

———. 1980. The mechanical properties of articular cartilage. In: *The joints and synovial fluid*, ed. L. Sokoloff, vol. 2, pp. 177–238. New York: Academic Press.

King, A. I., and Evans, F. G. 1967. Analysis of fatigue strength of human compact bone by the Weibull method. In: *Digest of the 7th international conference on medical and biological engineering*, ed. B. Jacobson, p. 514. Stockholm: Published by the organizing committee.

Knets, I. V., Krauya, U. E., and Vilks, Yu. K. 1975. Acoustic emission in human bone tissue subjected to longitudinal extension. *Mechanika polimerov* 11:685–90.

Korostoff, E., 1979. A linear piezoelectric model for characterizing stress generated potentials in bone. *Journal of biomechanics* 12:335–37.

Krenchel, H. 1964. *Fibre reinforcement*. Copenhagen: Akademisk Forlag.

Kriewall, T. J., McPherson, G. K., and Tsai, A. C. 1981. Bending properties and ash content of fetal cranial bone. *Journal of biomechanics* 14:73–79.

Lafferty, J. F. 1978. Analytical model of the fatigue characteristics of bone. *Aviation space and environmental medicine* 49:170–74.

Lafferty, J. F., and Raju, P.V.V. 1979. The influence of stress frequency on the fatigue strength of cortical bone. *Journal of biomedical engineering* 101:112–13.

Lakes, R., and Saha, S. 1979. Cement line motion in bone. *Science* 204:501–3.

Lanyon, L. E. 1974. Experimental support for the trajectorial theory of bone structure. *Journal of bone and joint surgery* 56B:160–66.

———. 1981. The measurement and biological significance of bone strain in vivo. In: Cowin, 93–105.

Lanyon, L. E., and Baggott, D. G. 1976. Mechanical function as an influence on the structure and form of bone. *Journal of bone and joint surgery* 58-B:436–43.

Lanyon, L. E., and Bourne, S. 1979. The influence of mechanical function on the development of remodeling of the tibia. *Journal of bone and joint surgery* 61-A:263–73.

Lanyon, L. E., Goodship, A. E., Pye, C. J., and MacFie, J. H. 1982. Mechanically adaptive bone remodelling. *Journal of biomechanics* 15:141–54.

Lanyon, L. E., and Hartman, W. 1977. Strain related electrical potentials recorded in vitro and in vivo. *Calcified tissue research* 22:315–27.

Lanyon, L. E., Magee, P. T., and Baggott, D. G. 1979. The relationship of functional stress and strain to the process of bone remodeling. An experimental study on the sheep radius. *Journal of biomechanics* 12:593–600.

Lanyon, L. E., and Rubin, C. T. 1983. Functional adaptation in skeletal structures. In: Hildebrand, ed. *Functional Vertebrate Morphology*. Harvard University Press.

Lawn, B. R., Hockey, B. J., and Wiederhorn, S. M. 1980. Atomically sharp cracks in brittle solids: an electron microscopy study. *Journal of materials science* 15:1207–23.

Lawn, B. R., and Wilshaw, T. R. 1975. *Fracture of brittle solids*. Cambridge: Cambridge University Press.

Lease, G. O'D., and Evans, F. G. 1959. Strength of human metatarsal bones under repetitive loading. *Journal of applied physiology* 14:49–51.

Lees, S., and Davidson, C. L. 1977. The role of collagen in the elastic properties of calcified tissues. *Journal of biomechanics* 10:473–86.

Leonard, F., Parsons, D. B., and Adams, J. P. 1980. Tensile strength of precalcified turkey tendon. *Journal of biomechanics* 13:731–32.

Lišková, M., and Heřt, J. 1971. Reaction of bone to mechanical stimuli. Part 2.

Periosteal and endosteal reaction of the tibial diaphysis in rabbit to intermittent loading. *Folia morphologica* 19:301–17.

Lovejoy, O., and Heiple, K. 1981. The analysis of fractures in skeletal populations with an example from the Libben site, Ottowa County, Ohio. *American journal of physical anthropology* 55:529–41.

Lowenstam, H. A. 1981. Minerals formed by organisms. *Science* 211:1126–31.

McBryde, A. M. 1975. Stress fractures in athletes. *Journal of sports medicine* 3:212–17.

McCutchen, C. W. 1975. Do mineral crystals stiffen bone by straightjacketing its collagen? *Journal of theoretical biology* 51:51–58.

———. 1981. Joint lubrication. Clinics in rheumatic diseases. In: *The biology of the joint*, vol. 7, no. 1, pp. 241–57. Philadelphia: W. B. Saunders.

McElhaney, J. H. 1966. Dynamic response of bone and muscle tissue. *Journal of applied physiology* 21:1231–36.

———. 1967. The charge distribution on the human femur due to load. *Journal of bone and joint surgery* 49A:1561–71.

McMahon, T. A. 1973. Size and shape in biology. *Science* 179:1201–4.

———. 1975. Allometry and biomechanics: limb bones in adult ungulates. *American naturalist* 109:547–63.

McPherson, G. K., and Kriewall, T. J. 1980a. The elastic modulus of fetal cranial bone: a first step towards an understanding of the biomechanics of fetal head molding. *Journal of biomechanics* 13:9–16.

———. 1980b. Fetal head molding: an investigation utilizing a finite element model of the fetal parietal bone. *Journal of biomechanics* 13:17–26.

Mack, R. W. 1964. Bone—a natural two-phase material. Technical memoirs of the biochemical laboratory, University of California at Berkeley.

Margaria, A. 1976. *Biomechanics and energetics of muscular exercise*. Oxford: Oxford University Press.

Marotti, G. 1961. Number and arrangement of osteons in corresponding regions of homotypic long bones. *Nature* 191:1400–01.

Maroudas, A. 1969. Studies on the formation of hyaluronic acid films. In: *Lubrication and wear in joints*, ed. V. Wright, pp. 124–33. London: Sector.

Martin, R. B., and Burr, D. B. 1982. A hypothetical mechanism for the stimulation of osteonal remodeling by fatigue damage. *Journal of biomechanics* 15: 137–39.

Mather, B. S. 1967. The symmetry of the mechanical properties of the human femur. *Journal of surgical research* 7:222–26.

Merrilees, M. J., and Flint, M. H. 1980. Ultrastructural study of tension and pressure zones in a rabbit flexor tendon. *American journal of anatomy* 157: 87–106.

Minaire, P., Meunier, P., Edourd, C., Bernard J., Coupron, P., and Bourret, J. 1976. Quantitative histological data on disuse osteoporosis. *Calcified tissue research* 17:57–73.

Minns, R. J., Bremble, G. R., and Campbell J. 1975. The geometrical properties of the human tibia. *Journal of biomechanics* 8:253–57.

Morey, E. R., and Baylink, D. J. 1978. Inhibition of bone formation during space flight. *Science* 201:1138–41.

Moss, M. L. 1961. Osteogenesis of acellular teleost fish bone. *American journal of anatomy* 108:99–109.

Murray, P.D.F. 1936. *Bones: a study of the development and structure of the vertebrate skeleton.* Cambridge: Cambridge University Press.

Murray, P.D.F., and Huxley, J. S. 1925. Self-differentiation in the grafted limb-bud of the chick. *Journal of anatomy* 59:379–84.

Muzik, K., and Wainwright, S. A. 1977. Morphology and habitat of five Fijian sea fans. *Bulletin of marine science* 27:308–37.

Nachtigall, W. 1971. *Biotechnik: Statische Konstruktionen in der Natur.* Heidelberg: Quelle und Meyer.

Netz, P., Eriksson, K., and Strömberg, L. 1980. Material reaction of diaphyseal bone under torsion. *Acta orthopaedica Scandinavica* 51:223–29.

Nichols, D., and Currey, J. D. 1968. The secretion, structure, and strength of echinoderm calcite. In: *Cell structure and its interpretation*, ed. S. M. McGee-Russell and K.F.A. Ross, pp. 251–61. London: Edward Arnold.

Niven, J.S.F. 1933. The development *in vivo* and *in vitro* of the avian patella. *Roux' Archiv* 128:480–501.

Norberg, U. M. 1981. Allometry of bat wings and legs and comparison with bird wings. *Philosophical transactions of the Royal society* 292B:359–98.

Nye, J. F. 1957. *Physical properties of crystals.* Oxford: Oxford University Press.

O'Connor, J. A., Goodship, A. E., Rubin, C. T., and Lanyon, L. E. 1981. The effect of externally applied loads on bone remodeling in the radius of the sheep. In: *Mechanical factors and the skeleton*, ed. I.A.F. Stokes, pp. 83–90. London: John Libbey.

Orava, S., Puranen, J., and Ala-Ketola, L. 1978. Stress fractures caused by physical exercise. *Acta orthopaedica Scandinavica* 49:19–27.

Ortner, D. J., and von Endt, D. W. 1971. Microscope and electron-microprobe characterization of sclerotic lamellae in human osteons. *Israel journal of medical science* 7:480–82.

Ørvig, T. 1967. Phylogeny of tooth tissues: evolution of some calcified tissues in early vertebrates. In: *Structural and chemical organization of teeth*, ed. A.E.W. Miles, vol. 1, pp. 45–110. New York: Academic Press.

Ostrom, J. H. 1966. Functional morphology and evolution of the ceratopsian dinosaurs. *Evolution* 20:290–308.

Oxnard, C. E. 1971. Tensile forces in skeletal structures. *Journal of morphology* 134:425–36.

Oxnard, C. E., and Yang, H. C. 1981. Beyond biometrics: studies of complex biological patterns. *Symposia of the zoological society of London* 46:127–67.

Paavolainen, P. 1978. Studies on mechanical strength of bone. I Torsional strength of normal rabbit tibio-fibular bone. *Acta orthopaedica Scandinavica* 49: 497–505.

Parkes, E. W. 1974. *Braced frameworks.* Oxford: Pergamon Press.

Pauwels, F. 1948. Die Bedeutung der Bauprinzipien des Stütz-und Bewegungsapparates für die Beanspruchung der Röhrenknochen. *Zeitschrift für Anatomie und Entwicklungsgeschichte* 114:129–66.

———. 1950. Die Bedeutung der Bauprinzipien der unteren Extremität für die

Beanspruchung des Beinskeletes. *Zeitschrift für Anatomie und Entwicklungsgeschichte* 114:525–38.

———. 1973. Kurzer Überblick über die mecanische Beanspruchung des Knochens und ihre Bedeutung für die funktionelle Anpassung. *Zeitschrift für Orthopadie und ihre Grenzgebiete* 111:681–705.

———. 1980. *Biomechanics of the locomotor apparatus.* Berlin: Springer-Verlag.

Peltokallio, P., and Peltokallio, V. 1968. Fractures of the humerus from muscular violence in sport. *Journal of sports medicine* 8:21.

Peterson, R. E. 1974. *Stress concentration factors.* New York: John Wiley.

Petit, G. 1955. Ordre des Siréniens. In: *Traité de Zoologie*, ed. P-P. Grassé, vol. 17, p. 926. Paris: Masson.

Piekarski, K. 1973. Analysis of bone as a composite structure. *International journal of engineering science* 11:557–65.

Piney, A. 1922. The anatomy of the bone marrow: with special reference to the distribution of the red marrow. *British medical journal* 2:792–95.

Prange, H. D., Anderson, J. F., and Rahn, H. 1979. Scaling of skeletal mass to body mass in birds and mammals. *American naturalist* 113:103–22.

Preuschoft, H. 1973. Functional anatomy of the upper extremity. In: *The chimpanzee*, ed. G. H. Bourne, vol. 6, pp. 34–120. Basel: Karger.

Puhl, J. J., Piotrowski, G., and Enneking, W. F. 1972. Biomechanical properties of paired canine fibulas. *Journal of biomechanics* 5:391–97.

Radin, E. L., and Paul, I. L. 1971. Importance of bone in sparing articular cartilage from impact. *Clinical orthopaedics and related research* 78:342–44.

Ramaekers, J. G. 1977. The dynamic shear modulus of bone in dependence on the form. *Acta morphologica Neerlando-Scandinavica* 15:185–201.

Reilly, D. T., and Burstein, A. H. 1974. The mechanical properties of cortical bone. *Journal of bone and joint surgery* 56A:1001–22.

———. 1975. The elastic and ultimate properties of compact bone tissue. *Journal of biomechanics* 8:393–405.

Reilly, D. T., Burstein, A. H. and Frankel, V. H. 1974. The elastic modulus for bone. *Journal of biomechanics* 7:271–75.

Ride, W.D.L. 1959. Mastication and taxonomy in the macropodine skull. *Systematics association publications* no. 3:33–59.

Roark, R. J. 1965. *Formulas for stress and strain.* New York: McGraw-Hill.

Robertson, D. M., and Smith, D. C. 1978. Compressive strength of mandibular bone as a function of microstructure and strain rate. *Journal of biomechanics* 11:455–71.

Rockey, K. C., Evans, H. R., Griffiths, D. W., and Nethercot, D. A. 1975. *The finite element method.* London: Crosby.

Roesler, H. 1981. Some historical remarks on the theory of cancellous bone structure (Wolff's law). In: Cowin, 27–42.

Rubin, C. T., and Lanyon, L. E. 1982. Limb mechanics as a function of speed and gait: a study of functional strains in the radius and tibia of horse and dog. Journal of experimental biology 101:187–211.

Ruth, E. B. 1953. Bone studies. II An experimental study of the Haversian-type vascular channels. *American journal of anatomy* 93:429–55.

Saha, S. 1982. The dynamic strength of bone and its relevance. In: *Osteoarthromechanics*, ed. D. N. Ghista, pp. 1–43. New York: McGraw-Hill.

Saha, S., and Hayes, W. C. 1976. Tensile impact properties of human compact bone. *Journal of biomechanics* 9:243–51.

Sammarco, G. J., Burstein, A. H., Davis, W. L., and Frankel, V. H. 1971. The biomechanics of torsional fractures: the effect of loading on ultimate properties. *Journal of biomechanics* 4:113–17.

Scapino, R. 1981. Morphological investigations into functions of the jaw symphysis in carnivorans. *Journal of morphology* 167:339–75.

Schultz, A. H. 1939. Notes on Diseases and healed fractures of wild apes. *Bulletin of the history of medicine* 7:571–82.

Shanley, F. R. 1957. *Weight-strength analysis of aircraft structures.* New York: McGraw-Hill.

Simon, B. R., ed. 1980. *Finite elements in biomechanics.* 2 vol. Tucson: University of Arizona Press.

Singer, F. L. 1963. *Strength of materials.* New York: Harper and Row.

Singh, I. 1978. The architecture of cancellous bone. *Journal of anatomy* 127: 305–10.

Smith, J. W. 1962a. The relationship of epiphysial plates to stress in some bones of the lower limb. *Journal of anatomy* 96:58–78.

———. 1962b. The structure and stress relationships of fibrous epiphysial plates. *Journal of anatomy* 96:209–225.

Sokoloff, L., ed. vol. 1, 1978; vol. 2, 1980. *The joints and synovial fluid.* New York: Academic Press.

Spengler, D. M., Morey, E. R., Carter, D. R., Turner, R. T., and Baylink, D. J. 1979. Effect of space flight on bone strength. *Physiologist* 22 (supplement): 75–76.

Starkebaum, W., Pollack, S. R., and Korostoff, E. 1979. Microelectrode studies of stress-generated potentials in four-point bending of bone. *Journal of biomedical materials research* 13:729–51.

Stern, J. T. 1974. Computer modeling of gross muscular dynamics. *Journal of biomechanics* 7:411–28.

Stokes, I.A.F., Hutton, W. C., and Stott, J.R.R. 1979. Forces acting on the metatarsals during normal walking. *Journal of anatomy* 129:579–90.

Stowell, E. Z., and Liu, T. S. 1961. On the mechanical behavior of fibre-reinforced crystalline materials. *Journal of the mechanics and physics of solids* 9:242–60.

Strömberg, L., and Dalén, N. 1976. Experimental measurements of maximum torque capacity of long bones. *Acta orthopaedica Scandinavica* 47:257–63.

Swanson, S.A.V., and Freeman, M.A.R. 1966. Is bone hydraulically strengthened? *Medical and biological engineering* 4:433–38.

Swanson, S.A.V., Freeman, M.A.R., and Day, W. H. 1971. The fatigue properties of human cortical bone. *Medical and biological engineering* 9:23–32.

Tattersall, H. G., and Tappin, G. 1966. The work of fracture and its measurement in metals, ceramics and other materials. *Journal of materials science* 1: 296–301.

Tavassoli, M. 1974. Differential response of bone marrow and extramedullary adipose cells to starvation. *Experientia* 30:424–25.

Taylor, C. R., Heglund, N. C., and Maloiy, G.M.O. 1982. Energetics and mechanics of terrestrial locomotion 1. Metabolic energy consumption as a function of speed and body size in birds and mammals. *Journal of experimental biology* 97:1–21.

Taylor, C. R., Shkolnik, A., Dmi'el, R., Baharav, D., and Borut, A. 1974. Running in cheetahs, gazelles, and goats: energy cost and limb configuration. *American journal of physiology* 227:848–50.

Thomas, R. A., Yoon, H. S., and Katz, J. L. 1977. Acoustic emission from fresh bovine femora. In: *Ultrasonic symposium proceedings*, cat. 77CH1264, pp. 237–41. New York: Institute of Electrical and Electronics Engineers.

Thompson, R. C., and Ballou, J. E. 1956. Studies of metabolic turnover with tritium as a tracer. V. The predominantly non-dynamic state of body constituents in the rat. *Journal of biological chemistry* 223:795–809.

Torzilli, P. A., Takebe, K., Burstein, A. H., Zika. J. M., and Heiple, K. G. 1982. The material properties of immature bone. *Journal of biomechanical engineering* 104:12–20.

Townsend, P. R., Raux, P., Rose, R. M., Miegel, R. E., and Radin, E. L. 1975. The distribution and anisotropy of the cancellous bone in the human patella. *Journal of biomechanics* 8:363–67.

Tschantz, P., and Rutishauser, E. 1967. La surcharge méchanique de l'os vivant. Les déformations plastiques initiales et l'hypertrophie d'adaptation. *Annales d'anatomie et pathologie* 12:223–48.

Unsworth, A. 1981. Cartilage and synovial fluid. In: *An introduction to the biomechanics of joints*, ed. D. Dowson and V. Wright, pp. 107–14. London: Mechanical Engineering Publications.

Utenkin, A. A. 1975. Anisotropy of compact bone tissue under impact loading. *Mechanika polimerov* 11:655–58.

Van Buskirk, W. C., and Ashman, R. B. 1981. The elastic moduli of bone. In: Cowin, 131–43.

Vasciaveo, F., and Bartoli, E. 1961. Vascular channels and resorption cavities in the long bone cortex. The bovine bone. *Acta anatomica* 47:1–33.

Vaughan, L. C., and Mason, B.J.E. 1975. *A clinico-pathological study of racing accidents in horses.* Dorking: Bartholomew Press.

Vincent, J.F.V. 1982. *Structural biomaterials.* London: Macmillan.

Vincent, J.F.V., and Currey, J. D. 1980. *The mechanical properties of biological materials.* Symposia of the Society for Experimental Biology, no. 34. Cambridge: Cambridge University Press.

Vogel, J. M., Whittle, M. W., Smith, M. C., and Rambaut, P. C. 1977. Bone mineral measurement—experiment MO 78. In: *Biomedical results from skylab*, ed. R. S. Johnson and L. R. Diethein, pp. 183–90. NASA SP 377.

Wainwright, S. A., Biggs, W. D., Currey, J. D., and Gosline, J. M. 1982. *Mechanical design in organisms.* Princeton: Princeton University Press.

Walker, P. S., Dowson, D., Longfield, M. D., and Wright, V. 1968. "Boosted lubrication" in synovial joints by fluid entrapment and enrichment. *Annals of rheumatic diseases* 27:512–18.

Waris, W. 1946. Elbow injuries of javelin-throwers. *Acta chirurgica Scandinavica* 93:563.

Weightman, B., and Kempson, G. E. 1979. Load carriage. In: Freeman, 291–331.

Weis-Fogh, T. 1973. Quick estimates of flight fitness in hovering animals, including novel mechanisms for lift production. *Journal of experimental biology* 59:169–230.

Wheeler, E. J., and Lewis, D. 1977. An X-ray study of the paracrystalline nature of bone apatite. *Calcified tissue research* 24:243–48.

White, A. A. III, Panjabi, M. M., and Hardy, R. J. 1974. Analysis of mechanical symmetry in rabbit long bones. *Acta orthopaedica Scandinavica* 45:328–36.

Whitehouse, W. J. 1975. Scanning electron micrographs of cancellous bone from the human sternum. *Journal of pathology* 116:213–24.

Whitehouse, W. J., and Dyson, E. D. 1974. Scanning electron microscope studies of trabecular bone in the proximal end of the human femur. *Journal of anatomy* 118:417–44.

Wolff, J. 1869. Über die Bedeutung der Architektur der spongiosen Substanz. *Zentralblatt fur die medizinische Wissenschaft*. VI Jahrgang. 223–34.

Wolff, J. D. 1892. *Das Gesetz der Transformation der Knochen*. Berlin: A. Hirschwald.

Woo, S. L-Y. 1981. The relationships of changes in stress levels on long bone remodeling. In: Cowin, 107–29.

Woo, S. L-Y., Kuei, S. C., Amiel, D., Gomez, M. A., Hayes, W. C., White, F. C., and Akeson, W. H. 1981. The effect of prolonged physical training on the properties of long bone: a study of Wolff's law. *Journal of bone and joint surgery* 63A:780–87.

Woodhead-Galloway, J. 1980. *Collagen: the anatomy of a protein*. London: Edward Arnold.

Wright, K.W.J., and Yettram, A. L. 1979. An analytical investigation into possible mechanical causes of bone remodelling. *Journal of biomedical engineering* 1:41–49.

Wright, T. M., and Hayes, W. C. 1976. Tensile testing of bone over a wide range of strain rates: effects of strain rate, microstructure and density. *Medical and biological engineering* 14:671–80.

Yoon, H. S., Caraco, B., and Katz, J. L. 1979. Further studies on the acoustic emission of fresh mammalian bone. In: *Ultrasonic symposium proceedings*, cat. 79 CH 1482, pp. 399–404. New York: Institute of Electrical and Electronics Engineers.

Yoon, H. S., and Katz, J. L. 1976a. Ultrasonic wave propagation in human cortical bone. I Theoretical considerations for hexagonal symmetry. *Journal of biomechanics* 9:407–12.

———. 1976b. Ultrasonic wave propagation in human cortical bone. II Measurements of elastic properties and microhardness. *Journal of biomechanics* 9:459–64.

Zienkiewicz, O. C. 1977. *The finite element method*. London: McGraw-Hill.

Zweben, C., and Rosen, B. W. 1970. A statistical theory of material strength with application to composite materials. *Journal of mechanics and physics of solids* 18:189–206.

Index

Library of Congress Cataloging in Publication Data

Currey, John D., 1932–
The mechanical adaptations of bones.

Bibliography: p.
Includes index.
1. Bones—Mechanical properties 2. Adaptation (Physiology) I. Title. [DNLM: 1. Bone and Bones. 2. Stress, Mechanical. 3. Adaptation, Physiological. WE 200 C976m]
QP88.2.C87 1984 596′.0471 84-42591
ISBN 0-691-08342-8